CHEMOMETRIC METHODS IN CAPILLARY ELECTROPHORESIS

CHEMOMETRIC METHODS IN CAPILLARY ELECTROPHORESIS

Edited by

GRADY HANRAHAN
FRANK A. GOMEZ

A JOHN WILEY & SONS, INC., PUBLICATION

Published by John Wiley & Sons, Inc., Hoboken, New Jersey
Published simultaneously in Canada

For general information on our other products and services or for technical support, please contact our Customer Care Department within the United States at (800) 762-2974, outside the United States at (317) 572-3993 or fax (317) 572-4002.

Wiley also publishes its books in a variety of electronic formats. Some content that appears in print may not be available in electronic formats. For more information about Wiley products, visit our web site at www.wiley.com.

Library of Congress Cataloging-in-Publication Data:

Chemometric methods in capillary electrophoresis / edited by Grady Hanrahan, Frank A. Gomez.
p. cm.
Includes index.
ISBN 978-0-470-39329-1 (cloth)
1. Capillary electrophoresis. 2. Chemometrics. I. Hanrahan, Grady. II. Gomez, Frank A.
TP248.25.C37C44 2010
660′.2972–dc22

2009014009

Printed in the United States of America

10 9 8 7 6 5 4 3 2 1

CONTENTS

PREFACE

The goal of this book is to present modern chemometric methods utilized in capillary electrophoresis (CE) to help alleviate the problems commonly encountered during routine analysis and method development. Its scope is to focus on current chemometric methods utilized in CE endeavors—techniques developed and routinely incorporated by research-active experts in the field.

The book begins with a thorough introduction to CE and chemometric-related concepts, followed by discussion on the need for modern chemometric methods in CE. Part 1 presents a unique blend of information from authors active in employing experimental design and optimization techniques in routine analyses. Useful information on differing types of screening design and response surface methodology is covered in application-based format. Part 2 presents vital discussion on various exploratory data analysis, prediction, and classification techniques utilized in CE-related studies. Part 3 houses two key chapters that provide practical information on modeling quantitative structure relationships. Finally, Part 4 explores transformation techniques, in particular, fundamental studies and applications of cross correlation and Hadamard Transform Electrophoresis.

All sections present timely chemometric methods and discuss how they are applied in a wide array of applications, including biological, medical, pharmaceutical, food, forensic, and environmental science. This book is not only highly significant to CE-based endeavors, but is also instructive for investigators active in other areas of separation science who could benefit from its informative content.

Los Angeles
March 2009

Grady Hanrahan
Frank A. Gomez

ACKNOWLEDGMENTS

The editors express strong gratitude to Bob Esposito, Michael Leventhal, and John Wiley & Sons, Inc. We are also grateful for the work and valuable expertise of the chapter contributors. We thank Jennifer Arceo and Sarah Muliadi for their kind efforts in formatting references for individual chapters. Finally, we thank our research students who have contributed significantly to the development of our own studies in capillary electrophoresis and chemometrics.

EDITOR BIOGRAPHIES

Grady Hanrahan, PhD, is the John Stauffer Endowed Professor of Analytical Chemistry at California Lutheran University. With experience in directing undergraduate and graduate research, he has taught in the fields of Environmental Science and Analytical Chemistry at California State University, Los Angeles (CSULA), and California Lutheran University (CLU). He is the author of *Environmental Chemometrics: Principles and Modern Applications* and *Modelling of Pollutants in Complex Environmental Systems*.

Frank A. Gomez, PhD, is the Director of the CSULA-Caltech Partnership for Research and Education in Materials (PREM) Collaborative. He is a Professor in the Department of Chemistry and Biochemistry at California State University, Los Angeles, and a Visiting Research Associate at the California Institute of Technology.

CONTRIBUTORS

Juliana Vieira Alberice, Instituto de Química de São Carlos, Universidade de São Paulo, São Carlos, SP-Brazil.

Silvia M. Albillos, Institute of Biotechnology IMBIOTEC, León, Spain.

Alison Beavis, Department of Chemistry and Forensic Science, University of Technology, Sydney, NSW 2007, Australia.

Lucas Blanes, Department of Chemistry and Forensic Science, University of Technology, Sydney, NSW 2007, Australia.

Susanne P. Boyle, School of Pharmacy, The Robert Gordon University, Schoolhill, Aberdeen AB10 1FR, UK.

María D. Busto, Department of Biotechnology and Food Science, University of Burgos, Burgos, Spain.

Emanuel Carrilho, Instituto de Química de São Carlos, Universidade de São Paulo, São Carlos, SP, Brazil.

Xiao-jia Chen, Institute of Chinese Medical Sciences, University of Macau, Macao SAR, China.

Alejandro Cifuentes, Department of Food Analysis, Institute of Industrial Fermentations (CSIC), Madrid, Spain.

Froseen Dahdouh, Department of Chemistry & Biochemistry, California State University, Los Angeles, Los Angeles, CA 90032.

Bieke Dejaegher, Department of Analytical Chemistry and Pharmaceutical Technology, Vrije Universiteit Brussel (VUB), Brussels, Belgium.

Luís G. Dias, Department of Chemistry, Faculty of Philosophy, Sciences and Languages of Ribeirão Preto (FFCLRP), University of Sao Paulo, Ribeirao Preto, SP, Brazil.

Philip Doble, Department of Chemistry and Forensic Science, University of Technology, Sydney, NSW 2007, Australia.

Melanie Dumarey, Department of Analytical Chemistry and Pharmaceutical Technology, Vrije Universiteit Brussel (VUB), Brussels, Belgium.

Alexandra Durand, Department of Analytical Chemistry and Pharmaceutical Technology, Vrije Universiteit Brussel (VUB), Brussels, Belgium.

David G Durham, School of Pharmacy, The Robert Gordon University, Schoolhill, Aberdeen AB10 1FR, UK.

João P.S. Farah, Institute of Chemistry, University of Sao Paulo, Sao Paolo, SP, Brazil.

Jessica L. Felhofer, Department of Chemistry, The University of Texas at San Antonio, San Antonio, TX 78249.

Maribel Elizabeth Funes-Huacca, Instituto de Química de São Carlos, Universidade de São Paulo, São Carlos, SP, Brazil.

Carlos D. Garcia, Department of Chemistry, The University of Texas at San Antonio, San Antonio, TX 78249.

Frank A. Gomez, Department of Chemistry & Biochemistry, California State University, Los Angeles, CA 90032.

Grady Hanrahan, Department of Chemistry, California Lutheran University, Thousand Oaks, CA 91360.

Javier Hernández-Borges, Department of Analytical Chemistry, Nutrition and Food Science, University of La Laguna (ULL), Tenerife, Canary Islands, Spain.

Mehdi Jalali-Heravi, Department of Chemistry, Sharif University of Technology, Tehran, Iran.

Takashi Kaneta, Department of Applied Chemistry, Graduate School of Engineering, Kyushu University, Motooka, Fukuoka, Japan; Division of Translational Research, Center of Future Chemistry, Kyushu University, Motooka, Fukuoka, Japan.

Hua Li, School of Chemistry and Material Science, Northwest University, Xi'an, 710069, China.

Shao-ping Li, Institute of Chinese Medical Sciences, University of Macau, Macao SAR, China.

Ann S. Low, School of Pharmacy, The Robert Gordon University, Schoolhill, Aberdeen AB10 1FR, UK.

Ruthy Montes, Department of Chemistry & Biochemistry, California State University, Los Angeles, CA 90032.

Edgar P. Moraes, Institute of Chemistry, University of Sao Paulo, Sao Paolo, SP, Brazil.

Natividad Ortega, Department of Biotechnology and Food Science, University of Burgos, Burgos, Spain.

Raymond G. Reid, School of Pharmacy, The Robert Gordon University, Schoolhill, Aberdeen AB10 1FR, UK.

Toni Ann Riveros, Department of Chemistry & Biochemistry, California State University, Los Angeles, CA 90032.

Miguel Ángel Rodríguez-Delgado, Department of Analytical Chemistry, Nutrition and Food Science, University of La Laguna, Tenerife, Canary Islands, Spain.

Javier Saurina, Department of Analytical Chemistry, University of Barcelona, 08028 Barcelona, Spain.

Gerhard K.E. Scriba, Department of Pharmaceutical Chemistry, Friedrich Schiller University of Jena, 07743 Jena, Germany.

Marina F.M. Tarvares, Institute of Chemistry, University of Sao Paulo, Sao Paolo, SP, Brazil.

Fernando G. Tonin, Department of Food Engineering, University of Sao Paulo, Pirassununga, SP, Brazil.

Amanda Van Gramberg, Department of Chemistry and Forensic Science, University of Technology, Sydney, NSW 2007, Australia.

Yvan Vander Heyden, Department of Analytical Chemistry and Pharmaceutical Technology, Vrije Universiteit Brussel (VUB), Brussels, Belgium.

Feng-qing Yang, Institute of Chinese Medical Sciences, University of Macau, Macao SAR, China.

Yaxiong Zhang, School of Chemistry and Material Science, Shan'xi Normal University, Linfen, 041004, China.

KEY ACRONYMS

ACE—affinity capillary electrophoresis
ANFIS—adaptive neuro-fuzzy inference system
ANOVA—analysis of variance
ANN—artificial neural networks
AZT—3′-azido-2′, 2′-dideoxythymidine
BBD—Box–Behnken design
BGE—background electrolyte
CAB—carbonic anhydrase
CARTs—classification and regression trees
CBSA—4-carboxybenzenesulfonamide
CC—cross correlation
CCD—central composite design
CDA—canonical discriminant analysis
CE—capillary electrophoresis
CEC—capillary electrochromatography
CE–DAD—capillary electrophoresis–diode-array detection
CGD—conjugate gradient descent
CMC—critical micellar concentration
COW—correlation optimized warping
CRF—chromatographic response function
CZE—capillary zone electrophoresis
DM—Doehlert matrices
FASS—field-amplified sample stacking
GRNN—generalized regression neural network
ED—experimental design
EFA—evolving factor analysis
EMMA—electrophoretically mediated microanalysis
EOF—electroosmotic flow
FSMW–EFA—fixed-size moving-window–evolving factor analysis
FT—Fourier transform
FTPFACE—flow-through partial-filling affinity capillary electrophoresis
G6P—glucose-6-phosphate
G6PDH—glucose-6-phosphate dehydrogenase

GC—gas chromatography
HCA—hierarchical cluster analysis
HELP—heuristic evolving latent projections
HHM—horse heart myoglobin
HPLC—high performance liquid chromatography
HPLC–DAD—high performance liquid chromatography–diode array detector
HT—Hadamard transform
IR—infrared spectroscopy
ITTFA—iterative target transformation factor analysis
kNN—k-nearest neighbors
LDA—linear discriminant analysis
LFER—linear free energy relationships
LGO—leave-group-out
LOO—leave-one-out
LSER—linear solvation energy relationship
MA—machine learning
MCDM—multicriteria decision-making
MCR–ALS—multivariate curve resolution based on alternating least squares
MEKC—micellar electrokinetic chromatography
MEKC–DAD—micellar electrokinetic chromatography–diode array detection
MLP—multilayer perceptron
MRLs—maximum residue limits
MS—mass spectrometry
MSC—multiplicative signal correction
NACE—nonaqueous capillary electrophoresis
NADH—nicotinamide adenine dinucleotide, reduced form
NJ—neighbor joining
OPA—orthogonal projection approach
ORM—overlapping resolution
OTU—operational taxonomic unit
OVAT—one-variable-at-a-time
PC—principal components
PCA—principal component analysis
PCO—principal coordinate analysis
PCR—principal component regression
PF—partial filling
PLS—partial least squares
PLSDA—partial least squares discriminant analysis
PNN—probabilistic neural network
PP—projection pursuit
PPFs—projection pursuit features
PRBS—pseudo-random binary sequence

QDA—quadratic discriminant analysis
QSMR—quantitative structure–mobility relationship
QSRR—quantitative structure–retention relationship
rPCA—robust principal component analysis
RAPD—random amplified polymorphic DNA
RBF—radial basis function
RMTR—relative migration time ratio
RP-HPLC—reverse-phase high performance liquid chromatography
RSM—response surface methodology
SCOFT—Shah convolution Fourier transform
SDA—stepwise discriminant analysis
SDS—sodium dodecylsulphate
SGE—slab gel electrophoresis
SIMCA—soft independent modelling of class analogy
SIMPLISMA—Simple-to-Use Interactive Self-modeling Mixture Analysis
SST—system suitability test
SVM—support vector machines
UPGMA—unweighted pair group method using arithmetic average
WFA—window factor analysis

PART I

EXPERIMENTAL DESIGN AND OPTIMIZATION CONSIDERATIONS

CHAPTER 1

INTRODUCTION

GRADY HANRAHAN[1] and FRANK A. GOMEZ[2]
[1]Department of Chemistry, California Lutheran University, Thousand Oaks, CA
[2]Department of Chemistry & Biochemistry, California State University, Los Angeles, CA

CONTENTS

1.1. CAPILLARY ELECTROPHORESIS (CE): AN OVERVIEW

Over the past two decades, CE has become the technique of choice in many analytical laboratories where analysis of small quantities of materials must be accurately, efficiently, and expeditiously assessed. It is a powerful separation technique that brings much needed speed, quantitation, reproducibility, and automation to the inherently highly resolving but labor-intensive methods of electrophoresis (1–5). CE comprises a family of techniques including:

1. capillary zone electrophoresis;
2. capillary gel electrophoresis;
3. isoelectric focusing; and
4. micellar electrokinetic capillary chromatography.

All employ narrow-bore (e.g. 20–200-μm i.d.) capillaries (Fig. 1.1) to perform high efficiency separations for the analysis of biological materials and is an unparalleled experimental tool for examining interactions in biologically relevant media. A generalized experimental setup for CE is presented in Figure 1.2. As shown, the instrumental configuration is relatively simple and includes

Chemometric Methods in Capillary Electrophoresis. Edited by Grady Hanrahan and Frank A. Gomez

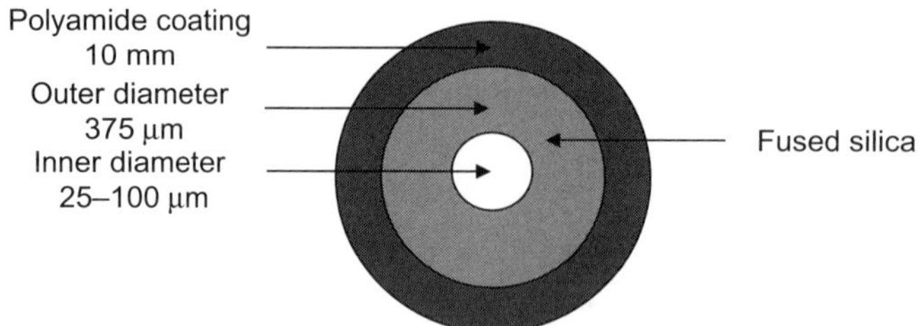

FIGURE 1.1. Fused silica capillary.

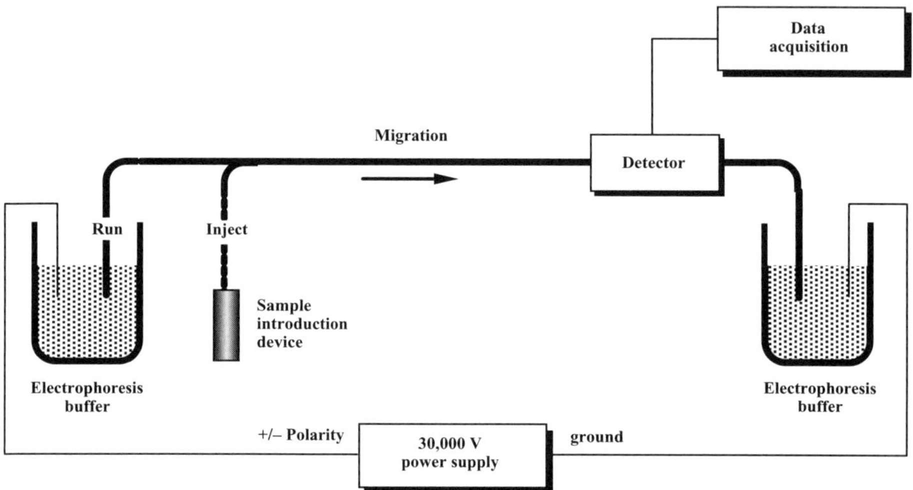

FIGURE 1.2. Generalized experimental setup for CE.

a narrow-bore capillary, a high-voltage power supply, two buffer reservoirs, a sample introduction device, and a selected detection scheme. Optical detection, typically absorbance (UV-visible) and laser-induced fluorescence, is employed. Signals are then transferred to a data acquisition module, which produces a representative electropherogram.

The underlying theory that governs electrophoresis is directly applicable to CE and can be explained by a variety of fundamental principles. CE differentiates charged species on the basis of mobility under the influence of an applied electric field gradient. Consequently, separation is reliant upon the difference in ion migration velocities expressed as:

$$v = \mu_e E \qquad \text{(Eq. 1.1)}$$

where v = the ion migration velocity (m/s), μ_e = the electrophoretic mobility (m^2/V/s), and E = the electric field potential (V/m). The latter is a function of the applied voltage divided by the total length of the chosen capillary. Electrophoretic mobility is a constant proportionality between the ion velocity and the electric field potential (6) expressed as:

$$\mu_e = \frac{q}{6\pi\eta r} \quad \text{(Eq. 1.2)}$$

where q = the energy of the ion, η = the solution's viscosity and r = the hydrodynamic radius of the ion. As evident in Equation 1.2, the differences in electrophoretic mobility are subject to differences in the charge-to-mass ratio of the analyte ions. For example, a higher charge and smaller ion mass will yield greater mobility. Due to the differences in mobility, it is possible to separate mixtures of different ions and solutes using electrophoresis (Fig. 1.3). Selectivity can be manipulated by the alteration of electrolyte properties including ionic strength, pH, electrolyte composition, or by incorporating electrolyte additives.

It is the high voltage source that facilitates separations, ultimately generating electroosmotic flow (EOF) of buffer solutions and ionic species within the capillary. EOF is defined by:

$$v_{eo} = \frac{\varepsilon\zeta}{4\pi\eta} \quad \text{(Eq. 1.3)}$$

where ε = the dielectric constant, η = the buffer viscosity, and ζ represents the zeta potential of the capillary wall. The latter is the potential difference measured at the plane of shear close to the liquid–solid interface (7).

The surface charges of the liquid–solid interface play crucial roles in the EOF phenomenon. When a buffer solution is introduced into the capillary, the negatively charged wall attracts the positively charged ions from solution,

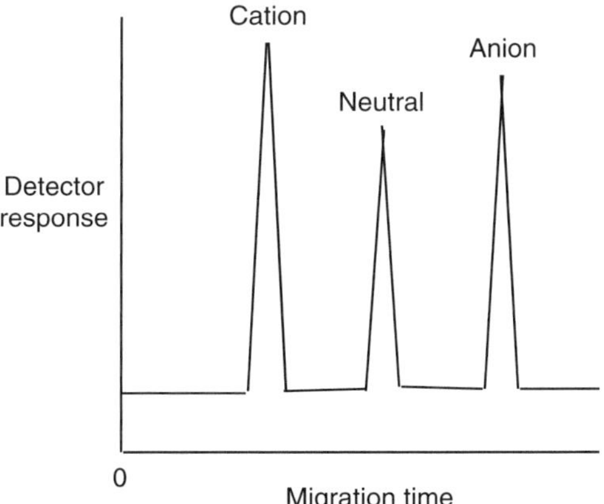

FIGURE 1.3. Separation of differing ions by CE.

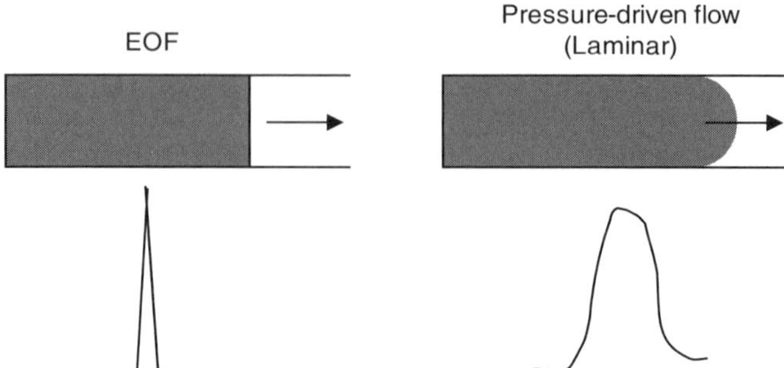

FIGURE 1.4. EOF and its generation of a flat flow profile alongside a parabolic laminar pressure-driven flow profile.

creating an electrical double layer (fixed and mobile) and a potential difference (zeta potential) close to the capillary wall. Accordingly, EOF mobility will vary with a change in the pH of the buffer solution. At pH > 7, the EOF mobility drives the net migration of the majority of ions toward the cathode (regardless of charge). As will be evident in subsequent chapters, the EOF must be controlled (or possibly suppressed) to run certain modes of CE. A beneficial feature of EOF is its generation of a flat flow profile alongside a parabolic laminar pressure-driven flow profile as typically seen in high performance liquid chromatography methods. This flat flow profile aids in minimizing zone broadening, ultimately allowing high separation efficiencies based on mobility differences as low as 0.05% (7). See representative diagram in Figure 1.4.

Indeed, there are a number of factors that must be considered for efficient and optimized separation, as well as in developing new methods to meet today's analytical challenges and routine laboratory needs. It is beyond the scope of this book to completely cover all theoretical aspects of CE. Complete coverage can be found in a variety of informative sources (6–9).

1.2. CHEMOMETRIC METHODS AND THEIR IMPORTANCE IN CE

CE offers a number of advantages as a separation technique: (i) it requires only small quantities of material; (ii) it is applicable to water-soluble, nonvolatile, high-molecular-weight species in aqueous buffer solution; (iii) it is readily automated and has good reproducibility; and (iv) various separation modes make it applicable for the analysis of a variety of biological and nonbiological species. Unfortunately, CE does suffer from a number of weaknesses. Adsorption of charged species to the capillary wall can occur in the absence of efforts to minimize adsorption and can change the magnitude of EOF. Overlapping peaks are a common occurrence, and methods devised to aid in separation are critical. The presence of Joule heating and other effects of using high voltage create variances in EOF, sometimes yielding irreproducible migration times for analytes, making comparison from run to run problematic. This disadvantage can be especially troubling in the pharmaceutical industry where quality control is a priority and where method development is critical in product manufacture, analysis, and marketing. Ultimately, the search for optimum separation conditions in CE is often time-consuming and tedious. Therefore, the development and utilization of robust chemometric techniques in CE are favorable and a definitive source of information is vital.

Fortunately, various chemometric-based techniques, including multivariate experimental design and data analysis techniques, have been devised to aid in optimizing the performance of systems and extend their separation capabilities. In broadest terms, chemometrics is a subdiscipline of analytical chemistry that uses mathematical, statistical, and formal logic to (10):

1. design and/or select optimal experimental procedures;
2. provide maximum relevant chemical information by analyzing chemical data; and
3. obtain knowledge about given chemical systems.

Although statistical methodologies such as "curve fitting" and "statistical control" were used in analytical chemistry throughout the 1960s, it was not until 1972 that Svante Wold coined the term "chemometrics." The broad definition described above was shaped by the evolution of this subdiscipline over the past 35 years. The first known paper with chemometrics in the title was subsequently written by Bruce Kowalski in 1975 (11), which presented the value of pattern recognition concepts to the analytical community. The 1980s brought about an era of enhanced computing capabilities and more sophisticated analytical instrumentation, including the development of more advanced CE methods. The deluge of data generated by these multielement and multicomponent instruments required the application of chemometric methods already established, as well as creating a need for higher-level methodologies. Such methods were expressed to the scientific community with the advent of two specialized journals: *Chemometrics and Intelligent Laboratory Systems*, established in 1986, and *Journal of Chemometrics* in 1987. An increased number of investigators began incorporating chemometrics into their research activities in the 1990s. Brown et al., in a 1996 comprehensive review of chemometrics, reported over 25,000 computer-generated citations for this broad topic (12). In a 1998 review, Wold and Sjöström presented an informative look at the acceptance and success of chemometrics in modern analytical research (13). This paper illustrated how analytical chemistry is driven by chemometrics and describes state-of-the-art methods including multivariate calibration, structure–(re)activity modeling, and pattern recognition, classification, and discriminant analysis. The twenty-first century has brought about even greater analytical sophistication allowing automated, high throughput capabilities with low reagent and sample use. In a 2008 review, Lavine and Workman describe the latest trends and acceptance of chemometrics in modern chemical analysis (14).

1.3. CURRENT AND FUTURE APPLICATION AREAS

In regard to CE, previous reviews and informative research papers provided systematic studies on early development efforts and use of experimental design methodology in CE (15–18). More recent papers have examined experimental design concepts and methods for data analysis in regard to CE applications in greater detail (19–25). The above list of citations is obviously not conclusive, but considering the information presented, it is obvious that chemometric methodologies are important tools in analytical chemistry, especially when considering modern CE applications. It is evident from the above papers and material presented in subsequent chapters that chemometric techniques are, and will continue to have, a profound effect on CE applications, including drug design, food technology, biomedical research, and environmental science. For example, microfluidics is one area where chemometrics has yet to be employed in earnest and where its integration will prove fruitful in the future. While the vast majority of papers in microfluidics have detailed elegant studies, optimization of parameters for a particular application has not been at the forefront.

REFERENCES

1. Guzman, N.A. (2004) *Anal Bioanal Chem*, **378**, 37–42.
2. Villareal, V., Azad, M., Zurita, C., Silva, I., Hernandez, L., Rudolph, M., Moran, J., and Gomez, F.A. (2003) *Anal Bioanal Chem*, **376**, 822–831.
3. Landers, J.P. (1997) *Handbook of Capillary Electrophoresis*, CRC Press, Boca Raton, FL.
4. Wiedmer, S., Cassely, A., Hong, M., Novotny, M.V., and Riekkola, M.-L. (2000) *Electrophoresis*, **21**, 3212–3219.
5. Riekkola, M.L., Jonsson, J.A., and Smith, R.M. (2004) *Pure Appl Chem*, **76**, 443–451.
6. Compton, S.W. and Brownlee, R.G. (1988) *Biotechniques*, **6**, 432–440.
7. Jorgenson, J.W. and Lukacs, K.D. (1981) *J Chromatogr*, **218**, 209–216.
8. Altria, K.D. (1996) Fundamentals of capillary electrophoresis theory, in *Capillary Electrophoresis Guidebook Principles, Operation, and Applications*, Vol. **52** (ed. K.D. Altria), Humana Press, Totowa, NJ, pp. 3–13.
9. Khaledi, M.G. (1998) *High-Performance Capillary Electrophoresis: Theory, Techniques, and Applications*, John Wiley & Sons, Hoboken, NJ.
10. Hopke, P.K. (2003) *Anal Chim Acta*, **500**, 365–377.
11. Kowalski, B.R. (1975) *J Chem Inf Comput Sci*, **15**, 201–203.

12. Brown, S.D., Sum, S.T., Despagne, F., and Lavine, B.K. (1996) *Anal Chem*, **68**, 21–61.
13. Wold, S. and Sjöström, M. (1998) *Chemom Intell Lab Syst*, **44**, 3–14.
14. Lavine, B. and Workman, J. (2008) *Anal Chem*, **80**, 4519–4531.
15. Alria, K.D., Clark, B.J., Filbey, S.D., Kelly, M.A., and Rudd, D.R. (1995) *Electrophoresis*, **16**, 2143–2148.
16. Vander Heyden, Y. and Massart, D.L. (1996). Review of robustness in analytical chemistry, in *Robustness of Analytical Chemical Methods and Pharmaceutical Technological Products* (eds. M.W.B. Hendriks, J.H. de Boer, and A.K. Smilde), Elsevier, Amsterdam, pp. 79–147.
17. Vargas, M.G., Vander Heyden, Y., Maftouh, M., and Massart, D.L. (1999) *J Chromatogr A*, **855**, 681–693.
18. Jimidar, M., Bourguignon, B., and Massart, D.L. (1996) *J Chrom A*, **740**, 109–117.
19. Siouffi, A.M. and Phan-Tan-Luu, R. (2000) *J Chromatogr A*, **892**, 75–106.
20. Sentellas, S. and Saurina, J. (2003) *J Sep Sci*, **26**, 875–885.
21. Duarte, A.C. and Capelo, S. (2006) *J Liq Chromatogr Related Technol*, **29**, 1143–1176.
22. Hanrahan, G., Montes, R., and Gomez, F.A. (2008) *Anal Bioanal Chem*, **390**, 169–179.
23. Maia, P.P., Amaya-Farfán, J., Rath, S., and Reyes, F.G.R. (2007) *J Pharm Biomed Anal*, **43**, 450–456.
24. Tran, A.T.K., Hyne, R.V., Pablo, F., Day, W.R., and Doble, P.A. (2007) *Talanta*, **71**, 1268–1275.
25. Hernández-Borges, J., Rodríguez-Delgado, M.A., García-Montelongo, F.J., and Cifuentes, A. (2005) *Electrophoresis*, **26**, 3799–3813.

CHAPTER 2

EXPERIMENTAL DESIGN IN METHOD OPTIMIZATION AND ROBUSTNESS TESTING

BIEKE DEJAEGHER, ALEXANDRA DURAND, and YVAN VANDER HEYDEN
Department of Analytical Chemistry and Pharmaceutical Technology, Vrije Universiteit Brussel (VUB), Brussels, Belgium

CONTENTS

Chemometric Methods in Capillary Electrophoresis. Edited by Grady Hanrahan and Frank A. Gomez

2.1. INTRODUCTION

Generally, in the development of a method aimed at analyzing one or more component(s) in a given matrix, different steps can be distinguished: method or technique selection, method optimization, and method validation. The different steps in method development and the possible approaches are presented in Figure 2.1.

To assay, for instance, drug compounds in different matrices, analytical techniques, such as high-performance liquid chromatography (HPLC) or capillary electrophoresis (CE), are frequently used. The selection of the method is mainly determined by the properties of the component to be analyzed and

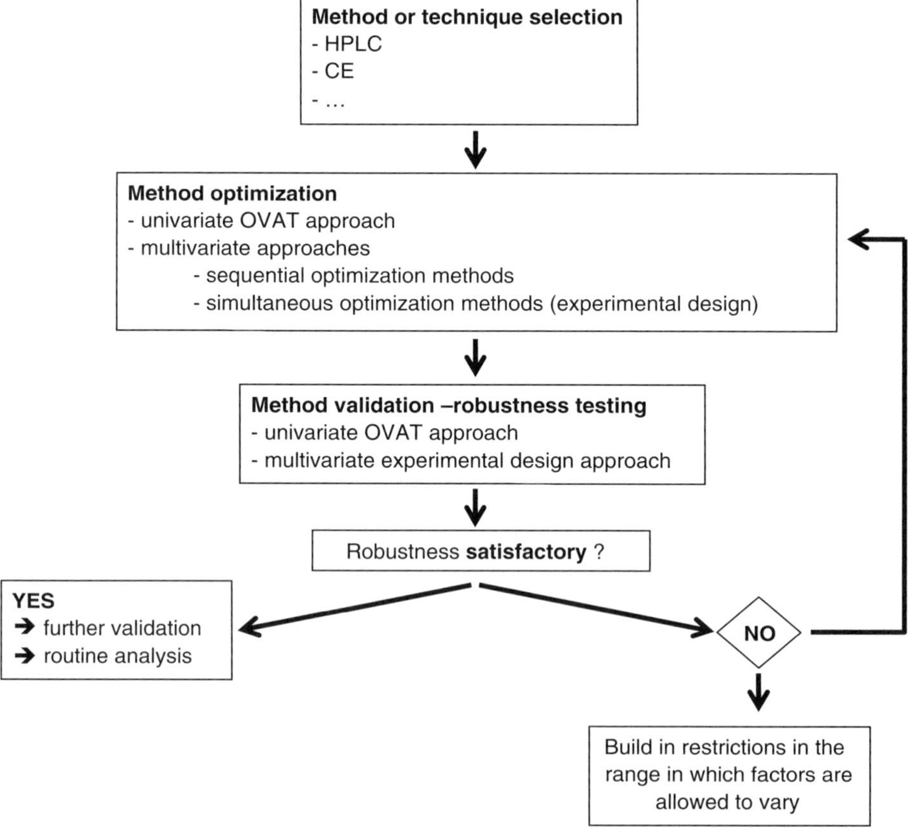

FIGURE 2.1. Different steps in method development and the possible approaches. OVAT = one-variable-at-a-time.

by the availability of the techniques in the development laboratory (*method or technique selection*).

After selecting the technique, the method should be developed and optimized (*method optimization*) (1), leading, for separation techniques, to the definition of the most optimal experimental conditions that allow a sufficient resolution of the relevant peaks as well as adequate and robust assay results in an acceptable analysis time.

Finally, the optimized method should be validated before being applied in routine analysis (*method validation*). This validation ensures the applicability and suitability of the analytical method for its intended purpose. Depending on the purpose of the method, certain validation issues are recommended to be considered (2, 3). A robustness test is a part of method validation and evaluates the effects of small but deliberate changes in some (method) parameters on the results (responses) of the method (2–5). Initially, such test is executed at the end of the validation procedure, just before an interlaboratory study to assess reproducibility, in order to identify potentially important factors, which could affect the results during such study (6). However, a method considered nonrobust should be adapted or redeveloped and revalidated, resulting in a waste of time and money. Therefore, nowadays, robustness is verified at a much earlier stage in the method lifetime, that is, at the end of development or at the beginning of validation (5). When the method robustness is considered satisfying, the method can be further validated and, when successful, applied routinely. Otherwise, the method should be adapted or reoptimized.

Different parameters or factors potentially can affect the results of a method. Several strategies can be applied to optimize (analytical) methods. When two or more factors need to be optimized, their influences on the response(s) can be examined by applying either univariate or multivariate approaches (7). A univariate method, such as the one-variable-at-a-time (OVAT) approach, varies only one factor at a time between consecutive experiments, while a multivariate procedure changes several factors simultaneously. However, the optimum found with the OVAT procedure may depend on the starting conditions of the optimization. Moreover, during this procedure, one might be trapped in a local optimum and never find the global. The approach also does not take into account interactions (see further) and is therefore only efficient when no interactions occur. When they are present, a given factor usually needs to be considered several times during the procedure in order to find the global optimum, while most frequently each factor is considered only once (Figure 2.2). Another drawback of the OVAT approach is that a high number of experiments may be required when the number of factors increases.

For these reasons, multivariate approaches seem better. The multivariate approaches, the topic of this chapter, can be further divided into sequential and simultaneous strategies (7–9). In sequential optimization strategies, initially only a few experiments are performed and their results are used to define the next experiment(s) (7, 8, 10). In simultaneous approaches, a predefined number of experiments are performed according to a well-defined experimen-

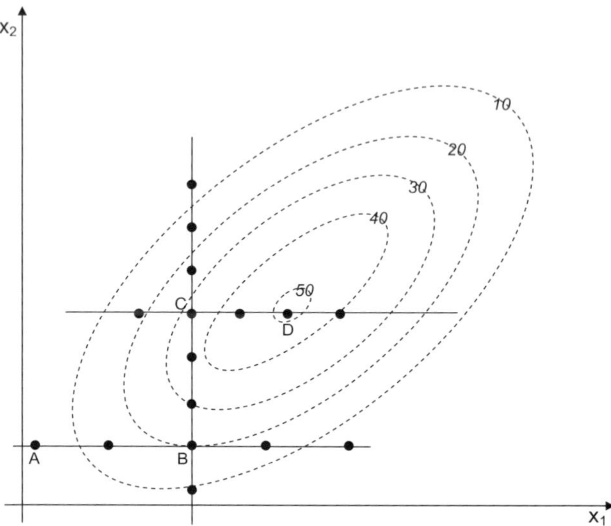

FIGURE 2.2. One-variable-at-a-time optimization procedure for two factors, x_1 and x_2, in the presence of an interaction effect between the factors. Dotted lines = hypothetical contour plot of response to optimize. A = starting point; B = best result after varying x_1 a first time; C = best result after varying x_2 a first time (= usually reported optimum); and D = best result after varying x_1 a second time (= real optimum).

tal setup, that is, an experimental design, in order to simultaneously examine a given number of factors (7).

Method optimization is often divided into a screening phase and an actual optimization phase (1, 11). During the screening phase, all factors potentially influencing the method are screened in a given range, in order to identify the most important. Thus, the experimental domain in which the optimum is probably situated is defined. In this phase, the so-called screening designs are applied (1, 11). The most important factors are then subsequently further optimized in the optimization phase, where the optimal experimental conditions are determined and the optimum is selected. In this latter phase, either response surface designs (1, 11) or sequential optimization methods (8, 11) are used. When further in the text the optimization step or phase is mentioned, the latter stage is meant.

To determine the robustness of a method, several approaches exist. Basically, the situation for robustness testing is similar to that for screening during optimization, except for the range within which the factors are examined. The influence of small but deliberate changes in parameters on the response(s) is evaluated using either an OVAT or an experimental design approach (12).

Robustness is sometimes also called ruggedness in the literature, while other sources define both as different validation items. For instance, Youden and Steiner (6) use the term ruggedness (for robustness), the United States Pharmacopoeia (USP) (13) distinguishes between both terms and provides different definitions, and the ICH (3) considers both as synonyms. We also adhere to the last approach.

Youden and Steiner (6) define a ruggedness test as a setup examining influences of minor but deliberate and controlled changes in the method parameters (factors) on the response(s), in order to detect those nonrugged factors with a large influence. Controlling the latter factors within specific limits should then avoid problems in a subsequent interlaboratory study.

The USP (13) defines ruggedness as follows: *"The ruggedness of an analytical method is the degree of reproducibility of test results obtained by the analysis of the same sample under a variety of normal test conditions, such as different laboratories, different analysts, different instruments, different lots of reagents, different elapsed assay times, different assay temperatures, different days, etc."* Here the method is performed under different test conditions without deliberately changing specific factors in a narrow interval. To evaluate the influences of the different test conditions, a nested design or a nested analysis of variance (ANOVA) can be used (4, 14). In fact the above definition is equivalent to that for either intermediate (within-laboratory) precision or reproducibility (between-laboratory variability), depending whether experiments are executed in one or several laboratories. For both the estimation of intermediate precision and reproducibility, ISO guidelines exist (14, 15).

The USP definition of robustness equals that of the ICH (3): *"The robustness of an analytical procedure is a measure of its capacity to remain unaffected by small, but deliberate variations in method parameters and provides an indication of its reliability during normal usage."* A robustness test is the experimental setup used to evaluate method robustness. It quantifies the insensitivity of the results for a method transfer to another laboratory or instrument. The ICH guidelines also state that *"One consequence of the evaluation of robustness should be that a series of system suitability parameters (e.g., resolution tests) is established to ensure that the validity of the analytical procedure is maintained whenever used"* (3).

In fact, the definition of ruggedness by Youden and Steiner equals the USP and ICH definitions of robustness. It is also the most widely applied definition. Further in this chapter, only consequences related to this definition are considered, and only the term robustness is used. In such type of robustness testing, usually screening designs are applied.

In this chapter, the use of multivariate approaches during method optimization and robustness testing is elaborated, discussed, and illustrated with examples.

2.2. AIMS/OBJECTIVES

2.2.1. Optimization

The goal of method optimization is to define (the best) experimental conditions that allow a sufficient resolution of the relevant peaks, and that provide satisfactory and robust results in an acceptable analysis time. Prior to method optimization, usually several factors (>3) can be selected or specified that potentially influence the method performance. Therefore, as already mentioned, method development is often divided into a screening and an optimization phase. In the screening phase, several (qualitative or quantitative) factors are examined in order to identify those most affecting the response(s). These latter factors are then further optimized in the optimization phase. The factors evaluated in robustness testing often are the same as those examined during the screening phase. However, in method development, normally the range in which the factors are studied is much larger than in robustness testing (12).

The responses of main interest also are different in method development and robustness testing. In development, the considered responses are related to the quality of the separation (1), such as, for electrophoretic methods, migration times, peak shapes, and the resolutions between neighboring peaks. When the separation is optimized and the method is validated, thus also in robustness testing, the responses of main interest are related to the quantitative aspects of the method, such as contents, concentrations, or recoveries. The responses considered during development occasionally are considered in a second instance, for example, as system suitability test (SST) parameters.

During the screening phase, screening designs are applied. These designs allow the examination of a relatively high number of factors in a rather small number of experiments. Usually the factors are evaluated at only two levels and two-level designs are applied. The results from screening designs are analyzed by estimating and interpreting the effects of the factors on the response(s) (4, 5, 7, 16) in order to determine those factors most influencing the outcome of the method. For the screening phase, the following steps can be distinguished:

(1) selection of the factors to be evaluated and their levels;
(2) selection of the screening design;
(3) definition of the responses;
(4) planning and execution of the entire experimental setup, and experimental determination of the responses;
(5) calculation of the (factor) effects on the responses;
(6) graphical and/or statistical interpretation of the estimated effects; and
(7) identification of the factors most influencing the method performance.

During the subsequent optimization phase, when only a limited number of variables (≤3) are evaluated, often response surface designs or sequential optimization methods are applied. When using a response surface design, the selected experimental domain, determined by the design geometry and the factor level ranges, is expected to contain the optimum. The design results are analyzed by building and interpreting a polynomial (usually quadratic) model describing the relation between the response(s) and the considered factors (1, 7, 17). When applying a response surface design methodology, the following steps are performed:

(1) selection of the level ranges of those factors most influencing the method;
(2) selection of the response surface design;
(3) definition of the responses;
(4) planning and execution of the entire experimental setup, and experimental determination of the responses;
(5) building the polynomial model(s) describing the relation between the response(s) and the factors;
(6) graphical and/or statistical evaluation of the model; and
(7) determination of the optimum.

On the other hand, in situations where the experimental region containing the optimum is not *a priori* known, a sequential optimization method, for example, a simplex approach, can be applied. Then, the following steps are considered:

(1) selection of the size and position (= levels) of the initial simplex for those factors most influencing the method;
(2) selection of the type of sequential method, for example, the type of simplex approach;
(3) definition of one response to optimize;
(4) experimental determination of the response;
(5) selection of the next simplex, that is, the subsequent experiment, based on a number of predefined rules and the results of the previous simplex;
(6) repeating steps (4) and (5) until the optimum is sufficiently approached; and
(7) determination of the optimum.

The different steps of the above-described approaches are discussed in more detail later and illustrated with an example taken from the literature.

2.2.2. Robustness Testing

The main goal of a robustness test is to examine potential sources (factors) causing variability in one or more responses of the method. To identify those sources, a number of factors, usually specified with a nominal level in the operating procedure of the method, are selected. These factors are then varied in an interval, representative for the fluctuations in the nominal factor levels, which can be expected when transferring a method between different instruments or laboratories (5, 18). The nominal level of a factor is the one described in the operating procedure or the level set during routine application. Preferably, the selected factors are evaluated simultaneously by means of a screening design. In a first instance, the considered responses describe quantitative aspects of the method, such as the estimated concentrations or percentage recoveries of the main and/or related compound(s). Second, also qualitative responses related to the separation, for example, responses for which SST limits should be defined, can be studied, such as, for electrophoretic methods, resolutions between neighboring peaks. After determining the response(s) for all design experiments, the factor effects on the response(s) are estimated. This allows determination of the factors with an important influence on the results and enables establishment of boundaries or limits to control the levels of these factors, if necessary.

A second goal from a robustness test can be to define SST limits. These SST limits can be determined in a systematic way based on the experimental data from the robustness test, although actually they are frequently chosen arbitrarily based on the experience of the analyst.

In general, in a robustness test, the following steps can be distinguished:

(1) selection of the factors to be evaluated and their levels;
(2) selection of the experimental design;
(3) definition of the responses;
(4) planning and execution of the entire experimental setup, and experimental determination of the responses;
(5) calculation of the (factor) effects on the responses;
(6) graphical and/or statistical interpretation of the estimated effects;
(7) drawing chemically relevant conclusions and, if necessary, taking precautions to improve the method performance;
(8) determining nonsignificance intervals for significant quantitative factors; and
(9) defining SST limits for certain qualitative responses.

The different steps are discussed in more detail and illustrated with an example taken from the literature.

2.3. FACTORS AND THEIR LEVELS

2.3.1. Selection of Factors

Before starting method development or robustness testing, the factors to be examined should be carefully defined and selected. Factors of CE methods can be divided into operational, environmental, and peak measurement/peak analysis factors. Operational parameters are those that after optimization are described in the operating procedure of the method, while environmental factors are not necessarily specified in that procedure, for example, room temperature. However, when such environmental factors have been examined in method development, then they normally also will be specified in the operating procedure.

Peak measurement/peak analysis parameters (12, 19) are related to the measurement of the signal at the detector, its treatment, and reporting. They affect the quality of responses, such as peak areas, peak heights, migration times, and resolutions. These latter factors can be found in the data-treatment software of an instrument, where often only their default settings are used by the analyst. However, except for the detection wavelength, the factors mentioned are usually not considered, although they can affect the electropherogram largely (Figure 2.3) (19).

In general, all factors potentially influencing the (quality of the) separation are chosen for screening, while for robustness testing, those factors that are most likely to vary when a method is transferred between different laboratories, analysts, or instruments are selected. Often, the same factors are concerned.

Table 2.1 presents an overview of factors that can potentially be considered for optimization and robustness testing of CE methods. Lists of commonly used electrolytes/buffers (20–23) or additives (20) and characteristic properties of frequently applied solvents and surfactants (20) can be found in the literature. Sample concentration (see Table 2.1) is a factor occasionally included. However, the aim of the analytical method is to estimate this concentration through the measured signal, from a calibration procedure. In method optimization, responses related to the quality of the separation, for example, resolutions, are considered, and in this situation one can verify whether the sample

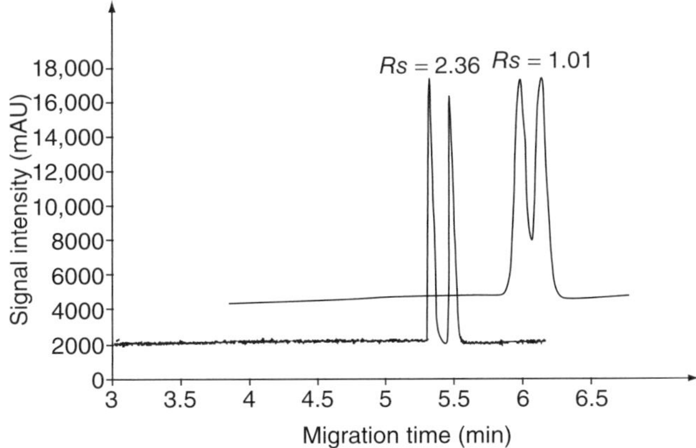

FIGURE 2.3. Two electropherograms, recorded with different settings for peak measurement/peak analysis parameters. Experimental conditions are identical. Adapted from Reference 19.

TABLE 2.1. Potential factors in the optimization or robustness testing of capillary electrophoretic methods

(1) Operational and environmental factors	*(2) Peak measurement/analysis parameters*
Additives concentrations	Detection
Chiral selectors	Detection wavelength (VIS, UV, or fluorimetric detection)
Inorganic salts	Reference wavelength
Organic solvents	Detection wavelength bandwidth
Surfactants	Reference wavelength bandwidth
Background electrolyte:	Integration: sensitivity
Electrolyte composition	Peak detection:
Electrolyte concentration	Peak width
Ionic strength of the buffer	Threshold
pH of the buffer	Signal processing:
Capillary	Data acquisition rate
Age	Type of filtering
Batch	Amount of filtering
Coating	
Internal diameter	
Length	
Manufacturer	
Capillary temperature	
Concentrations of rinsing liquids	
Rinse times	
Sample concentration[a] and composition	
Sample injection time	
Voltage	

[a]Comment: See text.
VIS = visible; UV = ultraviolet.

concentration has an influence on such responses. On the other hand, in robustness testing, in a first instance, quantitative responses are considered, and in this case one will thus evaluate the influence of the sample concentration on the sample concentration, which does not seem to be a good idea.

Another division of the factors can be made into mixture-related, quantitative (continuous), or qualitative (discrete) factors (4, 5, 16, 18, 24). A mixture-related factor in CE is usually related to a mixture of solvents, for example, the composition of the background electrolyte solution. A quantitative factor can vary on a continuous scale, for example, the buffer pH, the electrolyte concentration, the additive concentration, the capillary temperature, or the voltage. A qualitative factor, on the other hand, varies on a discrete nominal scale, for example, batch or manufacturer of a reagent, solvent, or capillary.

2.3.1.1. Mixture-Related Factors. Mobile phases in chromatography and electrolyte systems in electrophoresis are examples of frequently used solvent mixtures. A property of mixtures is that in a mixture of p components, only $p - 1$ components can be varied independently. Thus maximally $p - 1$ mixture-related variables can be examined in the types of experimental designs considered in this chapter. The value of the pth variable is determined by those of the other variables and used as adjusting component to complete the mixture. If one of the mixture components has an important effect on a response, then the composition of the whole mixture is important and should be strictly controlled (5, 18). To examine only mixture-related factors, so-called mixture designs are applied (1, 7, 17). However, they are not used in the optimization or robustness testing of CE methods.

Suppose the electrolyte system in CE consists of methanol/buffer 5:95 (V/V). When the methanol fraction is selected as factor to be varied in an experimental design approach, the buffer fraction will be used as adjusting component to sum the fractions to one.

2.3.1.2. Quantitative Factors. Quantitative factors are most often evaluated. They usually are described in the operating procedure and are used as such in the design, for example, capillary temperature. However, sometimes the selected factors by themselves do not represent a physicochemical property. These factors should preferably be defined in such a way that the effects can be linked to a physicochemical property. The following example illustrates this. A buffer can be defined either by the concentrations of its acid (C_a) and basic (C_b) compounds or by a given pH and ionic strength μ (5). The individual effects of C_a and C_b, that is, when considering the concentrations as factors, do not directly represent physically interpretable properties, and the significance of one factor (C_a or C_b) in a robustness test should lead to a strict control of both, as for mixture-related variables. They are linked and the variation of either one or both (C_a and/or C_b) might affect the properties of the background electrolyte, resulting in, for instance, a change in pH. To relate C_a and C_b in the definition of factors to pH and μ, they are combined, $\frac{C_a}{C_b}$, so that their effect corresponds to a change in pH and/or ionic strength μ (5). The latter approach might be preferred because it gives the analyst a better link between the physicochemical property and its estimated effect.

2.3.1.3. Qualitative Factors. For CE methods, also qualitative factors, such as the batch or manufacturer of the capillary, reagent, or solvent, can be selected. However, during method development, such factors are not frequently examined. Usually, initially a fused-silica capillary is selected, and only when for some reason the electroosmotic flow should be modified or the selectivity should be altered, a coated capillary can be used instead (20, 22). In the first phase of method development, where screening designs are applied, qualitative factors could, in principle, be included in the design. On the other hand, in the optimization phase, in the response surface designs, they cannot. The response, measured at the conditions defined by the design, is modeled as a function of the examined factors, in order to determine the (intermediate) optimum conditions. However, modeling a qualitative factor has no meaning because only discrete levels are possible and no intermediate values occur. Therefore, only mixture-related and quantitative factors are examined in the optimization step. Sequential optimization methods select successive experiments in the factor domain, which implies that again only mixture-related and quantitative factors can be examined.

On the other hand, qualitative factors are rather frequently considered in a robustness test. When evaluating the influence of such factors, the analyst should be aware that the estimated effects are only representative for the examined discrete levels and not for any other level of those factors, and certainly not for the whole population (4, 5). For example, when examining two capillaries, X and Y, then the estimated effect only allows drawing conclusions about these two capillaries and not about other capillaries available on the market. Such approach allows evaluation of whether capillary Y is an alternative for capillary X, used, for instance, to develop the method.

One also should be careful not to create situations that cannot be handled in the designs used (4, 5). For instance, the factors manufacturer and batch cannot be considered together. The designs used are two-level designs (see further), and it is impossible to define two levels for manufacturer and also two for batch in such a way that the two batches belong at the same time to both manufacturers.

2.3.2. Selection of Levels

In the screening phase of method development and in robustness tests, the factors usually are examined at two levels (−1, +1). On the other hand, in the response surface designs, applied in method optimization, the factors are examined at three or more levels, depending on the applied design (see further).

In method optimization, the range between the levels is much larger than in robustness tests. Often, the range selected for a factor in optimization represents the broadest interval in which the factor can be varied with the technique considered. In practice, the examined range is chosen based on earlier gathered knowledge and/or information from the literature.

In robustness tests, the selected range between the levels should represent the variability that can occur when transferring the method (4, 5, 16, 18, 25). However, specifications to estimate such variability are not given in regulatory documents, such as the ICH guidelines. Often the extreme levels are chosen based on personal experience, knowledge, or intuition. Sometimes they are defined as "nominal level $\pm x\%$." However, this approach based on relative variation is not appropriate because the absolute variation then depends on the value of the nominal level (18). Another systematic approach defines the levels based on the precision or the uncertainty with which they can be set (5, 18). The uncertainty can be estimated for the nominal factor level (18, 26). If the uncertainty or absolute error on a measured pH value is 0.01, this means that the true pH value is situated in the interval "measured pH ± 0.01" with 95% certainty. To define the extreme levels, the above interval is extended to simulate potential variability caused by transferring the method between instruments or laboratories, as well as to compensate for potential sources of variability that were neglected during the estimation of the uncertainty. For this purpose, the uncertainty is multiplied with a constant k, chosen arbitrarily, and usually $2 \leq k \leq 10$. Thus, the extreme factor levels are given by "nominal level ± k*uncertainty" (5, 18). The minimal k value should be 2 to enable a distinction between the factor levels, and often $k = 5$ is used as default value. The lower the k value for a factor, the smaller the examined interval, and the stricter that factor is to be controlled during later use, because only robustness in the narrow interval is verified. On the other hand, a higher k value increases the probability that a significant effect occurs in the examined interval, but allows a less strict control of the factor if no important effect is observed. Examples of the latter approach to select factor levels can be found in References 5 and 18.

In robustness testing, the extreme levels are most frequently chosen symmetrically around the nominal for mixture-related and quantitative factors. However, for some factors, an asymmetric interval might better represent the reality or better reflect the change in response occurring. A first example is the capillary temperature. Suppose a capillary temperature of 15 °C is prescribed. Symmetric levels, selected based on uncertainty are, for instance, 10 °C and 20 °C. However, many cooling systems do not allow temperatures of more than 10 °C below room temperature; therefore, 10 °C may not be attained accurately by the instrument. The lowest extreme level could then be taken equal to the nominal (15 °C).

A second example is the detection wavelength. Suppose a signal is measured at the maximum absorbance wavelength, λ_{max} or $\lambda_{nom,1}$ (see Figure 2.4). A small decrease in detection wavelength then often has a similar effect on the response as a small increase. This leads to an estimated effect, $E_{nom,1}$, close to zero, when evaluating the change between extreme levels chosen symmetrically around the nominal. Examining an asymmetric interval better reflects the change in response, and often one extreme level and the nominal are considered in the robustness test. On the other hand, when the nominal wavelength is in a slope of the spectrum, $\lambda_{nom,2}$ (see Figure 2.4), then a symmetric interval seems best because the response is continuously increasing or decreasing as a function of the factor levels, resulting in an effect estimation, $E_{nom,2}$, clearly representing the change in response.

For qualitative factors, only discrete values are possible, for example, capillaries X, Y, or Z. As already indicated, this means that only conclusions can be drawn about the examined capillaries and no extrapolation to other capillaries can be made. Most logic in a robustness test is to compare the nominal capillary with an alternative. Including two capillaries different from the nominal does not make sense because comparison with the nominal situation is no longer considered.

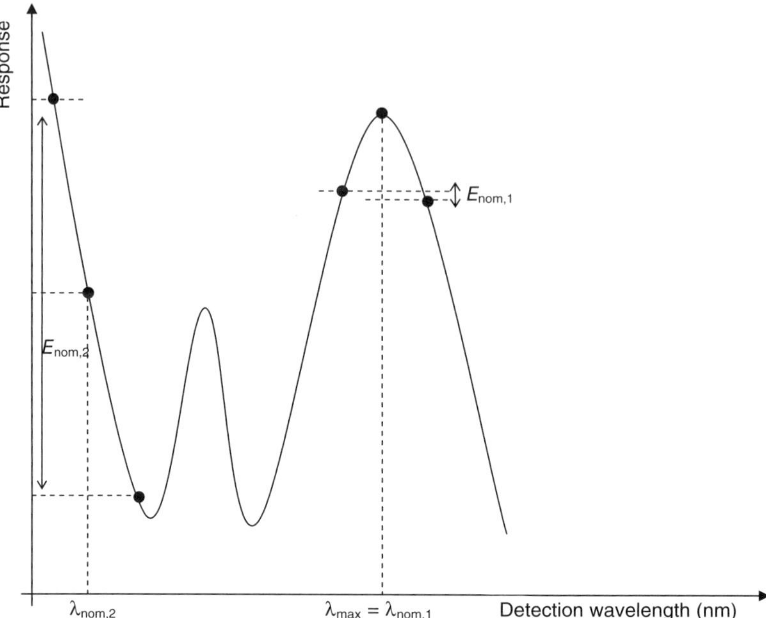

FIGURE 2.4. Response (e.g., signal intensity or absorbance) as a function of detection wavelength.

TABLE 2.2. Factors and their levels investigated during the screening phase in the development of a CE method to separate pronucleotide diastereoisomers of 3′-azido-2′,3′-dideoxythymidine in biological samples (27)

Factor	Levels	
	−1	+1
(A) Chiral additive concentration [CM-β-CD] (mM)	5	15
(B) Buffer concentration (mM)	50	100
(C) Percent MeOH (V/V)	0	10
(D) Injected volume (nL)	4.64	12.38
(E) Capillary length (cm)	31.2	51.2
(F) Voltage (V/cm)	0.50	0.80
(G) Capillary temperature (°C)	15	25

(−1) and (+1) = extreme levels.
CM-β-CD = carboxymethyl-β-cyclodextrine.

2.3.3. Examples of Factors and Their Levels from Some Case Studies

The factors and their levels examined during a screening phase in method development (27), an optimization phase in method development (28), and a robustness test (29) are presented in Tables 2.2, 2.3, and 2.4, respectively.

During a screening phase in method development, the seven factors in Table 2.2 were selected to develop a CE method to separate pronucleotide

TABLE 2.3. Factors and their levels investigated during the optimization phase in the development of a chiral enantioseparation method for a nonsteroidal anti-inflammatory drug (28)

Factor	Levels				
	−1.6818	−1	0	+1	+1.6818
(A) Chiral selector concentration (mM)	1	2.8	5.5	8.2	10
(B) pH	4	4.6	5.5	6.4	7
(C) Capillary temperature (°C)	14.9	18	22.5	27	30.1

TABLE 2.4. Factors and their levels investigated in a robustness test on a CE method to determine rufloxacin hydrochloride in coated tablets (29)

Factor	Levels		
	−1	0	+1
(A) Capillary temperature (°C)	26	27	28
(B) Voltage (kV)	17	18	19
(C) BGE concentration (M)	0.09	0.10	0.11
(D) pH	8.7	8.8	8.9

(−1) and (+1) = extreme levels.
BGE = background electrolyte.

diastereoisomers of 3′-azido-2′,3′-dideoxythymidine in biological samples (27). The examined factors were mixture-related (C) or quantitative (A, B, and D–G).

During an optimization phase in method development, the three factors in Table 2.3 were selected to develop the enantioseparation of a nonsteroidal anti-inflammatory drug (28). All examined factors were quantitative (A–C).

The four factors in Table 2.4 were selected from a robustness test on a CE method to determine rufloxacin hydrochloride in coated tablets (29). All factors were quantitative (A–D) and their extreme levels are situated symmetrically around the nominal.

2.4. TYPES OF EXPERIMENTAL DESIGNS

An experimental design is an experimental setup that allows a number of factors in a predefined number of experiments to be studied simultaneously. Several types of experimental designs are described in the literature. During the screening in method development and in robustness testing, so-called screening designs are most frequently used, while during the optimization phase, response surface designs or sequential optimization methods are applied.

2.4.1. Screening Designs

The aim of applying screening designs is to estimate the effect of the examined factors on the considered response(s) in order to determine the most important. Two-level screening designs (1, 4, 5, 17, 30, 31), such as fractional factorial (FF) or Plackett–Burman (PB) designs, are most often applied. Such designs allow evaluation of a relatively large number of factors f at $L = 2$ levels in a relatively small number of experiments ($N \geq f + 1$). The number of experiments required depends on the number of factors to be examined. In the literature, several FF and PB designs are described that allow including different numbers of factors. Also for a given number of factors, different designs, which differ in design properties and number of experiments, exist. The designs most frequently applied in separation science usually require the execution of 8, 12, or 16 experiments. The designs can be constructed manually by the analyst based on literature information (4, 5, 17) or by using (commercial) software packages (32–41).

2.4.1.1. Two-Level Full Factorial Designs. A two-level full factorial design contains all possible combinations between the f factors and their $L = 2$ levels. The number of experiments is $N = L^f = 2^f$. For example, to examine three factors, the full factorial design requires $N = 2^3$ experiments (Table 2.5). This design allows all main factor effects (E_A, E_B, E_C) and all interaction effects between the factors (E_{AB}, E_{AC}, E_{BC}, E_{ABC}) (see further) (1, 7, 17) to be estimated.

Occasionally, two-level full factorial designs are applied for screening purposes during method development (42, 43) or in robustness testing (44, 45) when the number of factors is low, that is, usually not more than four. For more factors, the required number of experiments is, in general, considered unfeasibly high because it increases exponentially. For example, to examine five factors with a two-level full factorial design, already $N = 2^5 = 32$ experiments need to be performed.

TABLE 2.5. Two-level full factorial design for three factors, and columns of contrast coefficients for the interactions

Experiment	Factors			Contrast Coefficients			
	A	B	C	AB	AC	BC	ABC
1	−1	−1	−1	1	1	1	−1
2	1	−1	−1	−1	−1	1	1
3	−1	1	−1	−1	1	−1	1
4	1	1	−1	1	−1	−1	−1
5	−1	−1	1	1	−1	−1	1
6	1	−1	1	−1	1	−1	−1
7	−1	1	1	−1	−1	1	−1
8	1	1	1	1	1	1	1

In some cases, a four-factor two-level full factorial design was used in optimization. Rarely, also 2^5 and 2^6 full factorial designs were applied for optimization purposes in the literature. Such designs are not recommended because of the large number of experiments required, that is, 32 and 64, respectively. The above full factorial designs examine the factors at two levels and allow only all main and interaction effects to be estimated but not quadratic effects; that is, they do not allow modeling of curvature. An intermediate optimum cannot be found because curvature in the response cannot be modeled from two-level design results.

2.4.1.2. Two-Level FF Designs. A two-level FF design contains only a fraction of the experiments from the full factorial design. In general, a two-level $2^{(f-v)}$ FF design examines f factors at two levels in $N = 2^{(f-v)}$ experiments, with $\frac{1}{2^v}$ representing the fraction of the full factorial ($v = 1, 2, 3, \ldots$) (1, 4, 5, 7, 17). In practice, half-fraction, quarter-fraction, eight-fraction, and even sixteenth-fraction factorial designs are frequently used in screening and robustness testing. The fact that a given number of factors are examined in a fraction of the number of experiments required by a full factorial has consequences regarding the information obtained. From an FF design, not all main and interaction effects can be individually estimated. Some effects are estimated together in a given design. It is said that these effects are confounded in that design.

The construction of FF designs has been thoroughly described in the literature; for more detailed information, refer to References 4, 5, and 17. To examine a given number of factors, different FF designs can be selected. These designs can either represent different fractions of the full factorial, or these designs can represent the same fraction of the full factorial design, but be constructed differently. All these designs differ in their so-called confounding pattern, that is, the different effects that are estimated together. For example, to examine five factors, a half-fraction factorial design requiring $2^{(5-1)} = 16$ experiments (Table 2.6) or a quarter-fraction factorial design with only $2^{(5-2)} = 8$ experiments (Table 2.7) is possible. Different $2^{(5-1)}$ and $2^{(5-2)}$ can be constructed, with different properties and confounding patterns, but we consider the discussion on their detailed differences outside the scope of this chapter. From the $2^{(5-1)}$ design, each estimated effect is a confounding of two effects, while from the $2^{(5-2)}$ design it is of four effects. However, in FF designs no confounding among the main effects occurs. The smallest fraction for which this does not occur is called a saturated FF design. In robustness testing, the interaction effects are considered negligible. Therefore, their estimated effects can be considered as a measure for the experimental error and used in the statistical evaluation of the estimated effects (see further). FF designs have been used for screening purposes during method development of CE methods in References 46–51 and during their robustness testing in References 52–54.

TABLE 2.6. A $2^{(5-1)}$ fractional factorial design

Experiment	Factors				
	A	B	C	D	E
1	–1	–1	–1	–1	1
2	1	–1	–1	–1	–1
3	–1	1	–1	–1	–1
4	1	1	–1	–1	1
5	–1	–1	1	–1	–1
6	1	–1	1	–1	1
7	–1	1	1	–1	1
8	1	1	1	–1	–1
9	–1	–1	–1	1	–1
10	1	–1	–1	1	1
11	–1	1	–1	1	1
12	1	1	–1	1	–1
13	–1	–1	1	1	1
14	1	–1	1	1	–1
15	–1	1	1	1	–1
16	1	1	1	1	1

TABLE 2.7. A $2^{(5-2)}$ fractional factorial design

Experiment	Factors				
	A	B	C	D	E
1	–1	–1	–1	1	1
2	1	–1	–1	–1	–1
3	–1	1	–1	–1	1
4	1	1	–1	1	–1
5	–1	–1	1	1	–1
6	1	–1	1	–1	1
7	–1	1	1	–1	–1
8	1	1	1	1	1

2.4.1.3. Two-Level PB Designs. PB designs are saturated factorial designs that allow examination of up to $N - 1$ factors in N (a multiple of four) experiments (4, 5, 7, 17, 55). PB designs are constructed by performing $N - 2$ cyclic permutations of the first row of the design, which is defined by Plackett and Burman (55), followed by adding a final row of –1 signs (see Tables 2.8 and 2.9). This construction has been thoroughly described in the literature (4, 5, 17). To examine a given number of factors, again different PB designs can be used. These designs differ in their dimensions and confounding patterns, although it is inherent in PB designs that both two-factor and higher-order interaction effects are confounded with the main effects (4, 5). For example,

TABLE 2.8. Plackett–Burman design to examine up to 11 factors in 12 experiments

Experiment	Factors										
	A	B	C	D	E	F	G	H	I	J	K
1	1	1	−1	1	1	1	−1	−1	−1	1	−1
2	−1	1	1	−1	1	1	1	−1	−1	−1	1
3	1	−1	1	1	−1	1	1	1	−1	−1	−1
4	−1	1	−1	1	1	−1	1	1	1	−1	−1
5	−1	−1	1	−1	1	1	−1	1	1	1	−1
6	−1	−1	−1	1	−1	1	1	−1	1	1	1
7	1	−1	−1	−1	1	−1	1	1	−1	1	1
8	1	1	−1	−1	−1	1	−1	1	1	−1	1
9	1	1	1	−1	−1	−1	1	−1	1	1	−1
10	−1	1	1	1	−1	−1	−1	1	−1	1	1
11	1	−1	1	1	1	−1	−1	−1	1	−1	1
12	−1	−1	−1	−1	−1	−1	−1	−1	−1	−1	−1

TABLE 2.9. Plackett–Burman design to examine up to seven factors in eight experiments

Experiment	Factors						
	A	B	C	D	E	F	G
1	1	1	1	−1	1	−1	−1
2	−1	1	1	1	−1	1	−1
3	−1	−1	1	1	1	−1	1
4	1	−1	−1	1	1	1	−1
5	−1	1	−1	−1	1	1	1
6	1	−1	1	−1	−1	1	1
7	1	1	−1	1	−1	−1	1
8	−1	−1	−1	−1	−1	−1	−1

to examine 5 factors, a PB with 12 experiments (Table 2.8), or one with only 8 experiments (Table 2.9), can be chosen. The selection of the larger design is then made to allow a given statistical interpretation of the effects (see further).

When the number of factors to be examined is lower than the number of factors that potentially can be examined in a PB design ($N - 1$), the remaining columns are defined as so-called dummy factors. A dummy factor is an imaginary variable and changing its levels does not correspond to any physical or chemical change. Therefore, its estimated effect can be considered as a measure for experimental error and used in the statistical evaluation of the estimated factor effects (see further). PB designs have been used for screening purposes during method development of CE methods in References 27, 56, and 57 and during their robustness testing in References 29, 56, and 58–62.

2.4.1.4. Three-Level Screening Designs. Although usually two-level screening designs are applied, occasionally it might be worthwhile investigating the factors at three levels, for example, in cases where it is expected that the effects between –1 and 0 considerably differ from those between 0 and +1. For example, this occurs when the response plotted as a function of the factor levels goes through an optimum in the interval [–1, +1], for example, the maximum absorbance wavelength for the factor detection wavelength (Figure 2.4). For such factor, it can be expected that in the region [–1, 0] the response will increase, while it will decrease in the region [0, 1]. In such situation, it is thus more informative to examine the factor at three levels (–1, 0, +1). When screening at only the extreme levels (–1, +1), the intermediate optimum is ignored.

A possible way to screen the factor(s) at three levels is by using so-called reflected designs (4, 32, 63–65). Reflected designs are duplicated two-level full factorial, FF, or PB designs. The latter designs are executed once with the factor levels (–1, 0) and once with (0, +1). As there is one common experiment (all factors at 0 level), this results in a reflected design examining f factors in $2N-1$ experiments. For a given number of factors, several reflected designs can be chosen. For example, to examine seven factors, a reflected PB design with 15 experiments (Table 2.10), or one with 23 experiments (Table 2.11), can be used. Reflected FF designs also can be constructed for this situation.

Reflected FF and PB designs were applied during robustness testing of CE methods in References 66–71. To screen the factor(s) at three levels, three-

TABLE 2.10. Reflected Plackett–Burman design to examine up to seven factors at three levels in 15 experiments

Experiment	Factors						
	A	B	C	D	E	F	G
1	1	1	1	0	1	0	0
2	0	1	1	1	0	1	0
3	0	0	1	1	1	0	1
4	1	0	0	1	1	1	0
5	0	1	0	0	1	1	1
6	1	0	1	0	0	1	1
7	1	1	0	1	0	0	1
8	0	0	0	0	0	0	0
9	–1	–1	–1	0	–1	0	0
10	0	–1	–1	–1	0	–1	0
11	0	0	–1	–1	–1	0	–1
12	–1	0	0	–1	–1	–1	0
13	0	–1	0	0	–1	–1	–1
14	–1	0	–1	0	0	–1	–1
15	–1	–1	0	–1	0	0	–1

TABLE 2.11. Reflected Plackett–Burman design to examine up to 11 factors at three levels in 23 experiments

Experiment	Factors										
	A	B	C	D	E	F	G	H	I	J	K
1	1	1	0	1	1	1	0	0	0	1	0
2	0	1	1	0	1	1	1	0	0	0	1
3	1	0	1	1	0	1	1	1	0	0	0
4	0	1	0	1	1	0	1	1	1	0	0
5	0	0	1	0	1	1	0	1	1	1	0
6	0	0	0	1	0	1	1	0	1	1	1
7	1	0	0	0	1	0	1	1	0	1	1
8	1	1	0	0	0	1	0	1	1	0	1
9	1	1	1	0	0	0	1	0	1	1	0
10	0	1	1	1	0	0	0	1	0	1	1
11	1	0	1	1	1	0	0	0	1	0	1
12	0	0	0	0	0	0	0	0	0	0	0
13	–1	–1	0	–1	–1	–1	0	0	0	–1	0
14	0	–1	–1	0	–1	–1	–1	0	0	0	–1
15	–1	0	–1	–1	0	–1	–1	–1	0	0	0
16	0	–1	0	–1	–1	0	–1	–1	–1	0	0
17	0	0	–1	0	–1	–1	0	–1	–1	–1	0
18	0	0	0	–1	0	–1	–1	0	–1	–1	–1
19	–1	0	0	0	–1	0	–1	–1	0	–1	–1
20	–1	–1	0	0	0	–1	0	–1	–1	0	–1
21	–1	–1	–1	0	0	0	–1	0	–1	–1	0
22	0	–1	–1	–1	0	0	0	–1	0	–1	–1
23	–1	0	–1	–1	–1	0	0	0	–1	0	–1

level PB designs were proposed in Reference 55. However, because these designs show a confounding of the main effects (65), they are useless. However, from the three-level designs proposed by Plackett and Burman, well-balanced three-level designs, that is, without confounded main effects, were constructed (65). A drawback is that only few designs requiring a feasible number of experiments are described, which also explains why they are not so frequently used (72). In References 17 and 73–75, so-called asymmetrical or mixed-level factorial designs were described or applied to screen different factors at different numbers of levels.

2.4.1.5. Examples of Applied Screening Designs. The applied screening design in the development of a CE method to separate pronucleotide diastereoisomers (27) was a 12-experiment PB design (Table 2.8). As the effects of seven selected factors (Table 2.2) were examined in 12 experiments, four dummies were included in the design (columns H–K).

The applied screening design in the robustness testing of a CE method to determine rufloxacin hydrochloride (29) was an 8-experiment PB design (Table 2.9). As the effects of four selected factors (Table 2.4) were examined in eight experiments, three dummies were included in the design (columns E–G). However, the four factors (Table 2.4) could as well have been examined using an 8-experiment $2^{(4-1)}$ FF design, for instance, as shown in Table 2.12. From the latter design, three interaction effects can be estimated.

2.4.2. Response Surface Designs

Three-level or more-level response surface designs, such as three-level full factorial, central composite (CCD), Box–Behnken, and Doehlert designs, have been selected for screening purposes in some publications, although they should not have been. These designs require many more experiments than the screening designs to examine a given number of factors. For example, to examine three factors, a three-level full factorial design requires $3^3 = 27$ experiments and a CCD of at least 15, while two-level screening designs with eight (FF or PB) or, theoretically, even four experiments (FF) (7) can be chosen. Moreover, during screening, usually (much) more than three factors are evaluated. Using three-level screening designs, such as reflected designs, up to seven factors can be examined in 15 experiments. Now, when more than three factors are examined, the number of experiments increases dramatically when response surface designs would be used. Moreover, analysis of the results of a response surface design is focused on building a mathematical model and the corresponding response surface, and much less on the estimation of the individual factor effects. It can be stated that response surface designs offer too much information when screening is done, and require too many experiments for the number of factors usually considered in screening.

In the optimization of method development, the main goal is to define (the best) experimental conditions that allow a sufficient resolution of the relevant peaks, and that provide robust results in an acceptable analysis time. The

TABLE 2.12. A $2^{(4-1)}$ fractional factorial design, and columns of contrast coefficients

Experiment	Factors				Contrast Coefficients		
	A	B	C	D	I_1	I_2	I_3
1	−1	−1	−1	−1	1	1	1
2	1	−1	−1	1	−1	−1	1
3	−1	1	−1	1	−1	1	−1
4	1	1	−1	−1	1	−1	−1
5	−1	−1	1	1	1	−1	−1
6	1	−1	1	−1	−1	1	−1
7	−1	1	1	−1	−1	−1	1
8	1	1	1	1	1	1	1

factors considered most important from the screening phase are further examined in this step.

In this method optimization phase, response surface designs or sequential optimization methods are applied. The main difference between the two is that for a response surface design the experimental domain enclosed by the design is expected to contain the optimum, while a sequential optimization method can be applied in situations where the experimental region containing the optimal result is not *a priori* known. Another difference is that the sequential methods allow optimization of only one response, while with response surface designs several responses can be considered simultaneously (see further).

In general, as already mentioned, the results from a response surface design are used to build a model, relating the response y to the considered x variable(s). In practice, response surface designs examine only a limited number of factors, that is, usually two or three important factors are evaluated. Response surface designs require at least three levels for each factor to enable modeling curvature in the response. The most frequently applied model is a quadratic polynomial. More information concerning the modeling is discussed later.

Response surface designs can be divided into symmetrical and asymmetrical designs (7). The first type examines the factors in a symmetrical experimental domain, while the second can be chosen when an asymmetrical experimental domain is to be examined.

2.4.2.1. Symmetrical Experimental Domain. The symmetrical experimental domain formed by the design experiments is sometimes (hyper)cubic, but usually (hyper)spherical (7). Examples of cubic designs are the three-level full factorial designs and the face-centered CCDs. Examples of spherical designs are the circumscribed CCDs, the Box–Behnken designs, and the Doehlert designs. These designs are discussed in more detail later.

Besides the design experiments, frequently additional experiments are performed (7). For example, to estimate the experimental error, the center point or one or several design experiment(s) can be replicated. To evaluate the prediction performance, additional points, different from the experimental design points, for example, the predicted optimum, can be measured.

2.4.2.1.1. Three-Level Full Factorial Designs. A three-level full factorial design contains all possible combinations between the f factors and their levels $L = 3$, and the number of experiments thus is $N = L^f = 3^f$. These three-level designs allow the coefficients of all factors, interactions, and quadratic terms to be estimated (1, 7, 17). An example of a three-level full factorial design to examine two factors in nine experiments is given in Table 2.13. Three-level full factorial designs have been used to optimize CE methods in References 76–79.

However, when the number of factors f increases, the number of required experiments N increases dramatically. Even for only three factors, already 27 experiments are to be executed. Therefore, these designs are not so frequently applied.

TABLE 2.13. Three-level full factorial design for two factors

Experiment	Factor	
	A	B
1	−1	−1
2	−1	0
3	−1	1
4	0	−1
5	0	0
6	0	1
7	1	−1
8	1	0
9	1	1

2.4.2.1.2. Central Composite Designs. CCDs are the most often used response surface designs (1, 7, 17). These designs are constructed by combining a two-level full factorial design (2^f experiments), a star design ($2f$ experiments), and a center point, which is often replicated a number of times. Thus, to examine f factors, at least $N = 2^f + 2f + 1$ experiments are required. For more than two factors, these designs are thus more economical in experiments and less time-consuming than the three-level full factorial designs.

The points of the full factorial design are situated at levels −1 and +1, those of the star design at levels $-\alpha$ and $+\alpha$, and the center point at level 0 (Figure 2.5). Depending on the α value, three types of CCDs are distinguished. An inscribed CCD has $|\alpha| < 1$, a face-centered CCD (FCCD) $|\alpha| = 1$, and a circumscribed CCD (CCCD) $|\alpha| > 1$. Usually the factors are examined at five levels ($-\alpha$, −1, 0, +1, $+\alpha$) and a CCCD is most often applied. Occasionally, an FCCD is used. In the latter design, the factors are varied at only three levels (−1, 0, +1).

To obtain a so-called rotatable circumscribed CCD, the levels of the star design ($-\alpha$, $+\alpha$) should fulfill the requirement $|\alpha| = (2^f)^{1/4}$. Then all experiments, except the center point, are situated on a circle or (hyper)sphere. Therefore, $|\alpha|$ is equal to 1.41, 1.68, 2.00, 2.38, and 2.83, for 2, 3, 4, 5, and 6 factors, respectively (7). As mentioned above, the center point is often replicated to evaluate experimental precision. In general, usually 3–5 center point replicates are performed.

In Table 2.14, an example is given of a CCD for three factors (at least 15 experiments). In Figure 2.5, this is graphically represented for a rotatable circumscribed CCD for three factors. CCDs are also frequently used during optimization of CE methods (28, 42, 43, 46, 48–51, 57, 80).

2.4.2.1.3. Box–Behnken Designs. As an alternative to CCDs, Box–Behnken designs can be applied (1, 7, 17, 81, 82). Box–Behnken designs are spherical. For three factors, minimally 13 experiments are required (Figure 2.6). This

TABLE 2.14. Central composite design for three factors

Experiment	Factors		
	A	B	C
1	−1	−1	−1
2	1	−1	−1
3	−1	1	−1
4	1	1	−1
5	−1	−1	1
6	1	−1	1
7	−1	1	1
8	1	1	1
9	−α	0	0
10	+α	0	0
11	0	−α	0
12	0	+α	0
13	0	0	−α
14	0	0	+α
15, etc.	0	0	0

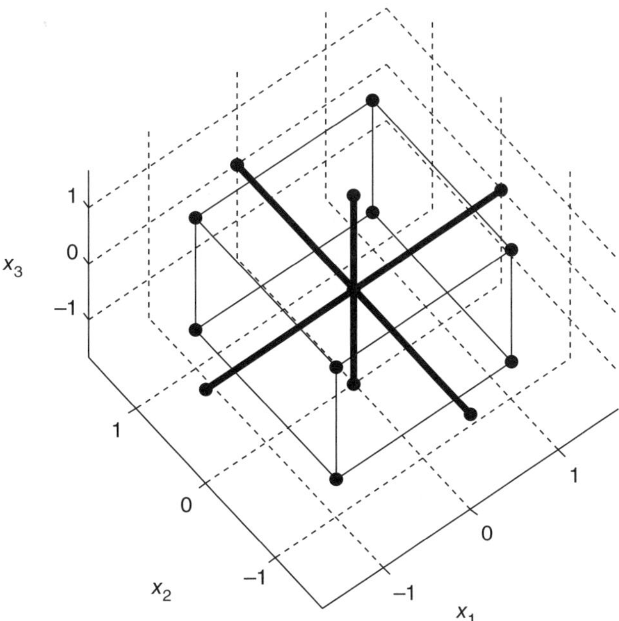

FIGURE 2.5. Circumscribed central composite design for three factors (at least 15 experiments).

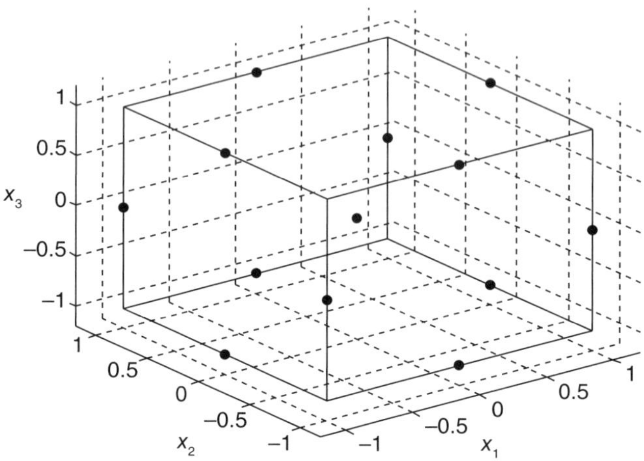

FIGURE 2.6. Box–Behnken design for three factors (at least 13 experiments).

TABLE 2.15. Box–Behnken design for three factors

Experiment	Factors		
	A	B	C
1	1	1	0
2	1	−1	0
3	−1	1	0
4	−1	−1	0
5	1	0	1
6	1	0	−1
7	−1	0	1
8	−1	0	−1
9	0	1	1
10	0	1	−1
11	0	−1	1
12	0	−1	−1
13, etc.	0	0	0

design is also the most frequently used Box–Behnken design. Concerning the required number of experiments, these designs are comparable to CCDs.

In Table 2.15, an example is given of a Box–Behnken design for three factors (at least 13 experiments). In Figure 2.6, the design is graphically represented. The Box–Behnken design consists of the middle points of the cube edges and the center point. As for the CCDs, this center point can be replicated. Because the design is spherical, part of the cubic domain is not covered by the model. Thus, predictions in these parts are obtained by extrapolation and should be interpreted with caution. Box–Behnken designs were applied during optimization of CE methods in References 83–87.

2.4.2.1.4. Doehlert Designs. Somewhat less known, but also useful response surface designs, are the Doehlert (uniform shell) designs (1, 7, 88). These designs also are spherical. The experiments are defined in such a way that uniformity in space filling is obtained. Thus, the distances between all neighboring experiments are equal. The Doehlert design for two factors consists of six points (vertices of a hexagon) with a center point (Figure 2.7), while for three factors it consists of a centered dodecahedron (Figure 2.8). The center point again can be replicated. Concerning the number of experiments to

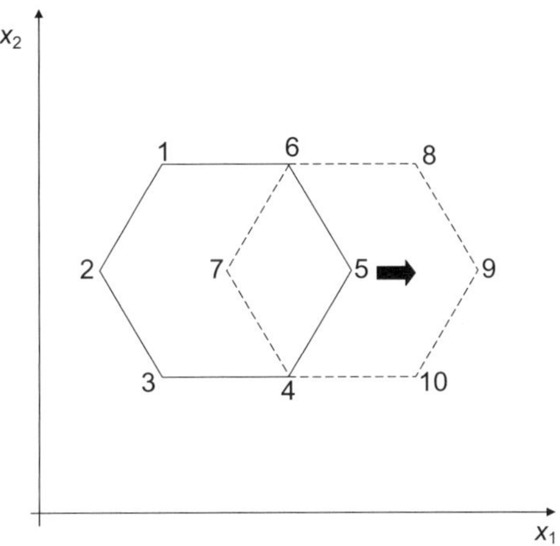

FIGURE 2.7. Doehlert design for two factors (at least seven experiments). Dotted line: possibility for sequentially moving the design in the direction of the arrow by executing three additional experiments.

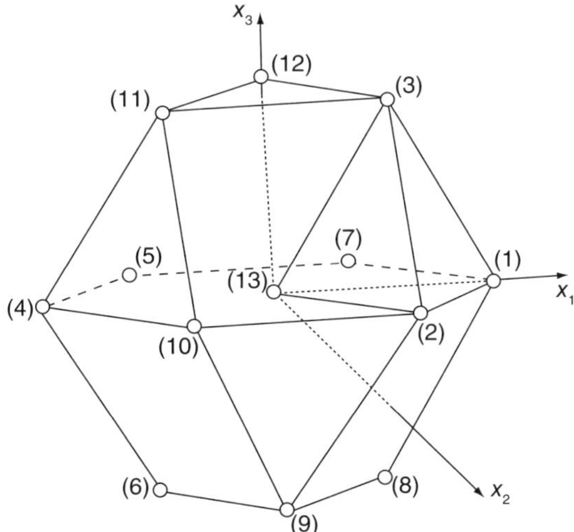

FIGURE 2.8. Doehlert design for three factors (at least 13 experiments).

examine a given number of factors, the Doehlert designs are more economical than CCDs.

In Table 2.16, Doehlert designs for two and three factors (at least 7 and 13 experiments, respectively) are given. In Figures 2.7 and 2.8, these designs are graphically represented. Contrary to the above response surface designs, the factors are varied at different numbers of levels in a Doehlert design, for example, one at three levels and one at five in the design for two factors.

An advantage of Doehlert designs is their potential for sequentiality. Suppose a Doehlert design for two factors was performed (e.g., points 1–7 in Figure 2.7). When further optimization would be needed in the direction of the arrow on Figure 2.7, four points from the initial Doehlert design (points 4–7) can be kept, and performing only three new experiments (points 8–10 in Figure 2.7) suffices to create a new Doehlert design. The same reasoning is

TABLE 2.16. Doehlert designs for (a) two and (b) three factors

(a)

Experiment	Factors	
	A	B
1	–0.5	0.866
2	–1	0
3	–0.5	–0.866
4	0.5	–0.866
5	1	0
6	0.5	0.866
7, etc.	0	0

(b)

Experiment	Factors		
	A	B	C
1	1	0	0
2	0.5	0.866	0
3	0.5	0.289	0.816
4	–1	0	0
5	–0.5	–0.866	0
6	–0.5	–0.289	–0.816
7	0.5	–0.866	0
8	0.5	–0.289	–0.816
9	0	0.577	–0.816
10	–0.5	0.866	0
11	–0.5	0.289	0.816
12	0	–0.577	0.816
13, etc.	0	0	0

valid in all directions. Doehlert designs were applied during optimization of CE methods in References 29, 47, 56, and 89. They were applied sequentially to optimize a spectrofluorimetric method in Reference 90.

2.4.2.2. Asymmetrical Experimental Domain. Sometimes irregular experimental domains can be found in both chromatography (91) and electrophoresis (92). In chromatography, for example, when optimizing pH and percentage organic modifier in the mobile phase, it is possible to determine with a few experiments the (asymmetric) area in which suitable retention (e.g., $1 < \text{retention factor } k < 10$) will occur for all compounds (91). In electrophoresis, for instance, when optimizing pH and concentration of sodium dodecylsulphate in the electrolyte, the area in which suitable migration and acceptable peak shapes occur for all compounds can be irregular (92). In Figure 2.9, an example is shown of a feasible experimental region obtained based on the migration behavior of three compounds. If the resulting area is irregular, it is recommended to use a nonsymmetrical design, which will cover the domain better than the symmetrical response surface designs do.

Also in situations where it is in practice impossible to perform one or more of the planned experiments from a symmetrical response surface design, irregular experimental areas remain and are to be explored. A situation similar to Figure 2.10a (see further) is obtained. For example, when considering the variables pH and percentage organic modifier in the mobile phase or the background electrolyte, it can happen that one of the compounds to be analyzed does not dissolve anymore and/or that conditions are created where no elution occurs.

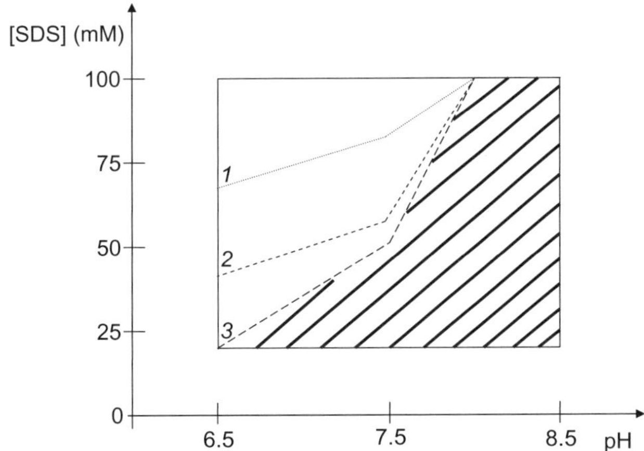

FIGURE 2.9. Migration boundary map obtained by the migration behavior of three components. The feasible experimental region is indicated. SDS = sodium dodecylsulphate.

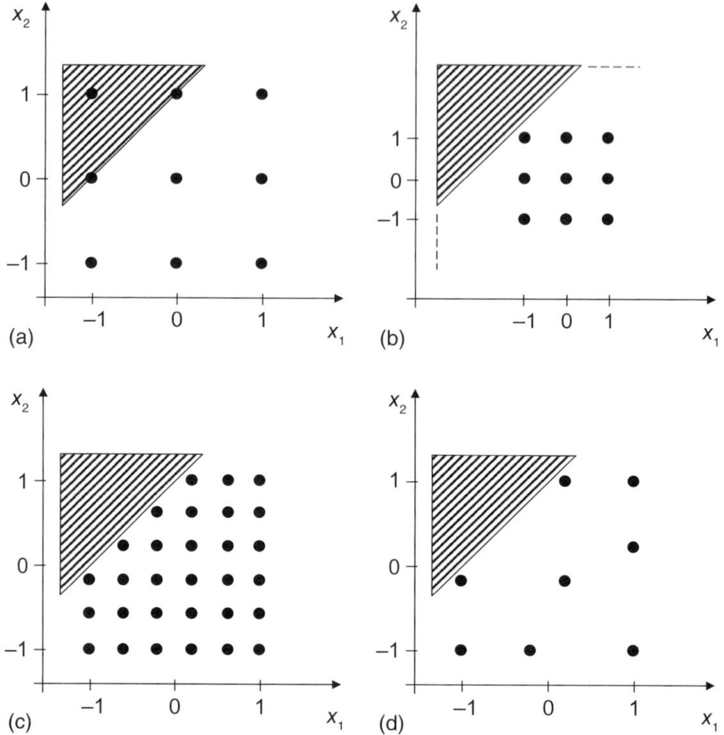

FIGURE 2.10. Mapping properties (●) of (a) a 3^2 full factorial design in a rectangular symmetrical domain; (b) a 3^2 full factorial design in a restricted rectangular symmetrical domain; (c) the candidate points of the grid in the asymmetrical domain; and (d) the selected points forming the 8-experiment *D*-optimal design.

Two types of response surface designs, applicable in an asymmetrical experimental domain, are discussed, that is, *D*-optimal designs and designs constructed with the Kennard and Stone algorithm (93).

2.4.2.2.1. D-optimal Designs. *D*-optimal designs are a first possibility to apply in an irregular experimental domain (1, 7, 94). *D*-optimality (see below) is a desirable characteristic of many symmetrical designs (7, 94), but it can also be applied to construct a design with an asymmetrical domain. Suppose that the desired experimental domain to examine is rectangular with +1 and −1 as scaled boundaries for the factors x_1 and x_2. One could choose to perform a 3^2 full factorial design, as shown in Figure 2.10a. However, suppose that the experiments at the conditions (−1, 1), (−1, 0), and (0, 1) are impossible in practice. Then the feasible experimental domain is as shown on Figure 2.10a, and the indicated 3^2 design is impossible because three experiments cannot be performed. When fitting a 3^2 full factorial design in the remaining experimental domain, as shown in Figure 2.10b, a large part of the area is not covered. Therefore, the experimental domain is represented by defining a number of candidate points forming a grid over the feasible domain (Figure 2.10c). From these candidate points, some will be selected in such a way that the whole domain is covered. The points are selected according to the D-optimality criterion, and form an asymmetrical design (Figure 2.10d).

When constructing a D-optimal design that covers an asymmetrical experimental domain, first the model that will be built is defined. A given model requires a minimal number of experiments, N_{min}, to be able to estimate the coefficients, and the analyst defines the number of experiments, N, that will be performed ($N \geq N_{min}$). Then the N experiments forming the D-optimal design are selected from all possible combinations to select N experiments from all candidate points of the grid. The D-optimal design is the selection for which the determinant of $\mathbf{X}^T\mathbf{X}$ is maximal (= D-optimality), with $\mathbf{X}^T$ the transpose of the model matrix $\mathbf{X}$. The situation of the selected points in the domain depends on N. A selection with one experiment more ($N + 1$) will result in a different selection, and not the N previous experiments plus one. Several software packages allow construction of D-optimal designs (35–41), although not always for asymmetrical experimental domains.

As for the symmetrical designs and in agreement with the philosophy of experimental designs, the experimental domain is mapped as well as possible. This explains why, except for a central point, often all experiments of the D-optimal design are situated toward the boundaries of the experimental domain (Figure 2.10d). During method optimization, D-optimal designs with a symmetrical experimental domain were applied in References 19, 60, and 95, and with an asymmetrical experimental domain in Reference 92.

2.4.2.2.2. Designs Constructed with the Kennard and Stone Algorithm. A second approach to examining an asymmetrical experimental domain is by applying the so-called uniform mapping algorithms, such as the algorithm of Kennard and Stone (1, 7, 93). This approach does not require the *a priori* specification of a model.

Using a uniform mapping algorithm ensures that the experiments cover the experimental domain as uniformly as possible, and that the experiments are situated as far as possible from each other. Another benefit is that the number of experiments can be sequentially increased. Here, the selection of $N + 1$ experiments equals the N previous plus a new one, in contrast to the D-optimal designs. Another advantage is the flexibility. Besides allowing irregular experimental domains that are neither spherical nor cubic to be examined, these designs also allow that certain obligatory conditions are included in the design. For example, when certain earlier performed experiments are available and should be included, this forms the starting point for the selection of new conditions. These new experiments are then chosen to be as different as possible from those already performed.

The Kennard and Stone algorithm maximizes the minimal Euclidean distance of a new point to those previously selected. The Euclidean distance between two points i and j, d_{ij}, is calculated with Equation 2.1:

$$d_{ij} = \sqrt{\sum_{v=1}^{w} (x_{iv} - x_{jv})^2} \quad \text{(Eq. 2.1)}$$

where v corresponds to the variables or factors ($v = 1, 2, \ldots, w$).

The algorithm can be initiated in two ways. In the first situation, no earlier performed experiments or *a priori* selected conditions are included. In the second situation, one or some are included. When no experiments need to be included, the distances between all pairs of points are calculated (Eq. 2.1) and the largest is selected (Eq. 2.2), which determines the first two points.

$$d_{selected} = \max(d_{ij}) \quad \text{(Eq. 2.2)}$$

To define the following point k to be included, the two distances between a remaining point k and the already selected points are calculated and the smallest distance is retained ($\min_i(d_{ik})$). This is done for all points k, and consecutively that point k that maximizes the minimal distance to the closest point already selected is chosen.

$$d_{selected} = \max_k \left[\min_i (d_{ik}) \right] \quad \text{(Eq. 2.3)}$$

In Figure 2.11, the consecutively selected points by the algorithm of Kennard and Stone are shown. In Figure 2.11a, no requirements were set, and in Figure 2.11b it was required that a central point be the first selected point. However, these designs, to our knowledge, were so far not used during the optimization of CE methods. In chromatographic method optimization, they were already applied (96, 97).

2.4.2.3. Example of an Applied Response Surface Design. In the optimization phase of the development of a CE method for the chiral enantioseparation of a nonsteroidal anti-inflammatory drug (28), a circumscribed CCD was performed. The applied symmetrical response surface design is as shown in Table 2.14, with $|\alpha| = (2^f)^{1/4} = 1.68$. The center point (experiment 15 in Table 2.14) was replicated five times (experiments 15–19).

2.4.3. Simplex Approaches

In the optimization phase, sequential optimization methods (7–11) can also be applied, instead of response surface designs. As already mentioned, in response surface designs, the experimental domain enclosed by the design is expected to contain the optimum, while a sequential optimization method can be applied in situations where one *a priori* has no idea about the situation of the optimum in the experimental domain.

In sequential methods, only a restricted number of experiments, that is, usually one more than the number of selected factors, are initially performed. From the obtained results, the next experiment is then defined. The result of

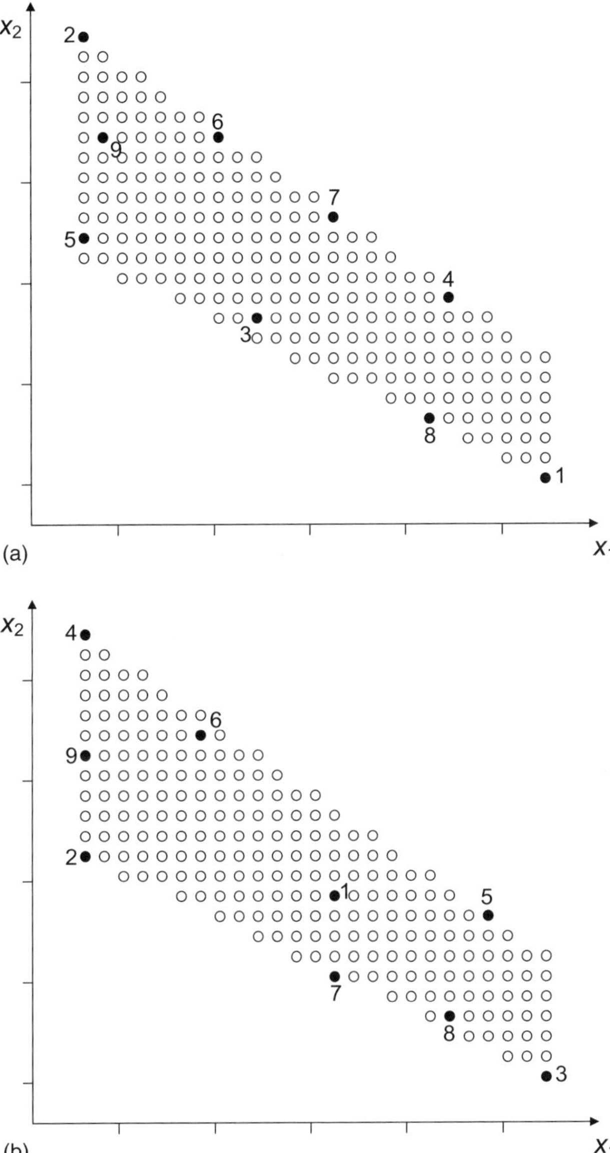

FIGURE 2.11. Selected experimental points by the uniform mapping algorithm of Kennard and Stone: (a) without requirements; and (b) with the requirement that the center point was the first selected point.

the new experiment together with some of the previous results is then used to select the next experiment to be performed, etc.

Different sequential optimization methods can be distinguished, of which the simplex approaches are most commonly applied. They can be further divided into the basic simplex procedure, the variable-size or modified simplex procedure, and the super modified simplex procedure. For a detailed overview of the sequential optimization methods, we refer to Reference 8. In this chapter, only the basic simplex and the modified simplex procedures will be discussed.

In general, a simplex for f factors is a geometric figure in the f-dimensional factor space, defined by $f + 1$ points or vertices, that is, one more than the number of factors. During optimization, the simplex sequentially moves through the experimental domain in the direction of the optimum. The next simplex to be performed is based on the results of the previous, and is defined according to specific rules.

In the following, the basic and modified simplex procedures are discussed for the optimization of two factors. The simplex is then a triangle ($f + 1 = 3$ vertices). For the basic procedure, it is an equilateral triangle, while for the modified procedure, it does not necessarily have to be.

In the *basic simplex procedure*, proposed by Spendley et al. (98), the first three experiments are performed according to the conditions of the initial simplex, called BNW (Figure 2.12). B, N, and W correspond to the vertices with the best, next-to-best, and worst responses, respectively. The best response is usually either the highest or the lowest, depending on what is the most desired situation. The size of the initial simplex is arbitrarily chosen by the analyst. B, N, and W can be represented by the vectors **b**, **n**, and **w**, that is, $\mathbf{b} = [x_{1b}, x_{2b}]$, $\mathbf{n} = [x_{1n}, x_{2n}]$, and $\mathbf{w} = [x_{1w}, x_{2w}]$. Depending on the obtained results, the next experiment will be selected.

The basic simplex procedure is further described by four rules (9, 10, 98, 99).

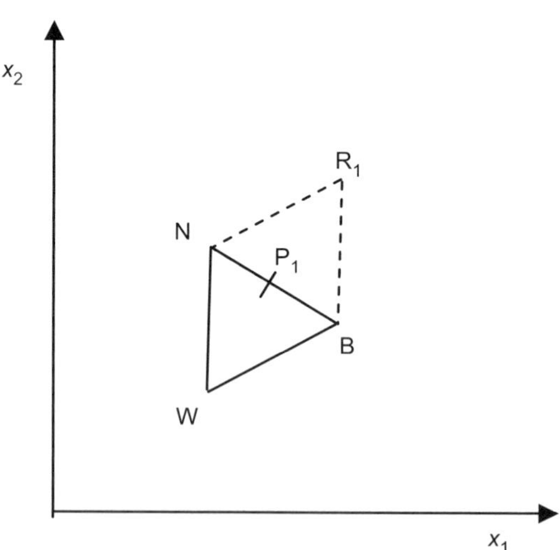

FIGURE 2.12. Basic simplex BNW: B = vertex with best response, W = vertex with worst response, N = vertex with next-to-best response, P_1 = centroid of the line segment BN, and R_1 = reflected vertex.

Rule 1: The new simplex is created by keeping the two vertices from the former simplex with the best results (B, N), and replacing the rejected vertex, that is, the one with the worst result (W), with its mirror image (R_1) across the line defined by the two remaining vertices (BN).

If the centroid P_1 of the line segment BN is represented by the vector $\mathbf{p_1}$ (Eq. 2.4), the coordinates of the new vertex R_1 are obtained by adding the vector ($\mathbf{p_1} - \mathbf{w}$) to $\mathbf{p_1}$. Thus, the vector $\mathbf{r_1}$ representing point R_1 is defined as in Equation 2.5:

$$\mathbf{p}_1 = \frac{1}{2}(\mathbf{n}+\mathbf{b}) = \frac{[(x_{1n}+x_{1b}),(x_{2n}+x_{2b})]}{2} \quad \text{(Eq. 2.4)}$$

$$\mathbf{r_1} = \mathbf{p_1} + (\mathbf{p_1} - \mathbf{w}) = 2\mathbf{p_1} - \mathbf{w} \quad \text{(Eq. 2.5)}$$

In the first steps of the procedure, the new experiment R_1 will usually lead to better results than at least one of the two other vertices because the simplexes tend to move toward the optimum. Nevertheless, in case the new experiment R_1 does not yield better results, that is, the simplex does not move toward the optimum anymore, a change in the progression axis is required. Applying rule 1 is useless because it rejects R_1 and reflects it back to the point W from the former simplex. Therefore, a second rule is applied.

Rule 2: In case the new vertex in a simplex corresponds to the worst result, rule 1 is not applied. The vertex corresponding to the next-to-worst response (N) is now eliminated from the latter simplex and its mirror image (R) across the line defined by the two remaining vertices (BW) is defined as the new vertex.

Applying rule 2 changes the direction of progression toward the optimum. This occurs most often in the region around the optimum. If a vertex in the vicinity of the optimum has been obtained, all new vertices are situated further from the optimum, and circle around it. This indicates that one is as near to the optimum as one can get with the initially chosen simplex size and starting from the initially chosen start conditions. Nevertheless, in practice, when the response surface is unknown, the optimum found may be only a local one.

When circling around a given set of conditions, rule 3 is applied.

Rule 3: When a certain vertex is retained in three (f + 1) successive simplexes, its response is redetermined.

If the new obtained result is the best compared with all vertices from the last three (f + 1) simplexes, it is considered the best optimum that can be obtained with the chosen simplex size. On the other hand, if the new obtained result is not the best, the simplex has become stuck into a false optimum, and then it is better to start again.

Rule 4 is related to what to do when the experimental conditions, defined as new vertex, are situated outside the feasible experimental domain.

Rule 4: If a vertex falls outside the boundaries of the feasible domain, an artificially worst response should be assigned to it and one should proceed further with rules 1-3. This will force the simplex back into the boundaries.

In Figure 2.13, an example is given of the basic simplex procedure. Consider the imaginary response surface of a method, representing the response as a function of two factors (x_1 and x_2) and shown as contour plot (dotted lines). Suppose the highest response value is considered to be the optimum.

First three experiments (points 1, 2, and 3) will be performed, according to the conditions defined by the initial simplex (S_1). By applying rule 1, the vertex with the worst response (point 1) is rejected and reflected to create point 4. Points 2, 3, and 4 then form the new simplex (S_2). An experiment is then run at the conditions defined by point 4, and the procedure is repeated. For simplexes 2–7, all defined according to rule 1, the new experiment always yielded better results than at least one of the two remaining experiments of the preceding simplex. From simplex 7, point 7 is considered to be the worst, rejected, and reflected to point 10 (S_8). However, this vertex falls outside the boundaries of the feasible domain, and an undesirable response is assigned to point 10.

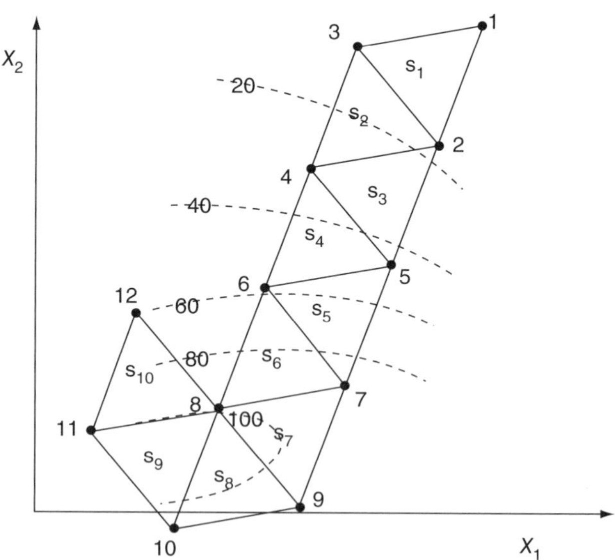

FIGURE 2.13. Example of the (basic) simplex procedure.

In this case, the new vertex corresponds to the worst result and rule 2 is applied. Thus, the next-to-best point (9) is reflected and replaced by point 11, leading to simplex 9. As point 8 is retained in three successive simplexes (S_6-S_7-S_8), first rule 3 is applied, and the response at point 8 is redetermined and evaluated. In case its result is confirmed, simplex 9 is considered. In the next step, rule 1 is again applied, and the worst point 10 is replaced by point 12, resulting in simplex 10.

It is observed that the simplexes circle around the optimum and point 8 is the closest the real optimum can be reached by the simplex used. The number of experiments or simplexes required to approach the optimum depends on the size of the simplex. A larger simplex will require fewer experiments than a smaller simplex. However, a smaller simplex will allow approaching the real optimum closer than a larger one. From this need to find a compromise between speed of moving through the domain and approachability of the optimum, the variable-size or modified simplex procedure has been developed.

In the basic simplex method, the simplex thus can only be reflected to obtain the next experiment, and the simplex size remains the same throughout the procedure. In the modified simplex method, suggested by Nelder and Mead (100), the simplex can be reflected, expanded, or contracted to define the next experiment. Thus, in case the simplex is expanded or contracted, the simplex size changes. More information about the simplex procedures can be found in References 7, 9, 10, and 98–102.

Let us now consider the *variable-size or modified simplex procedure*, proposed by Nelder and Mead (100). Whereas in the basic procedure, the size is fixed and determined by the initially chosen simplex, the size in the modified simplex procedure is variable. Besides the rules of the basic procedure, the modified procedure additionally allows expansion or contraction of simplexes. In favorable search directions, the simplex size is expanded to accelerate finding the optimum, while in other circumstances, the simplex size is contracted, for example, when approaching the optimum (Figure 2.14).

Similarly to the basic procedure, the points of the initial simplex BNW are represented by the vectors **b**, **n**, and **w**. In the following, it is assumed that the best response is the highest. By applying rule 1 of the basic procedure, the vertex W is rejected and reflected through the centroid P_1, represented by the vector $\mathbf{p_1}$ (Eq. 2.4), to obtain point R_1, represented by the vector $\mathbf{r_1}$ (Eq. 2.5). Consecutively, the experiment is performed according to the experimental conditions of R_1, and three situations are possible. In the first, the response is higher at R_1 than at B, in the second it is between those at B and at N, and in the third it is lower at R_1 than at N.

In the first situation (response at R_1 > response at B), the simplex seems to move in a favorable direction and the simplex is expanded by generating the expansion vertex E_1, represented by vector $\mathbf{e_1}$, and defined as

$$\mathbf{e}_1 = \mathbf{p}_1 + \gamma(\mathbf{p}_1 - \mathbf{w}) \qquad \text{(Eq. 2.6)}$$

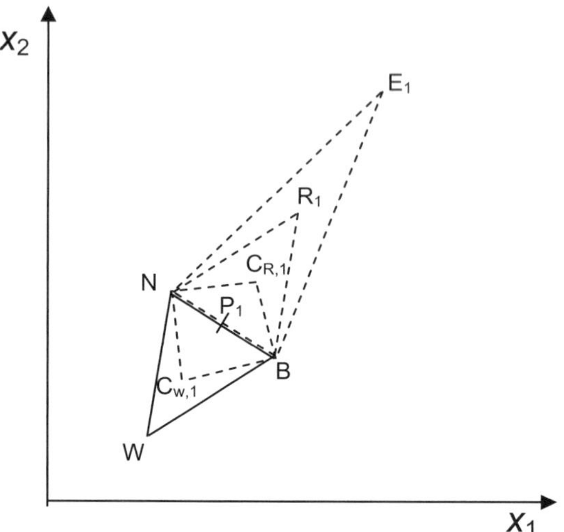

FIGURE 2.14. Modified simplex BNW: B = vertex with best response, W = vertex with worst response, N = vertex with next-to-best response, P_1 = centroid of the line segment BN, R_1 = reflected vertex, E_1 = expansion vertex ($\gamma = 2$), $C_{R,1}$ = contraction vertex on the reflection side ($\beta = 0.5$), and $C_{W,1}$ = contraction vertex on the worst side ($\beta = 0.05$).

where γ is the expansion coefficient, representing the expansion of the simplex ($\gamma > 1$). The larger the value of γ, the more the simplex is expanded. Usually γ is chosen equal to 2, as in Figure 2.14.

When response at $E_1 \geq$ response at B, the expansion is considered successful and the simplex BNE_1 is used to select the next experiment. On the other hand, if response at $E_1 <$ response at B, the expansion has failed and the simplex BNR_1 is taken to define the next experiment. Instead of comparing E_1 with B, a small and logic modification to the above was introduced in References 101 and 103, where the results at E_1 and R_1 are compared and the one with the most desirable result is retained. Further, the usual procedure is followed, that is, the worst vertex of the new simplex is rejected, reflected, etc.

In the second situation (response at N $\leq$ response at $R_1 \leq$ response at B), neither expansion nor contraction is considered. The next experiment to be performed is determined from BNR_1, using the classic rules.

In the third situation (response at $R_1 <$ response at N), it seems the simplex has moved too far, and it should be contracted. Two possibilities are distinguished. In the first, response at $R_1 \geq$ response at W, and the new vertex $C_{R,1}$, represented by vector $\mathbf{c}_{\mathbf{R},1}$ (Eq. 2.7) and situated nearer to R_1 than to W, is selected.

$$\mathbf{c}_{\mathbf{R},1} = \mathbf{p}_1 + \beta(\mathbf{p}_1 - \mathbf{w}) \qquad \text{(Eq. 2.7)}$$

β is the contraction coefficient, representing the contraction of the simplex ($0 < \beta < 1$). The smaller the value of β, the more the simplex will be contracted. Usually, β is chosen equal to 0.5, as in Figure 2.14.

Consequently, the response at this vertex is determined. When response at $C_{R,1} \geq$ response at R_1, the contraction is considered successful and the simplex $BNC_{R,1}$ is used to determine the following experiment. Otherwise (response at $C_{R,1} <$ response at R_1), the contraction is considered a failure, and BNR_1 remains the simplex to define the next experiment, which is done according to the classic procedure.

The second possibility is that response at $R_1 <$ response at W, and then the new vertex $C_{W,1}$, represented by vector $\mathbf{c}_{W,1}$ and situated nearer to W (Eq. 2.8, is selected.

$$\mathbf{c}_{W,1} = \mathbf{p}_1 - \beta(\mathbf{p}_1 - \mathbf{w}) \qquad \text{(Eq. 2.8)}$$

Consequently, the response at this vertex is determined. When response at $C_{W,1} \geq$ response at R_1, the contraction is found successful and the simplex for further use is $BNC_{W,1}$. On the other hand, when response at $C_{W,1} <$ response at R_1, the contraction is said to have failed, and BNR_1 remains the simplex to define the next experiment. Further, the usual procedure is followed.

Furthermore, rule 3 of the basic procedure is applied if a certain point is retained in $f + 1$ successive simplexes. A difficulty is to define a criterion to stop the (modified) simplex procedure. In Reference 8, different possibilities are discussed.

In Figure 2.15, an example is given of the modified simplex procedure for the determination of fluticasone propionate with flow injection analysis (104). The initial simplex is formed by points 1, 2, and 3. Points 4–14 represent the sequentially selected vertices. Point 6 seems to be situated close to the optimum because it is maintained in many simplexes. It is observed that again, as in the classic procedure, the simplexes circle around the optimum, but here also their size decreases as the procedure continues. To optimize three or more factors, the simplex procedures can be generalized, as described in Reference 8.

2.5. RESPONSES

During method optimization, initially qualitative responses, related to the quality of the separation, are considered. On the other hand, during robustness testing, first quantitative responses are studied. Nevertheless, all types of responses can be evaluated during both method optimization and robustness testing.

2.5.1. Qualitative or SST Responses

During method optimization, in a first instance, qualitative responses, related to the quality of the separation and providing information on the qualitative

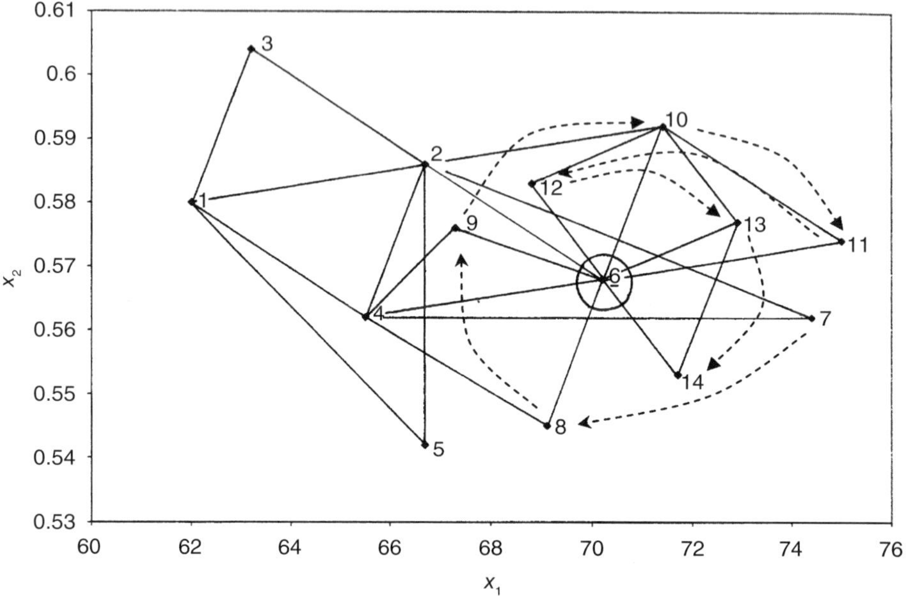

FIGURE 2.15. Example of the modified simplex procedure for the determination of fluticasone propionate with flow injection analysis, based on Reference 104. 1, 2, 3 = initial simplex, and 4, 5, ... , 14 = sequentially selected vertices.

aspects of the method, are considered. In a robustness test, these qualitative responses, for which occasionally SST limits can be defined, can also be examined (5, 16, 105). Regardless of whether or not the method is considered robust concerning its quantitative aspect, these SST responses often are "nonrobust"; that is, they contain significant effects (105).

In electrophoresis, qualitative responses, such as the migration time, the peak shape, the selectivity factor, the resolution between neighboring peaks, and the signal-to-noise ratio, can potentially be considered. When evaluating the robustness of a separation, responses describing the quality of the separation are studied, such as the selectivity factor or the resolution.

2.5.2. Quantitative Responses

During robustness testing, in a first instance, the considered responses usually represent quantitative aspects of the method (5, 16). An analytical method is considered robust if no significant effects are found on the response(s) describing the quantitative aspect of the method. Although during method optimization usually quantitative responses are initially not considered, they can, however, be studied.

Examples of quantitative responses are the concentrations or the percentage recoveries of the main and/or related compound(s), and occasionally also peak area or peak area/migration time.

TABLE 2.17. Responses determined from the 12-experiment Plackett–Burman design (Table 2.8) applied during the screening phase of CE method development in Reference 27: selectivity factor *S*, resolution *Rs*, and analysis time *t*

Experiment	Responses		
	S	*Rs*	*t*
1	1.12	1.91	4.02
2	1.05	1.69	10.15
3	1.05	1.17	14.50
4	1.12	4.30	26.70
5	1.06	1.45	6.53
6	1.05	1.76	22.85
7	1.06	2.76	19.41
8	1.13	1.10	4.49
9	1.11	1.81	12.31
10	1.10	2.33	7.74
11	1.05	1.79	7.52
12	1.10	2.05	8.71

2.5.3. Examples of Responses Studied

In Reference 27, the responses from the 12-experiment PB design (Table 2.8) applied during the screening phase of a CE method development were all qualitative, that is, the selectivity factor *S*, the resolution *Rs*, and the analysis time *t* (Table 2.17).

The responses considered from the circumscribed CCD (Table 2.14) applied during the optimization phase of the development of a chiral enantioseparation method in Reference 28 were also all qualitative, that is, migration time of the first and the second enantiomer (t_{m1} and t_{m2}), and resolution between the two enantiomers *Rs* (Table 2.18).

In Reference 29, the response studied in the 8-experiment PB design (Table 2.9) during the robustness testing of a CE method was quantitative, that is, peak area/migration time ratio A/t_m (Table 2.19).

2.6. PLANNING AND EXECUTION OF EXPERIMENTAL SETUP

At this point, the required experiments can be defined. For this purpose, the levels (e.g., $-\alpha, -1, 0, +1, +\alpha$) in the theoretical experimental design (e.g., Tables 2.8, 2.14, and 2.9) are replaced by the real factor levels (e.g., Tables 2.2–2.4, respectively). This results in the experimental conditions for each experiment. The dummy factor columns in PB designs can be ignored at this point. Often a number of replicated experiments at nominal or center point conditions are added to the setup (see above).

TABLE 2.18. Responses studied in the circumscribed central composite design (Table 2.14 with |α| = 1.68, five center point replicates (exp 15–19)) applied during the optimization phase of the development of a chiral enantioseparation method in Reference 28: migration time of the first and the second enantiomer (t_{m1} and t_{m2}), and resolution between the two enantiomers *Rs*

Experiment	Responses		
	t_{m1}	t_{m2}	*Rs*
1	13.67	13.96	0.85
2	14.60	15.24	1.86
3	6.84	7.14	1.55
4	8.55	9.12	2.58
5	12.08	12.35	0.74
6	12.85	13.45	1.72
7	6.04	6.29	1.19
8	7.36	7.86	2.32
9	6.43	6.52	0.49
10	9.00	9.65	2.50
11	13.58	14.03	0.91
12	6.06	6.41	1.70
13	7.94	8.39	2.07
14	6.78	7.14	1.80
15	7.90	8.32	1.92
16	7.58	8.01	1.92
17	7.61	8.02	1.90
18	7.62	8.06	1.92
19	7.62	8.06	1.90

TABLE 2.19. Response studied in the 8-experiment Plackett–Burman design (Table 2.9) during the robustness testing of a CE method in Reference 29: peak area/migration time ratio A/t_m

Experiment	Response
	A/t_m
1	2784
2	2707
3	2667
4	2762
5	2692
6	2733
7	2751
8	2586

It is often advisable to perform the experiments in a *random sequence* in order to minimize uncontrolled influences on the estimated effects (4, 5). A time effect reflects response changes, which are larger than the experimental error, over time, when measured at a set of fixed conditions. A special case of a time effect is called drift and occurs when the response continuously increases or decreases as a function of time. Randomization does not avoid biased effect estimates when a time effect is present. Depending on the executed sequence of the experiments, some estimated effects still will be influenced by the time effect (106).

Using the so-called *antidrift screening designs* (5, 107) might solve the time effect problem in some cases. These designs are just regular screening designs but executed in a particular sequence. In antidrift screening designs, the experiments are executed in such sequence that the main effects are not or minimally confounded with the drift effect, while the columns of the interaction or dummy terms in FF and PB designs, respectively, are most confounded with the drift effect (106). It should be noticed that the estimated interaction or dummy effects then cannot be used any longer in the statistical evaluation of effects (see further).

Another approach that allows correcting for the problem is the execution of *replicated* (*nominal*) *experiments* between the design experiments (4, 5, 16, 106). This approach can be applied with all types of designs. The replicated experiments are performed before, at regular times between (e.g., every *n* design experiments) and after the design experiments. They allow verification of the method performance before and at the end of the experimental design, and checking and correction for time effects (5, 16, 106). A drift plot, visualizing a drift or a time effect, can be drawn by plotting the replicated response, usually measured at nominal levels, as a function of time (Figure 2.16).

The magnitude of the drift can be expressed as follows (16),

$$\%\text{Drift} = \frac{y_{repl,end} - y_{repl,begin}}{y_{repl,begin}} \times 100 \qquad \text{(Eq. 2.9)}$$

where $y_{repl,begin}$ and $y_{repl,end}$ are the replicated (nominal) responses measured before and after the design experiments, respectively. For each response, it could be verified whether a time effect occurs.

When such time effect is present, the design responses are corrected relative to the (nominal) experiment performed at the beginning of the experimental design (Eq. 2.10) (Figure 2.16) (5, 16, 106). These corrected responses are then used to estimate the factor effects from screening designs or to build the model from response surface designs (see further). From both the estimated effects and the model coefficients then the time effect has been removed:

$$y_{i,corrected} = y_{i,measured} + y_{repl,begin} - \left(\frac{(p+1-i)\, y_{repl,before} + i y_{repl,after}}{p+1} \right) \qquad \text{(Eq. 2.10)}$$

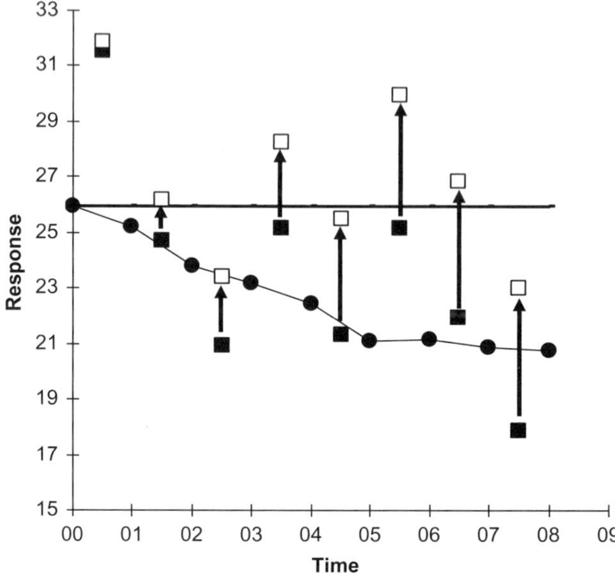

FIGURE 2.16. Drift plot (●): replicated responses measured as a function of time. Design responses (■) and corresponding corrected design responses (□) are also shown. The horizontal line (——) represents the initial replicated response value.

In Equation 2.10, $i = 1, 2, \ldots, p$, and p is the number of design experiments between two consecutive replicated (nominal) experiments. $y_{i,corrected}$ is a corrected design response, $y_{i,measured}$ the corresponding measured design response, $y_{repl,begin}$ the replicated (nominal) response at the beginning of the design experiments, and $y_{repl,before}$ and $y_{repl,after}$ the replicated (nominal) responses measured before and after the $y_{i,measured}$ that is being corrected, respectively.

Although it is not recommended, frequently, for practical reasons, experiments are *blocked* or *sorted* by one or more factors (4, 5, 16, 106). Then all experiments at one level of the factor are executed first, followed by all at the other level. The experiments are usually performed in a random sequence within one block. At least before and after each block, a check for drift can be recommended in such situation. These latter experiments permit observation and correction of occasional block effects.

2.7. DATA HANDLING

The results from screening designs are analyzed by estimating and interpreting the effects of the selected factors on the response(s), in order to determine those factors most influencing the method. On the other hand, the response surface design results are usually analyzed by building and interpreting a polynomial model describing the relation between the response(s) and the considered factors.

2.7.1. Screening Designs

The calculation of the factor effects on the considered responses is followed by a graphical and/or statistical interpretation of the estimated effects.

2.7.1.1. Estimation of Effects. Depending on the absence or presence of drift in the considered response, effects are estimated from the measured or corrected responses, respectively. The effect of factor X, E_X, on a response Y is calculated with Equation 2.11 (4, 5, 7),

$$E_X = \frac{\sum Y(+1) - \sum Y(-1)}{N/2} \tag{Eq. 2.11}$$

where $\sum Y(+1)$ and $\sum Y(-1)$ represent the sums of the responses where factor X is at (+1) and (–1) level, respectively, and N is the number of design experiments. Sometimes, the normalized effect of factor X, $E_X(\%)$, is also calculated (4, 5).

$$E_X(\%) = \frac{E_X}{\overline{Y}} \times 100\% \tag{Eq. 2.12}$$

In the absence of drift, $\overline{Y}$ is the average nominal result or the average design result. On the other hand, when drift is present, it is recommended to estimate the factor effects from the corrected responses (Eq. 2.10), and $\overline{Y}$ represents the replicated response value measured before the design experiments (see Figure 2.16) (5, 16).

Instead of calculating effects, some authors estimate the coefficients of the following regression model (7, 17),

$$y = \beta_0 + \sum_{i=1}^{f} \beta_i x_i \tag{Eq. 2.13}$$

where y is the response, β_0 the intercept, and β_i the main coefficient. The true β-coefficients are then estimated by the b-coefficients using least squares. In fact, the coefficients are related to the effects (Eq. 2.14). Effects reflect the change in response when changing the factor level from –1 to +1, while coefficients reflect the change between the levels 0 and +1:

$$E_X = 2b_X \tag{Eq. 2.14}$$

2.7.1.2. Interpretation of Effects. Consecutively, a graphical and/or statistical interpretation of the estimated effects usually is performed to determine the effects significance. We recommend combining a graphical with a statistical evaluation of the estimated effects.

The graphical interpretation consists of drawing normal probability (Figure 2.17a) or half-normal probability (Figure 2.17b) plots (4, 5, 7, 17). The normal probability plot presents the expected values from a normal distribution as a function of the estimated effects, while the half-normal probability plot, also called Birnbaun plot, shows the absolute values of the estimated effects as a function of so-called rankits, derived from a normal distribution. In these plots, the nonsignificant effects are found on a straight line through zero, while the significant effects deviate from this line. However, when many significant effects occur, it is no longer easy to distinguish the straight line of the nonsignificant. Drawing the least squares line through all effects also might obscure the significant effects (Figure 2.17a).

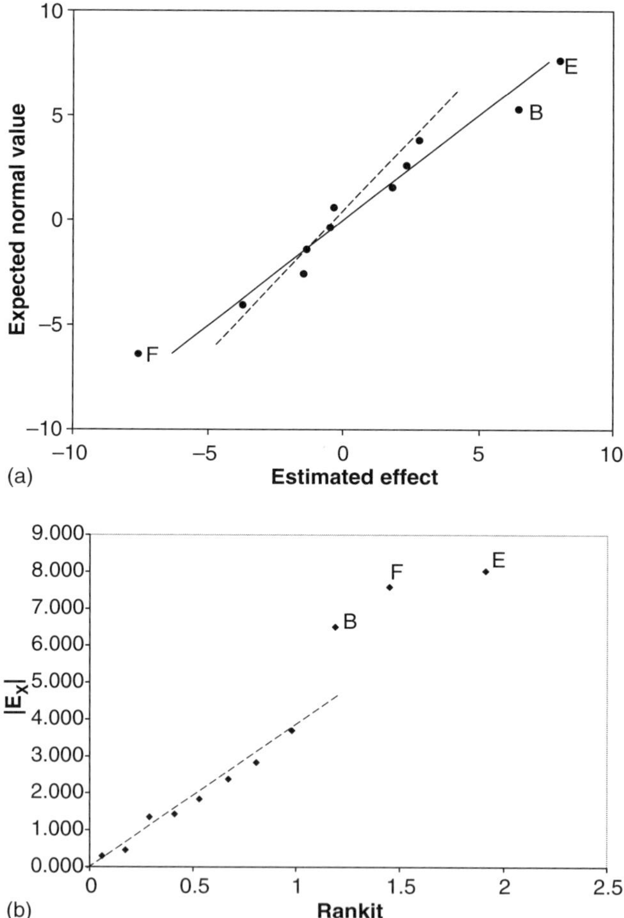

FIGURE 2.17. (a) Normal and (b) half-normal probability plot, for 11 effects on the response analysis time *t*, estimated from a 12-experiment Plackett–Burman design (27). The significant effects are identified. (——) represents least squares line through all effects, and (------) represents assumed line through nonsignificant effects.

The statistical interpretations usually apply the t-test statistic (Eq. 2.15) (4, 5). Occasionally, an ANOVA approach with F-tests is used, which in fact is equivalent to the t-test approach (4, 30, 108):

$$t = \frac{|E_X|}{(SE)_e} \Leftrightarrow t_{critical} \qquad \text{(Eq. 2.15)}$$

The calculated t-value (Eq. 2.15), based on the effect of factor X, E_X, and on the standard error of an effect, $(SE)_e$, is compared with a (tabulated) critical t-value, $t_{critical}$. The $t_{critical}$ depends on the number of degrees of freedom (d.f.) associated with the estimation of $(SE)_e$ and is usually determined at a significance level $\alpha = 0.05$. All effects with a t-value larger than or equal to $t_{critical}$ are considered significant.

The t-test statistic can be rewritten in such a way that a critical effect, $E_{critical}$ (Eq. 2.16), instead of a t-value is used (4, 5, 16). All effects that in absolute value are larger than or equal to this $E_{critical}$ are then considered significant:

$$|E_X| \Leftrightarrow E_{critical} = t_{critical} \times (SE)_e \qquad \text{(Eq. 2.16)}$$

$(SE)_e$ can be estimated in different ways, that is, from the variance of replicated experiments, for instance, at the nominal or center point level, from *a priori* declared negligible effects or from *a posteriori* defined negligible effects (4, 5, 7, 16, 24, 31, 74, 105, 106, 109–114).

Using the first approach, $(SE)_e$ is derived from the variance of replicated experiments, s^2, and estimated as follows:

$$(SE)_e = \sqrt{\frac{2s^2}{n}} \qquad \text{(Eq. 2.17)}$$

with n the number of experiments performed at each factor level.

In Equation 2.17, the variance of R replicates at the nominal or center point level, or the variance of duplicated design experiments ($s^2 = \frac{\sum d_i^2}{2n}$, with d_i the difference between the duplicated experiments), can be used, with n equal to $N/2$ or N, respectively, and the number of d.f. for $t_{critical}$ R − 1 or N, respectively (4, 5).

When using replicates, it is required that they are measured at intermediate precision conditions, and not at repeatability conditions. The latter leads to an underestimation of $E_{critical}$ and consequently most effects will be erroneously considered significant (110). Second, it is recommended to have at least three d.f. available to estimate $(SE)_e$.

In the second approach, $(SE)_e$ is obtained from n_N *a priori* declared negligible effects, E_N, such as two-factor interaction effects in robustness testing or higher-order interaction effects in screening during method development, or dummy factor effects in both situations, from FF and PB designs, respectively (Eq. 2.18) (4, 5, 7). Similar to the first approach, it is recommended that at least three negligible effects (d.f. = n_N = 3) are available to estimate $(SE)_e$ (5):

$$(SE)_e = \sqrt{\frac{\sum E_N^2}{n_N}} \qquad \text{(Eq. 2.18)}$$

In robustness testing, the two-factor interactions and the dummy factor effects in FF and PB designs, respectively, can indeed be considered negligible (5) and thus be used to estimate $(SE)_e$. On the other hand, during the screening phase of method development, this negligibility is not *a priori* assumed anymore. Therefore, when using this approach to estimate the critical effect, one should carefully consider the two-factor interactions or dummy factor effects, prior to inclusion in the estimation of $(SE)_e$. If available, preferably higher-order interaction estimates are used.

The third approach computes $(SE)_e$ from *a posteriori* defined negligible effects by using the algorithms of Lenth (111) or Dong (5, 112). They start from the idea of effect sparsity, that is, the hypothesis that in a screening design or robustness testing, no or only few effects are important. The algorithm of Dong leads to practically more relevant $E_{critical}$ values than that of Lenth (24, 74, 112, 113).

Dong's algorithm estimates from an initial error estimate s_0 (Eq. 2.19) the final error estimate $(SE)_e$ (Eq. 2.20), based on the m effects, E_k, that are not considered important, that is, those that fulfill the requirement $|E_k| \leq 2.5^* s_0$. The estimated critical effect (Eq. 2.16) from the algorithm of Dong is also called the margin of error:

$$s_0 = 1.5 \times median|E_X| \qquad \text{(Eq. 2.19)}$$

$$(SE)_e = \sqrt{\frac{\sum E_k^2}{m}} \qquad \text{(Eq. 2.20)}$$

Nevertheless, in situations where the effect sparsity principle is violated and the number of significant effects approaches 50%, the algorithm of Dong becomes incapable of detecting the significant effects correctly. This leads to a number of effects incorrectly considered nonsignificant, that is, to false negative results (114, 115). For these situations, an adaptation to the algorithm of Dong was suggested in Reference 114, where it is recommended to apply the 75% lowest absolute factor effects for the initial error estimation s_0, that is, $s_0 = 1.5 \times median|E_{75\%}|$ instead of using Equation 2.19.

2.7.1.3. Examples of Data Handling from Screening Design Results. For the two examples, described in References 27 and 29, the estimated effects on the responses (Tables 2.17 and 2.19) are given in Tables 2.20 and 2.21,

TABLE 2.20. Effects on the responses of Table 2.17(27) and critical effects according to the different graphical and statistical interpretation methods

Factor	Effects on the Responses		
	S	*Rs*	*t*
A	−0.058	−0.48	2.83
B	0.009	0.31	6.50
C	0.010	0.35	2.38
D	0.001	−0.81	−1.35
E	−0.006	0.90	8.02
F	0.002	−0.24	−7.59
G	0.003	−0.67	−3.71
H = d_1	−0.003	−0.13	−1.45
I = d_2	0.000	0.11	−0.46
J = d_3	0.000	0.12	−0.30
K = d_4	−0.001	0.46	1.84
Graphical interpretation	Number of Important Effects		
Normal or half-normal probability plots	1	0	3
Method to estimate $(SE)_e$	Critical Effects (number of significant effects)		
Variance of replicated experiments at $\alpha = 0.05$	/	/	/
Dummies at $\alpha = 0.05$	0.005 (4)	0.70 (2)	3.34 (4)
Algorithm of Dong at $\alpha = 0.05$	0.007 (3)	1.09 (0)	9.37 (0)
Adapted algorithm of Dong at $\alpha = 0.05$	0.007 (3)	1.09 (0)	4.82 (3)

/ = not possible to calculate from reported setup.

respectively. Their significance according to different graphical and statistical interpretation methods was determined.

From Table 2.20 (27), different numbers of effects are considered important for response *S* when evaluating the graphical and statistical methods. From the plots, only one clearly deviating effect was observed. It is nevertheless clear that factor *A*, responsible for the effect, should be examined further. For response *Rs*, usually the same number of effects is considered important, except for the approach based on dummies, where the critical effect seems somewhat underestimated. For response *t*, all approaches lead to the same number of significant effects, except the algorithm of Dong, which leads to an overestimation of the critical effect, probably caused by a violation of the effect sparsity principle (about half of effects are important).

2.7.1.4. Nonsignificance Intervals for Significant Quantitative Factors in Robustness Testing. When significant effects are indicated on the response(s) describing the quantitative aspect of the method, the results from the robustness test can be used to set restrictions on the levels of significant continuous factors. When factor *X* has a significant effect, the initially exam-

TABLE 2.21. Effects on the response of Table 2.19 (29) and critical effects according to different graphical and statistical interpretation methods

Factor	Effects on the Response
	A/t_m
A	94.5
B	46.5
C	25.0
D	23.0
E = d_1	32.0
F = d_2	26.5
G = d_3	1.0
Graphical interpretation	Number of Important Effects
Normal or half-normal probability plots	1
Method to estimate $(SE)_e$	Critical Effects (number of significant effects)
Variance of replicated experiments at $\alpha = 0.05$	/
Dummies at $\alpha = 0.05$	76.4 (1)
Algorithm of Dong at $\alpha = 0.05$	105.7 (0)
Adapted algorithm of Dong at $\alpha = 0.05$	70.9 (1)

/ = not possible to calculate from reported setup.

ined interval is reduced and the nonsignificance interval limits are estimated as follows (5):

$$\left[X_{(0)} - \frac{|X_{(+1)} - X_{(-1)}| \times E_{critical}}{2 \times |E_X|},\ X_{(0)} + \frac{|X_{(+1)} - X_{(-1)}| \times E_{critical}}{2 \times |E_X|}\right] \qquad \text{(Eq. 2.21)}$$

For example, the effect of factor A on response A/t_m at $\alpha = 0.05$ was found significant when using the dummy effects to estimate the critical effect (Table 2.21) (29). Factor A has 26, 27, and 28 °C as extreme low, nominal, and extreme high levels, respectively (Table 2.4), and an effect of 94.5 on response A/t_m, with the critical effect equal to 76.4 (Table 2.21). The nonsignificance interval for this factor is then estimated as [26.2 °C, 27.8 °C]. Thus, when restricting the levels of A to this interval, the quantitative aspect of the method is considered robust. It can be noticed that the interval is symmetrically around the nominal level and Equation 2.21 is meant for factors examined with their extreme levels symmetrically around the nominal.

2.7.1.5. Determination of SST Limits from the Results of a Robustness Test. An SST is an integral part of many analytical methods (3). It verifies the suitability and the efficacy of the instrument and/or the setup for the intended purpose of the method. Occasionally, SST limits for some responses are derived from the method optimization and validation results, but quite often they are based on the experience of the analyst.

Alternatively, SST limits can be determined from the results of a robustness test (5, 12), as recommended by the ICH (3). Using the worst-case situation for a given qualitative response allows definition of SST limits for it. The most extreme results are thus considered, obtained under experimental conditions resulting in acceptable quantitative determinations, that is, when the method is considered robust concerning its quantitative aspect in the entire examined experimental domain. The worst-case conditions can be derived from the estimated effects (5). The worst-case situation is that combination of factor levels resulting in the worst result, for example, the lowest resolution. Only the effects significant at a significance level $\alpha = 0.10$ are considered, while all other effects are considered to solely represent experimental error and are kept at nominal level in the worst-case conditions setting (5, 12). Consequently, SST limits can be derived either mathematically or experimentally. In the first situation, the limits are calculated as follows (5, 12):

$$Y = b_0 + \left(\frac{E_1}{2} \times F_1\right) + \left(\frac{E_2}{2} \times F_2\right) + \cdots + \left(\frac{E_k}{2} \times F_k\right) \qquad \text{(Eq. 2.22)}$$

where Y is the calculated SST limit, b_0 the average design result for the considered response, E_i the effect of factor i, and F_i the level of factor i. Significant factors (at $\alpha = 0.10$) have $F_i = -1$ or $+1$, that is, the level leading to the worst result, while nonsignificant factors are at their nominal level ($F_i = 0$).

Second, the SST limits can also be experimentally determined from measurements at the worst-case conditions (n measurements with standard deviation s) (5, 12). The SST limit is then defined as the lower (Eq. 2.23) or upper (Eq. 2.24) limit of the one-sided 95% confidence interval around the worst-case average result (7). For example, for resolution, the lower limit will be considered, while for peak asymmetry it would be the upper limit:

$$\left[\bar{Y}_{worst-case} - t_{\alpha,n-1}\left(\frac{s}{\sqrt{n}}\right), +\infty\right] \qquad \text{(Eq. 2.23)}$$

$$\left[-\infty, \bar{Y}_{worst-case} + t_{\alpha,n-1}\left(\frac{s}{\sqrt{n}}\right)\right] \qquad \text{(Eq. 2.24)}$$

$\bar{Y}_{worst\text{-}case}$ is the average of n replicated measurements with standard deviation s executed at the worst-case conditions. The t-value is determined by the significance level α and the number of d.f. for s, here $n - 1$.

2.7.2. Response Surface Designs

As mentioned earlier, the response surface design results are analyzed by building and interpreting a polynomial model describing the relation between the response(s) and the considered factors.

2.7.2.1. Estimation of Model. Two types of models can be built: mechanistic and empirical models. Usually, empirical models are applied in an experimental design context (1, 7). Most frequently, a second-order polynomial quadratic model is built. Such model includes an intercept, the main effect terms, the interaction effect terms, and the quadratic effect terms. Occasionally, not all possible terms are included in the model; that is, the nonsignificant terms can be deleted.

In general, the model for f factors can be written as follows:

$$y = \beta_0 + \sum_{i=1}^{f} \beta_i x_i + \sum_{1 \le i \le j}^{f} \beta_{ij} x_i x_j + \sum_{i=1}^{f} \beta_{ii} x_i^2 \qquad \text{(Eq. 2.25)}$$

where y is the response, β_0 the intercept, β_i the main coefficients, β_{ij} the interaction coefficients, and β_{ii} the quadratic coefficients (7). Usually, the interaction effect terms are restricted to two-factor interactions.

The experimental design results allow an estimation of the β-coefficients, that is, of the so-called b-coefficients. For two variables, x_1 and x_2, this results, for example, in the calculated model:

$$\hat{y} = b_0 + b_1 x_1 + b_2 x_2 + b_{12} x_1 x_2 + b_{11} x_1^2 + b_{22} x_2^2 + \varepsilon \qquad \text{(Eq. 2.26)}$$

where $\hat{y}$ is the predicted response from the model, b_0 the intercept, b_1 and b_2 the main coefficients, b_{12} the interaction coefficient, b_{11} and b_{22} the quadratic coefficients, and ε the residual (1, 7).

Regression leads to a model estimating the relation between the $N \times 1$ response vector **y**, and the $N \times t$ model matrix **X** (7, 17, 116) (Eq. 2.27). N is the number of design experiments, and t the number of terms included in the model. For example, in Equation 2.26, the number of terms equals six, since one intercept, two main effect terms, one interaction term, and two quadratic effect terms were included. The model matrix **X** is obtained by adding a row of ones before the $N \times (t - 1)$ design matrix, which consists of the coded factor levels and columns of contrast coefficients, as defined by the chosen experimental design.

$$\mathbf{y} = (\mathbf{X}\boldsymbol{\beta}) + \boldsymbol{\varepsilon} \qquad \text{(Eq. 2.27)}$$

$\boldsymbol{\beta}$ is the $t \times 1$ vector of regression coefficients and $\boldsymbol{\varepsilon}$ is an $N \times 1$ error vector. The regression coefficients **b** (e.g., b_0, b_1, b_2, b_{12}, b_{11}, b_{22} in Eq. 2.26) are calculated with the least squares estimation as follows:

$$\mathbf{b} = (\mathbf{X}^\mathrm{T}\mathbf{X})^{-1}\mathbf{X}^\mathrm{T}\mathbf{y} \qquad \text{(Eq. 2.28)}$$

where $\mathbf{X}^\mathrm{T}$ is the transpose of matrix **X**.

Besides the least squares estimation, also other regression techniques can be used to obtain the b-coefficients of the model (7). However, this is rarely done.

The surface representing the model is called the response surface. Graphically, the response surface can be visualized by drawing 2D contour plots or 3D response surface plots (7). A 2D contour plot shows the isoresponse lines as a function of the levels of two variables, while a 3D response surface plot represents the response, on a third dimension, as a function of the levels of two variables. An example of a 2D contour plot and a 3D response surface plot is shown in Figure 2.18. When more than two factors

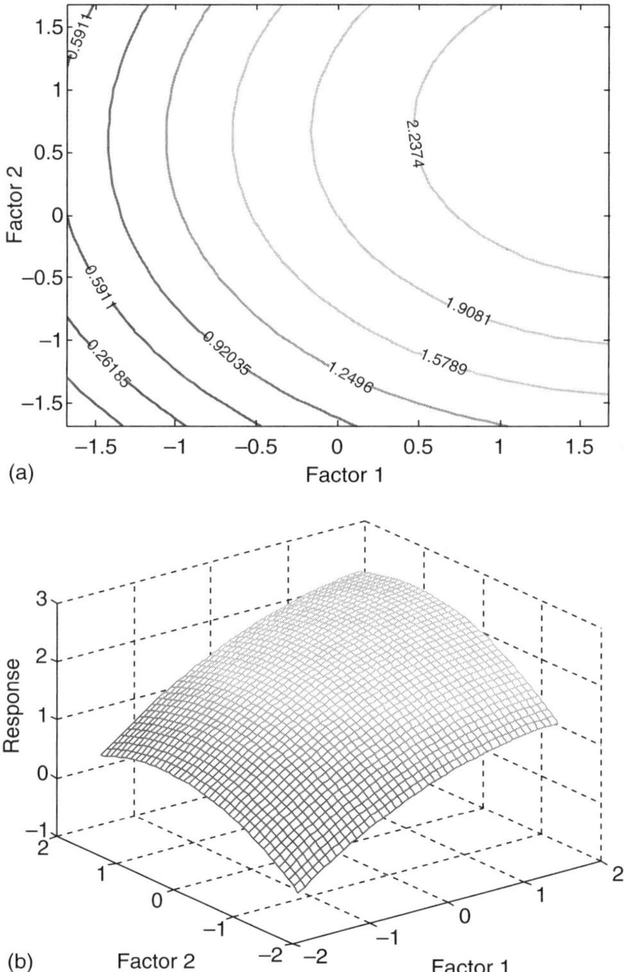

FIGURE 2.18. Graphical representation of the response resolution of Table 18 (28) as (a) 2D contour plot; and (b) 3D response surface plot. The response is presented as a function of factors 1 (A) and 2 (B), while factor 3 (C) is fixed at nominal level (Table 2.3).

are examined, all but two factors need to be fixed at a given level to draw response surfaces.

The number of experiments N in response surface designs is larger than the number of b-coefficients that needs to be estimated. The obtained model then can be used to predict the response for given experimental conditions. It should be emphasized that only predictions within the examined experimental domain are recommended. Extrapolations should be avoided because the model may not be correct anymore and the prediction error will increase (7). However, most frequently the model is used to determine the optimum, and this is selected from the graphical representation (Figure 2.18), rather than using the model for predictive purposes.

Higher-order models are rarely applied. In many cases, the true response surface can be sufficiently well approximated by the second-order model. Occasionally, higher-order models can be used when quadratic models are clearly inadequate, for example, when a sigmoid-like relation between the response and a variable is observed (7). Then, either a third-order model, an appropriate transformation, a mechanistic physical model, nonlinear modeling techniques, or neural networks can be applied (1, 7).

2.7.2.2. Model Validation. In a next step, the fit of the model to the experimental data can be evaluated. This can be done by the approaches summarized below. However, in an optimization context, such evaluation is not always performed. The reason is that the model often only needs to predict a value (the optimum) once and is then not used anymore. The goodness of prediction is then usually experimentally verified, and often method optimization stops here.

To evaluate the fit of the calculated model, usually ANOVA is applied (1, 7, 17, 116). ANOVA will evaluate the data set variation. Often a test for the significance of regression and a lack-of-fit test are performed (7, 17, 116). A model is then considered adequate and well fitted to the data when both a significant regression and a nonsignificant lack-of-fit occur.

Another possibility to evaluate the model is by performing a residual analysis (1, 7, 17, 116). Here, the experimental response and the response predicted by the model are compared for each experimental design point. Large residuals or tendencies in the residuals indicate that the model is not adequate and should be revised.

To evaluate the predictive properties, which is rarely done in method optimization, an external validation can be made (1, 7). This requires an external test set, which consists of experiments at other conditions than those of the experimental design. Again the experimental and the predicted responses are compared and the residuals evaluated.

2.7.2.3. Example of Data Handling from Response Surface Design Results. With the resolution results of the response surface design applied in Reference 28, a second-order polynomial model was built. The model is

$$\hat{y} = 1.91 + 0.55x_1 + 0.28x_2 - 0.10x_3 - 0.14x_1^2 - 0.21x_2^2 + 0.02x_3^2 + 0.02x_1x_2 + 0.01x_1x_3 - 0.05x_2x_3,$$

where x_1 is factor A, x_2 factor B, and x_3 factor C (Table 2.3).

The elution order of the enantiomers was the same for all experiments. Thus, a modeling of the resolution is meaningful. The 2D contour plot and the 3D response surface plot for this response *Rs* are shown in Figure 2.18.

2.7.3. Multicriteria Decision-Making (MCDM) Methods

MCDM methods are applied when at least two responses need to be optimized simultaneously. Different approaches can be distinguished, for example, window programming, threshold approaches, utility functions, Derringer's desirability functions, Pareto optimality methods, Electre outranking relationships, and Promethee (7). In this chapter, only the Pareto optimality methods (7, 117, 118) and Derringer's desirability functions (7, 119, 120) will be discussed.

A first MCDM approach is Pareto optimality. An experiment is considered Pareto-optimal when no other experiment exists with a better result on one criterion without having a worse result on another. This method mostly is used when only two responses are examined, because of the easy graphical interpretation. Theoretically, it can also be applied for more than two responses, although the (graphical) interpretation then is less straightforward. Moreover, the more responses are examined, the more unlikely it becomes that one experiment will dominate another for all considered responses, which makes this method less useful. It also should be noticed for the two-response case that a Pareto-optimal point is not always representing a practically suitable optimum.

Let us consider the data set of Figure 2.19 to explain the Pareto-optimality principle. Suppose 12 experiments are performed and two responses (y_1 and y_2) are measured for each experiment. Suppose the first response corresponds to resolution, and the second to migration time. The first response thus should be maximized, while the second minimized. The line connecting the experiments 1, 2, and 7 links the Pareto-optimal points for this situation. When comparing, for instance, experiments 1 and 4, experiment 1 is considered Pareto-optimal because it dominates. Similarly, experiment 2 dominates experiment 6. In both cases, migration time is shorter for a similar resolution.

A second MCDM approach is the use of Derringer's desirability functions. In this approach, all responses are transformed on the same scale and combined to one response, *D*, which then should be maximized. Each response is transformed on a scale between 0, representing the most undesirable outcome, and 1, representing the most desirable situation. The values of the transformed responses are called desirabilities. Different transformations are used, depend-

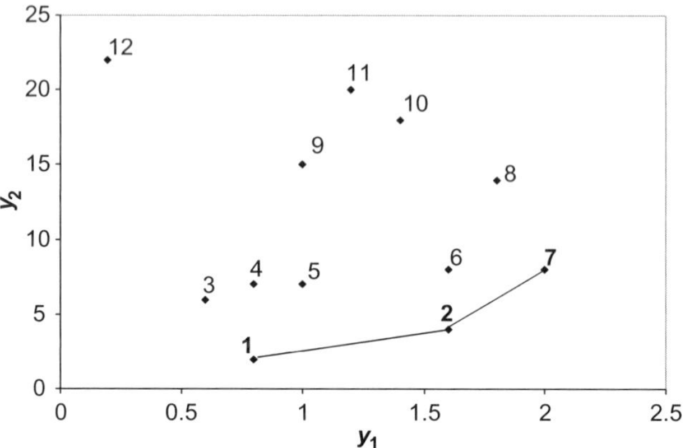

FIGURE 2.19. Pareto-optimality. Response y_1 (e.g., resolution) is to be maximized, while response y_2 (e.g., migration time) is to be minimized.

ing on whether the response is optimal when it is maximized, minimized, or at a predefined value. These three different transformations are represented in Figure 2.20.

In case a response needs to be maximized, Equations 2.29–2.31 are applied. In case a response needs to be minimized, Equations 2.32–2.34 are used.

$$\begin{cases} d_i = 0 & \text{for } y_i < y_{\min} & \text{(Eq. 2.29)} \\ d_i = 1 & \text{for } y_i > y_{\max} & \text{(Eq. 2.30)} \\ d_i = \left[\dfrac{y_i - y_{\min}}{y_{\max} - y_{\min}}\right]^r & \text{for } y_{\min} \le y_i \le y_{\max} & \text{(Eq. 2.31)} \end{cases}$$

$$\begin{cases} d_i = 0 & \text{for } y_i > y_{\max} & \text{(Eq. 2.32)} \\ d_i = 1 & \text{for } y_i < y_{\min} & \text{(Eq. 2.33)} \\ d_i = \left[\dfrac{y_{\max} - y_i}{y_{\max} - y_{\min}}\right]^r & \text{for } y_{\min} \le y_i \le y_{\max} & \text{(Eq. 2.34)} \end{cases}$$

In Equations 2.29–2.34, d_i represents the desirability value, y_i the measured response for experiment i, $y_{\min}$ and $y_{\max}$ either the smallest and largest measured or user-defined responses, and r a coefficient defined by the analyst. When $r = 1$, linear transformations are performed. In Figure 2.20a,b, transformation is performed with $r = 1$. A third transformation is the one shown in Figure 2.20c, where the optimal response is at a well-defined value. The transformation is composed of both earlier applied transformations. Also here $r = 1$ was applied.

The combined response or the global desirability D is then calculated as the geometric mean of the R individual desirabilities and given by Equation 2.35. Consecutively, this combined response D should be maximized.

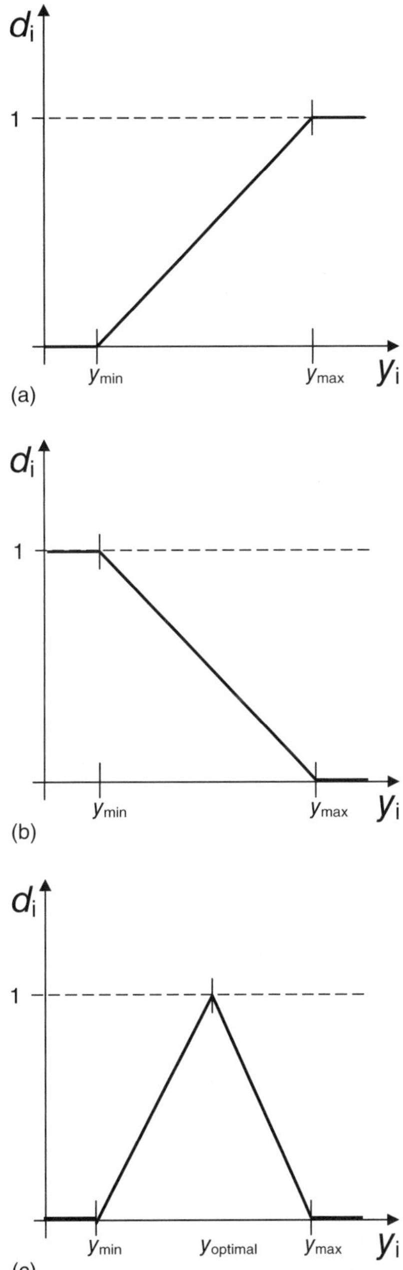

FIGURE 2.20. Derringer's desirability functions: the response is optimal when (a) maximized, (b) minimized, and (c) at a given value.

$$D = \left(\prod_{i=1}^{R} d_i \right)^{1/R} \tag{Eq. 2.35}$$

The d_i values in Equation 2.35 occasionally can be given different weights by raising them to a given power (now all powers $p = 1$), either $p > 1$ or $0 < p < 1$.

Determining the maximal D value can be done either by selecting the design experiment with the largest D value or by modeling D, following the approach discussed above for the responses of a response surface design.

2.8. SUMMARY AND CONCLUSIONS

In this chapter, the use of experimental design approaches during method development and robustness testing was discussed and illustrated with examples.

Method development is often divided into a screening and an actual optimization phase. During the screening phase, all factors potentially influencing the method should be examined, in order to determine the most important. Screening designs are applied in this phase. The results from such designs are analyzed by estimating the factor effects on the response(s), followed by a graphical and/or statistical interpretation of the estimated effects.

During the optimization phase, the most important factors are further examined. Here response surface designs or sequential optimization methods are applied. The results from response surface designs are analyzed by building a model relating the response(s) to the variables, occasionally followed by a validation of the model. Instead of using response surface designs, sequential optimization methods, such as simplex approaches, can also be applied. In a first instance, responses examined during method development are qualitative responses, related to the separation quality, and providing information concerning the qualitative aspects of the method.

Robustness testing is a part of method validation and evaluates the effects of small deliberate changes in some (method) parameters on the results (responses) of the method. All factors potentially influencing the method are examined. For this purpose, again screening designs are applied. A first difference with their application during method development is that the ranges in which the factors are examined are much smaller during robustness testing. A second difference is that the responses initially examined during robustness testing are quantitative, that is, related to the quantitative aspects of the method. However, the analysis of the results is similar to that in method development.

ACKNOWLEDGMENTS

Bieke Dejaegher is a post-doctoral fellow of the Fund for Scientific Research (FWO)—Vlaanderen, Belgium.

REFERENCES

1. Vander Heyden, Y., Perrin, C., and Massart, D.L. (2000) Optimization strategies for HPLC and CZE, in *Handbook of Analytical Separations*, Vol. 1, *Separation Methods in Drug Synthesis and Purification* (ed. K. Valkó), Elsevier, Amsterdam, pp. 163–212.
2. United States Food and Drug Administration (FDA), Department of Health and Human Services (1994) Validation of chromatographic methods, CMC3, http://www.fda.gov/ (accessed November 19, 2008).
3. Guidelines prepared within the International Conference on Harmonisation of Technical Requirements for the Registration of Pharmaceuticals for Human Use (ICH) (2005) Validation of analytical procedures: Text and methodology, Q2(R1), pp. 1–13, http://www.ich.org/ (accessed November 19, 2008).
4. Vander Heyden, Y. and Massart, D.L. (1996) Review of robustness in analytical chemistry, in *Robustness of Analytical Chemical Methods and Pharmaceutical Technological Products* (eds. M.W.B. Hendriks, J.H. de Boer, and A.K. Smilde), Elsevier, Amsterdam, pp. 79–147.
5. Vander Heyden, Y., Nijhuis, A., Smeyers-Verbeke, J., and Massart, D.L. (2001) *J Pharm Biomed Anal*, **24**, 723–753.
6. Youden, W.J. and Steiner, E.H. (1975) *Statistical Manual of the Association of Official Analytical Chemists*, The Association of Official Analytical Chemists, Washington, DC.
7. Massart, D.L., Vandeginste, B.G.M., Buydens, L.M.C., De Jong, S., Lewi, P.J., and Smeyers-Verbeke, J. (1997) *Handbook of Chemometrics and Qualimetrics: Part A*, Elsevier, Amsterdam.
8. Dejaegher, B. and Vander Heyden, Y. (2009) Chapter 17: Sequential optimization methods, in *Comprehensive Chemometrics*, Vol. 1 (eds. S. Brown, R. Tauler, and B. Walczak), Elsevier, Oxford, pp. 547–575.
9. Massart D.L., Dijkstra A., and Kaufman L. (1978) *Evaluation and Optimization of Laboratory Methods and Analytical Procedures, A Survey of Statistical and Mathematical Techniques*, Elsevier, Amsterdam.
10. Walters, F.H., Parker, L.R., Morgan, S.L., and Deming, S.N. (1991) *Sequential Simplex Optimization, A Technique for Improving Quality and Productivity in Research, Development, and Manufacturing*, CRC Press, Boca Raton, Florida.
11. Gabrielsson, J., Lindberg, N.-O., and Lundstedt T. (2002) *J Chemometrics*, **16**, 141–160.
12. Dejaegher, B. and Vander Heyden, Y. (2008) Robustness tests of CE methods, in *Capillary Electrophoresis Methods for Pharmaceutical Analysis* (eds. M. Jimidar and S. Ahuja), Elsevier, Amsterdam, pp. 185–224.
13. United States Pharmacopoeia 29th ed., National Formulary 24th ed., (2006) United States Pharmacopoeial Convention, Rockville, Maryland.
14. International Organization for Standardization (ISO) (1994(E)) Statistical methods for quality control, Accuracy (trueness and precision) of measurement methods and results—Part 3: Intermediate measures of the precision of a standard measurement method, ISO, Geneva, Vol. 2, 4th ed., 5725-3, pp. 75–104.
15. International Organization for Standardization (ISO) (1994(E)) Statistical methods for quality control, Accuracy (trueness and precision) of measurement

methods and results—Part 2: Basic method for the determination of repeatability and reproducibility of a standard measurement method, ISO, Geneva, Vol. 2, 4th ed., 5725-2, pp. 30–74.

16. Vander Heyden, Y., Questier, F., and Massart, D.L. (1998) *J Pharm Biomed Anal*, **17**, 153–168.
17. Montgomery, D.C. (1997) *Design and Analysis of Experiments*, 4th ed., John Wiley, New York.
18. Vander Heyden, Y., Questier, F., and Massart, D.L. (1998) *J Pharm Biomed Anal*, **18**, 43–56.
19. Perrin, C., Fabre, H., Massart, D.L., and Vander Heyden, Y. (2003) *Electrophoresis*, **24**, 2469–2480.
20. Poole, C.F. (2003) Chapter 8: Capillary-electromigration separation techniques, in *The Essence of Chromatography*, Elsevier, Amsterdam, pp. 619–717.
21. Li, S.F.Y. (1992) *Capillary Electrophoresis: Principles, Practice and Applications*, Journal of Chromatography Library vol. 52, Elsevier: Amsterdam.
22. Altria, K.D. (1996) Methods in molecular biology, Vol. 52, in *Capillary Electrophoresis Guidebook, Principles, Operation, and Applications* (ed. K.D. Altria), Humana Press, Totowa, New Jersey.
23. Altria, K.D. (1998) Chromatographia CE series, in *Analysis of Pharmaceuticals by Capillary Electrophoresis* (ed. K.D. Altria), Vieweg, Wiesbaden, Germany.
24. Nijhuis, A., van der Knaap, H.C.M., de Jong, S., and Vandeginste, B.G.M. (1999) *Anal Chim Acta*, **391**, 187–202.
25. van Leeuwen, J.A., Buydens, L.M.C., Vandeginste, B.G.M., Kateman, G., Schoenmakers, P.J., and Mulholland, M. (1991) *Chemometrics Intell Lab Syst*, **10**, 337–347.
26. Eurachem (1995) *A focus for Analytical Chemistry in Europe, Quantifying Uncertainty in Analytical Measurement*, 1st ed., Berlin, Germany.
27. Perrin, C., Coussot, G., Lefebvre, I., Périgaud, C., and Fabre, H. (2006) *J Chromatogr A*, **1111**, 139–146.
28. Ficarra, R., Cutroneo, P., Aturki, Z., Tommasini, S., Calabrò, M.L., Phan-Tan-Luu, R., Fanali, S., and Ficarra, P. (2002) *J Pharm Biomed Anal*, **29**, 989–997.
29. Furlanetto, S., Orlandini, S., La Porta, E., Coran, S., and Pinzauti, S. (2002) *J Pharm Biomed Anal*, **28**, 1161–1171.
30. Morgan, E. (1991) *Chemometrics: Experimental Design, Analytical Chemistry by Open Learning*, Wiley, Chichester.
31. Box, G.E.P., Hunter, W.G., and Hunter, J.S. (1978) *Statistics for Experimenters, An Introduction to Design, Data Analysis, and Model Building*, John Wiley, New York, pp. 306–418.
32. van Leeuwen, J.A., Buydens, L.M.C., Vandeginste, B.G.M. Kateman, G., Schoenmakers, P.J., and Mulholland, M. (1991) *Chemometrics Intell Lab Syst*, **11**, 37–55.
33. van Leeuwen, J.A., Buydens, L.M.C., Vandeginste, B.G.M., Kateman, G., Cleland, A., Mulholland, M., Jansen, C., Maris, F.A., Hoogkamer, P.H., and van den Berg, J.H.M. (1991) *Chemometrics Intell Lab Syst*, **11**, 161–174.
34. Questier, F., Vander Heyden, Y., and Massart, D.L. (1998) *J Pharm Biomed Anal*, **18**, 287–303.

35. Mathieu, D., Nony, J., and Phan-Tan-Luu, R. NEMROD (New Efficient Methodology for Research using Optimal Designs) software, LPRAI, Marseille, France, http://www.nemrodw.com/ (accessed November 28, 2008).
36. Modde, Umetrics, Umea, Sweden. http://www.umetrics.com/ (accessed November 28, 2008).
37. Statgraphics, Manugistics Inc., Rockville, USA. http://www.statgraphics.com/ (accessed November 28, 2008).
38. Design Ease and Design Expert, Stat-Ease Inc., Minneapolis, USA. http://www.statease.com/ (November 28, 2008).
39. Minitab, Minitab Inc., Pennsylvania, USA. http://www.minitab.com/ (November 28, 2008).
40. Unscrambler, Camo software Inc., Woodbridge, USA. http://www.camo.com/ (accessed November 28, 2008).
41. SAS, SAS Institute, North Carolina, USA Inc. http://www.sas.com/ (accessed November 28, 2008).
42. Bianchi, F., Careri, M., and Corradini, C. (2005) *J Sep Sci*, **28**, 898–904.
43. Maia, P.P., Amaya-Farfán, J., Rath, S., and Reyes, F.G.R. (2007) *J Pharm Biomed Anal*, **43**, 450–456.
44. Sänger-van de Griend, C.E., Wahlström, H., Gröningsson, K., and Widahl-Näsman, M. (1997) *J Pharm Biomed Anal*, **15**, 1051–1061.
45. Tobback, K., Li, Y,-M., Pizarro, N.A., De Smedt, I., Smeets, T., Van Schepdael, A., Roets, E., and Hoogmartens, J. (1999) *J Chromatogr A*, **857**, 313–320.
46. Mamani, M.C.V., Farfán, J.A., Reyes, F.G.R., and Rath, S. (2006) *Talanta*, **70**, 236–243.
47. Lara, F.J., García-Campaña, A.M., Alés-Barrero, F., Bosque-Sendra, J.M., and Garciá-Ayuso, L.E. (2006) *Anal Chem*, **78**, 7665–7673.
48. Schappler, J., Guillarme, D., Prat, J., Veuthey, J.-L., and Rudaz, S. (2007) *Electrophoresis*, **28**, 3078–3087.
49. Lu, C.-C., Jong, Y.-J., Ferrnace, J., Ko, W.-K., and Wu, S.-M. (2007) *Electrophoresis*, **28**, 3290–3295.
50. González, L., Akesolo, U., Jiménez, R.M., and Alonso, R.M. (2002) *Electrophoresis*, **23**, 223–229.
51. Ronda, F., Rodríguez-Nogales, J.M., Sancho, D., Oliete, B., and Gómez M. (2008) *Food Chem*, **108**, 287–296.
52. Altria, K.D., Frake, P., Gill, I., Hadgett, T., Kelly, M.A., and Rudd, D.R. (1995) *J Pharm Biomed Anal*, **13**, 951–957.
53. Altria, K.D., Bryant, S.M., and Hadgett, T.A. (1997) *J Pharm Biomed Anal*, **15**, 1091–1101.
54. Mardones, C., Vizioli, N., Carducci, C., Rios, A., and Valcárcel, M. (1999) *Anal Chim Acta*, **382**, 23–31.
55. Plackett, R.L. and Burman, J.P. (1946) *Biometrika*, **33**, 302–325.
56. Gotti, R., Furlanetto, S., Andrisano, V., Cavrini, V., and Pinzauti, S. (2000) *J Chromatogr A*, **875**, 411–422.
57. Brunnkvist, H., Karlberg, B., Gunnarsson, L., and Granelli, I. (2004) *J Chromatogr A*, **813**, 67–73.

58. Fabre, H. and Mesplet, N. (2000) *J Chromatogr A*, **897**, 329–338.

59. Owens, P.K., Wikström, H., Nagard, S., and Karlsson, L. (2002) *J Pharm Biomed Anal*, **27**, 587–598.

60. Orlandini, S., Fanali, S., Furlanetto, S., Marras, A.M., and Pinzauti, S. (2004) *J Chromatogr A*, **1032**, 253–263.

61. Gotti, R., Furlanetto, S., Pinzauti S., and Cavrini, V. (2006) *J Chromatogr A*, **1112**, 345–352.

62. Berzas-Nevado, J.J., Villaseñor-Llerena, M.J., Guiberteau-Cabanillas, C., and Rodríguez-Robledo, V. (2006) *Electrophoresis*, **27**, 905–917.

63. Mulholland, M. and Waterhouse, J. (1987) *J Chromatogr*, **395**, 539–551.

64. Mulholland, M. (1988) Trends *Anal Chem*, **9**, 383–389.

65. Vander Heyden, Y., Khots, M.S., and Massart, D.L. (1993) *Anal Chim Acta*, **276**, 189–195.

66. Toasaksiri, S., Massart, D.L., and Vander Heyden, Y. (2000) *Anal Chim Acta*, **416**, 29–42.

67. Rodríguez-Flores, J., Berzas Nevado, J.J., Contento Salcedo, A.M., and Cabello Díaz, M.P. (2005) *J Chromatogr A*, **1068**, 175–182.

68. Berzas Nevado, J.J., Guiberteau Cabanillas, C., Villaseñor Llerena, M.J., and Rodríguez Robledo, V. (2005) *J Chromatogr A*, **1072**, 249–257.

69. Rodríguez Flores, J., Berzas Nevado, J.J., Contento Salcedo, A.M., and Cabello Díaz, M.P. (2005) *Talanta*, **65**, 155–162.

70. Berzas Nevado, J.J., Castañeda Peñalvo, G., and Rodríguez Dorado, R.M. (2005) *Anal Chim Acta*, **533**, 127–133.

71. Berzas Nevado, J.J., Rodríguez Flores, J., Castañeda Peñalvo, G., and. Guzmán Bernardo, F.J. (2006) *Anal Chim Acta*, **559**, 9–14.

72. Vargas, M.G., Vander Heyden, Y., Maftouh, M., and Massart, D.L. (1999) *J Chromatogr A*, **855**, 681–693.

73. Addelman, S. (1962) *Technometrics*, **4**, 21–46.

74. Hund, E., Vander Heyden, Y., Haustein, M., Massart, D.L., and Smeyers-Verbeke, J. (2000) *Anal Chim Acta*, **404**, 257–271.

75. Hund, E., Vander Heyden, Y., Haustein, M., Massart, D.L., and Smeyers-Verbeke, J. (2000) *J Chromatogr A*, **874**, 167–185.

76. Hillaert, S., Vander Heyden, Y., and Van den Bossche, W. (2002) *J Chromatogr A*, **978**, 231–242.

77. Hillaert, S. and Van den Bossche, W. (2002) *J Chromatogr A*, **979**, 323–333.

78. Hillaert, S., Snoeck, L., and Van den Bossche, W. (2004) *J Chromatogr A*, **1033**, 357–362.

79. Capella-Peiró, M.E., Bossi, A., and Esteve-Romero, J. (2006) *Anal Biochem*, **352**, 41–49.

80. Galeano-Díaz, T., Acedo-Valenzuela, M.-I., Mora-Díez, N., and Silva-Rodríguez, A. (2005) *Electrophoresis*, **26**, 3518–3527.

81. Box, G.E.P. and Behnken, D.W. (1960) *Ann Math Stat*, **31**, 838–864.

82. Ferreira, S.L.C., Bruns, R.E., Ferreira, H.S., Matos, G.D., David, J.M., Brandão, G.C., da Silva, E.G.P., Portugal, L.A., dos Reis, P.S., Souza, A.S., and dos Santos, W.N.L. (2007) *Anal Chim Acta*, **597**, 179–186.

83. Martinez-Gomez, M.A., Villanueva-Camañas, R.M., Sagrado, S., and Medina-Hernández, M.J. (2005) *Electrophoresis*, **26**, 4116–4126.

84. Montes, R.E., Gomez, F.A., and Hanrahan, G. (2008) *Electrophoresis*, **29**, 375–380.

85. Luces, C.A., Fakayode, S.O., Lowry, M., and Warner, I.M. (2008) *Electrophoresis*, **29**, 889–900.

86. Montes, R.E., Hanrahan, G., and Gomez, F.A. (2008) *Electrophoresis*, **29**, 3325–3332.

87. Dahdouh, F.T., Clarke, K., Salgado, M., Hanrahan, G., and Gomez, F.A. (2008) *Electrophoresis*, **29**, 3779–3785.

88. Doehlert, D.H. (1970) *Appl Statist*, **19**, 231–239.

89. Lara, F.J., García-Campaña, A.M., Gámiz-Gracia, L., Bosque-Sendra, J.M., and Alés-Barrero, F. (2006) *Electrophoresis*, **27**, 2348–2359.

90. García-Campaña, A.M., Rodríguez, L.C., González, A.L., Alés-Barrero, F., and Ceba, M.R. (1997) *Anal Chim Acta*, **348**, 237–246.

91. Bourguignon, B., de Aguiar, P.F., Khots, M.S., and Massart, D.L. (1994). *Anal Chem*, **66**, 893–904.

92. Jimidar, M., de Aguiar, P.F., Pintelon, S., and Massart, D.L. (1997) *J Pharm Biomed Anal*, **15**, 709–728.

93. Kennard, R.W. and Stone, L.A. (1969) *Technometrics*, **11**, 137–148.

94. de Aguiar, P.F., Bourguignon, B., Khots, M.S., Massart, D.L., and Phan-Than-Luu, R. (1995) *Chemometrics Intell Lab Syst*, **30**, 199–210.

95. Fradi, I., Servais, A.-C., Pedrini, M., Chiap, P., Iványi, R., Crommen, J., and Fillet, M. (2006) *Electrophoresis*, **27**, 3434–3442.

96. de Aguiar, P.F., Bourguignon, B., Khots, M.S., and Massart, D.L. (1997) *Anal Chim Acta*, **356**, 7–18.

97. Torres-Lapasió, J.R., Massart, D.L., Baeza-Baeza, J.J., and García-Alvarez-Coque, M.C. (2000) *Chromatographia*, **51**, 101–110.

98. Spendley, W., Hext, G.R., and Himsworth, F.R. (1962) *Technometrics*, **4**, 441–461.

99. Burton, K.W.C. and Nickless, G. (1987) *Chemometrics Intell Lab Syst*, **1**, 135–149.

100. Nelder, J.A. and Mead, R. (1965) *Computer J*, **7**, 308–313.

101. Morgan, E., Burton, K.W., and Nickless, G. (1990) *Chemometrics Intell Lab Syst*, **7**, 209–222.

102. Morgan, S.L. and Deming, S.N. (1974) *Anal Chem*, **46**, 1170–1181.

103. Aberg, E.R. and Gustavsson, A.G.T. (1982) *Anal Chim Acta*, **144**, 39–53.

104. Vannecke, C., Nguyen Minh Nguyet, A., Bloomfield, M.S., Staple, A.J., Vander Heyden, Y., and Massart, D.L. (2000) *J Pharm Biomed Anal*, **23**, 291–306.

105. Vander Heyden, Y., Jimidar, M., Hund, E., Niemeijer, N., Peeters, R., Smeyers-Verbeke, J., Massart, D.L., and Hoogmartens, J. (1999) *J Chromatogr A*, **845**, 145–154.

106. Vander Heyden, Y., Bourgeois, A., and Massart, D.L. (1997) *Anal Chim Acta*, **347**, 369–384.

107. Goupy, J. (1993) *Methods for Experimental Design, Principles and Applications for Physicists and Chemists*, Elsevier, Amsterdam.
108. Draper, N.R. and Smith, H. (1981) *Applied Regression Analysis*, 2nd edn, Wiley, New York.
109. Mulholland, M. (1996) Ruggedness tests for analytical chemistry, in *Robustness of Analytical Chemical Methods and Pharmaceutical Technological Products* (eds. M.W.B. Hendriks, J.H. de Boer, and A.K. Smilde), Elsevier, Amsterdam, pp. 191–232.
110. Vander Heyden, Y., Luypaert, K., Hartmann, C., Massart, D.L., Hoogmartens, J., and De Beer, J. (1995) *Anal Chim Acta*, **312**, 245–262.
111. Lenth, R.V. (1989) *Technometrics*, **31**, 469–473.
112. Dong, F. (1993) *Stat Sin*, **3**, 209–217.
113. Haaland, P.D. and O'Connell, M.A. (1995) *Technometrics*, **37**, 82–93.
114. Dejaegher, B., Durand, A., and Vander Heyden, Y. *J Chromatogr B* doi:10.1016/j.jchromb.2008.10.019 (in press).
115. Dejaegher, B., Capron, X., Smeyers-Verbeke, J., and Vander Heyden, Y. (2006) *Anal Chim Acta*, **564**, 184–200.
116. Bezerra, M.A., Santelli, R.E., Oliveira, E.P., Villar, L.S., and Escaleira, L.A. (2008) *Talanta*, **76**, 965–977.
117. Smilde, A.K., Knevelman, A., and Coenegracht, P.M.J. (1986) *J Chromatogr*, **369**, 1–10.
118. Keller, H.R., Massart, D.L., and Brans, J.P. (1991) *Chemometrics Intell Lab Syst*, **11**, 175–189.
119. Derringer, G. and Suich, R. (1980) *J Qual Technol*, **12**, 214–219.
120. Bourguignon, B. and Massart, D.L. (1991) *J Chromatogr A*, **586**, 11–20.

CHAPTER 3

CHEMOMETRICAL EXPERIMENTAL DESIGN-BASED OPTIMIZATION STUDIES IN CAPILLARY ELECTROPHORESIS APPLICATIONS

RUTHY MONTES,[1] TONI ANN RIVEROS,[1] FROSEEN DAHDOUH,[1] GRADY HANRAHAN,[2] and FRANK A. GOMEZ[1]
[1]Department of Chemistry & Biochemistry, California State University, Los Angeles, CA
[2]Department of Chemistry, California Lutheran University, Thousand Oaks, CA

CONTENTS

3.1. INTRODUCTION

Over the past two decades, capillary electrophoresis (CE) has emerged as a powerful and versatile separation tool due to its high sensitivity, resolution, and ability to detect minute quantities of samples (1–11). It is an excellent tool for many types of bioanalyses and is an unparalleled experimental tool for biophysical studies of interactions in biologically relevant media. CE differentiates charged species on the basis of mobility differences under the influence of an applied electric field. Selectivity can be manipulated by the alteration of

Chemometric Methods in Capillary Electrophoresis. Edited by Grady Hanrahan and Frank A. Gomez

electrolyte properties such as pH, ionic strength, and electrolyte composition, or by the incorporation of electrolyte additives.

In our laboratory, work related to CE and its applications has focused on two techniques: affinity capillary electrophoresis (ACE) and electrophoretically medicated microanalysis (EMMA). Both techniques have proven to be quite useful in providing insight into the physicochemical properties of biological materials. Since the first papers in 1992 (12–15) documenting its use in measuring affinity parameters between biological species, ACE has become a staple in many laboratories in probing a variety of receptor–ligand interactions including protein–drug, protein–DNA, peptide–peptide, peptide–carbohydrate, carbohydrate–drug, and antibody–antigen (12–30). In ACE, the resolving power of CE is used to distinguish between free and bound forms of a receptor as a function of the concentration of free ligand in the electrophoresis buffer. In a typical form of ACE, a sample of receptor and standard(s) is exposed to an increasing concentration of ligand in the running buffer, causing a shift in the migration time of the receptor relative to the standard(s). In EMMA, differential electrophoretic mobility is utilized to merge distinct zones of analyte and analytical reagent(s) under the influence of an applied electric field. The reaction is allowed to proceed within the region of reagent overlap either in the presence or absence of an applied potential, and the resultant product is transported to the detector under the influence of an electric field (31–45). Many studies have detailed the use of EMMA in examining a plethora of enzyme systems resulting in the development of an excellent compliment to traditional biological assay techniques.

Several multivariate chemometric-based techniques including response surface methodology (RSM) have been developed to aid in the optimization of a given system's performance. The use of chemometrics in high performance liquid chromatography, mass spectrometry, atomic absorption, and other techniques is well documented (46–50). Whereas most work detailing the use of chemometrics in CE has focused on peak separation and how to best optimize the resolution of overlapped species, few studies have examined specific applications where solutions might be more universally applied in the examination of small molecules or macromolecular species (51–53). Herein, this chapter will describe our work in optimizing experimental conditions in ACE (flow-through partial-filling ACE [FTPFACE] and competitive binding FTPFACE [CBFTPFACE]) and EMMA.

3.2. RESULTS AND DISCUSSION

3.2.1. FTPFACE

In the first study, we used chemometrics RSM to predict extent of protein–ligand binding in FTPFACE (51). In FTPFACE, the capillary is partially filled with ligand (or receptor), and a sample plug of receptor (or ligand) is introduced into the capillary and electrophoresed (20). During electrophoresis, zones of sample overlap, then one flows through the other but not before equilibrium is established prior to detection. As long as the time of contact between ligand and receptor is sufficient for equilibrium to result, a value for K_d can be estimated. In this work, the value for K_d was estimated using one noninteracting standard, which relates changes in the electrophoretic mobility of carbonic anhydrase B (CAB, E.C.4.2.1.1) on complexation with 4-carboxybenzenesulfonamide present in the electrophoresis buffer. Experimental factors including injection time, capillary length, and applied voltage were selected and tested at three levels in a Box–Behnken design. Statistical analysis results were used to create a mathematical model for response surface prediction via contour and surface plots at a given target response of $K_d = 1.19 \times 10^{-6}$ M. The adequacy of the model was validated by experimental runs with the predicted model solution (capillary length = 47 cm, voltage = 11 kV, injection time = 0.01 min).

The design matrix (including actual and model predicted responses) generated for the Box–Behnken study is shown in Table 3.1. Here, three center point experiments were incorporated to compute an estimate of the error term that does not depend on the fitted model. Figure 3.1a shows the whole model leverage plot of actual-versus-predicted responses (based on all effects) with the quality of fit expressed by the coefficient of determination (r^2). This coefficient is variation in the response around the mean that can be attributed to terms in the model rather than to random error.

Typically, points on the leverage plot are actual data coordinates, and the horizontal line, the sample mean of the response. Here we have multiple effects, with the horizontal line representing a partially constrained model instead of a model fully constrained to a single mean value. As shown, the confidence curves (dashed lines) cross the horizontal line, thus the test is considered significant at the 5% level. Overall, an r^2 value of 0.89 was obtained with a mean response of 1.57. Analysis of variance for a linear regression partitions the total variation of a sample into components. Effect test results (Table 3.2) revealed that injection time and capillary length had significant single effects on the target response. The only significant interactive effect was capillary length × injection time. Here, Prob > F is the significance probability for the F-ratio.

Figure 3.1b shows the contour profiles of injection time-versus-capillary length. Two others (not shown) include voltage-versus-capillary length and

TABLE 3.1. Experimental factors and levels used in the Box–Behnken design (reprinted with permission from Reference 51)

Factor	Level (–)	Level (0)	Level (+)
Capillary length (cm)	37	47	57
Voltage (kV)	5.0	12.5	20
Injection time (min)	0.01	0.11	0.20

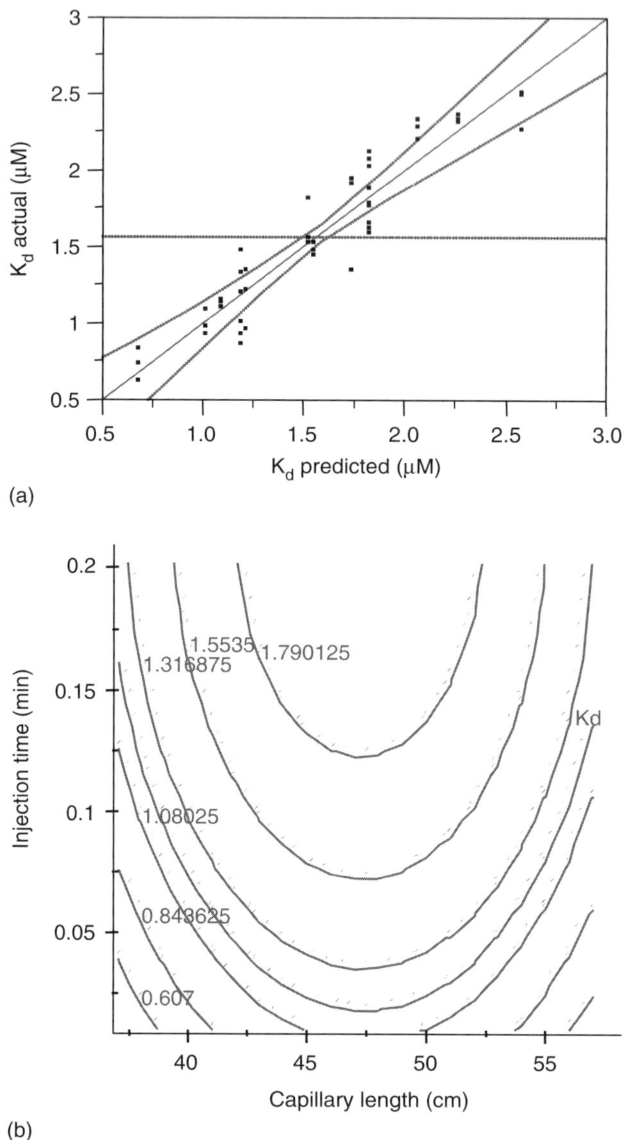

FIGURE 3.1. (a) Whole model leverage plot of actual-versus-predicted responses and (b) model generated contour plots showing injection time-versus-capillary length. (Reprinted with permission from Reference 51.)

voltage-versus-injection time. Here, we have assessed how the predicted values change with respect to changing each factor, two at a time. As before, a target value of $K_d = 1.19 \times 10^{-6}$ M was set, and the adjusted response surface glider moved along the axes of each combination of factors until the levels of factors reached the target response. As expected, there were a number of

TABLE 3.2. Effect test results for the Box–Behnken design (reprinted with permission from Reference 51)

Term	Estimate	Sum of Squares	*F*-Ratio	Prob > *F*
Capillary length	1	0.4180	10.803	0.0023
Voltage	1	0.0000	0.0000	0.9654
Injection time	1	2.1195	54.765	<0.0001
Capillary length × voltage	1	0.1279	3.306	0.0773
Capillary length × injection time	1	0.5146	13.296	0.0008
Voltage × injection time	1	0.0005	0.8200	0.3664

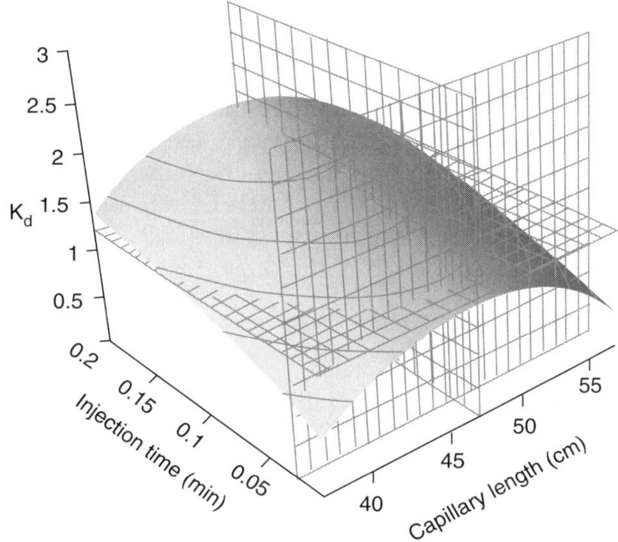

FIGURE 3.2. Response surface generated plot showing main interaction injection time-versus-capillary length. (Reprinted with permission from Reference 51.)

predicted solutions that reached our target response based on the significance of each factor at appropriate levels. This is very important in situations where one or more factors cannot be varied at a large range of levels (as in the case of capillary length in the above studies). Here, we were limited to set capillary lengths of 37, 47, and 57 cm due to the nature of the commercial instrument setup.

Representative resolution response surfaces in function of one of the chosen factors and levels (from the contour plot analysis) that reached our predicted response are depicted in Figure 3.2. Here, a control changes to a drop-down list of predefined resolutions for density grids in the JMP software. Too coarse a resolution means a function with a sharp change might not be represented as well, but setting the resolution high makes evaluating and displaying the surface slower. Grids parallel to each axis were generated to further enhance the response surface effects for interpretation purposes.

The generated model was validated experimentally by a representative series of electropherograms of CAB in capillaries partially filled with increasing concentrations of (0–25 μM) of **1** run at optimized conditions (Fig. 3.3). CAB is a zinc protein of the lyase class that catalyzes the equilibration of dissolved carbon dioxide and carbonic acid. It is strongly inhibited by sulfonamide-containing molecules.

At the point of detection, separate peaks for CAB, horse heart myoglobin (HHM), and mesityl oxide (MO) are observed. The complex that forms between CAB and **1** is more negatively charged than CAB uncomplexed and, hence, the peak for the complex shifts to longer migration time on increasing the concentration of **1** partially filled in the capillary column. A fourth peak (designated with an asterisk [*]) appears under the original CAB peak and is designated as inactive CAB as a result of using an older sample of CAB in some of our studies. This inactive CAB does not affect the measurement of a binding constant. The zone of ligand, typically seen in FTPFACE when the ligand is chromophoric, was observed after the maximal value of the x-axis shown in Figure 3.3. CAA(+) is an isozyme of CAB and gives values of K_d indistinguishable from CAB. A binding constant of 1.29×10^{-6} M was obtained, an 8.4% discrepancy difference from the target response (1.19×10^{-6} M).

3.2.2. CBFTPFACE

In the second study, we used RSM to optimize conditions for CBFTPFACE (52). In this technique, the capillary was first partially filled with a negatively charged ligand, a sample containing CAB and two noninteracting standards, and a neutral ligand, then electrophoresed (Fig. 3.4). Upon application of a voltage, the sample plug migrates into the plug of negatively charged ligand (L_-), resulting in the formation of a CAB-L_- complex. Continued electrophoresis results in mixing between the neutral ligand (L_o) and the CAB-L_- complex. L_0 successfully competes out L_- to form the new CAB-L_o complex. Analysis of the change in migration time relative to the standards yields a value for K_d.

For this study, three factors (injection time, voltage, and [L_0]) were chosen and tested at three levels in a Box–Behnken response surface design (Table 3.3). The design matrix (including actual [experimental] and model predicted responses) generated for the Box–Behnken study is shown in Table 3.4. Here, three center point experiments were incorporated to compute an estimate of the error term that does not depend on the fitted model. A whole model levarage plot (not shown) was generated to show actual values of the response plotted againt the model predicted values with the quality of fit expressed by the r^2. This coefficient is the variation in the response around the mean that can be attributed to terms in the model rather than to random error.

In the present work, a dual marker form of analysis, called the relative migration time ratio (RMTR), was used to obtain a value for K_b between a

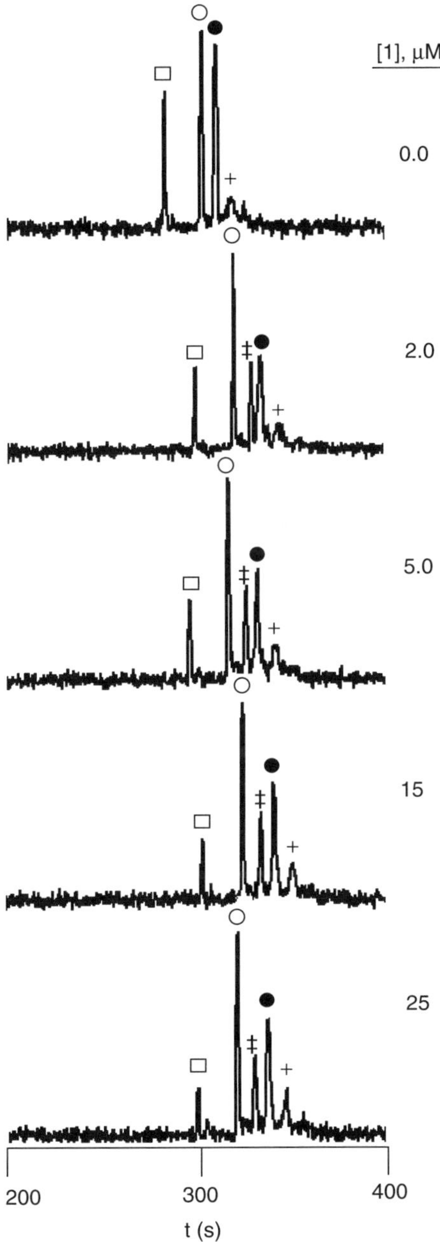

FIGURE 3.3. A representative set of electropherograms of CAB (darkened circle) in 192 mM glycine-25 mM Tris buffer (pH 8.3) containing various concentrations of **1** using the flow-through partial-filling affinity capillary electrophoresis (FTPFACE) technique. The total analysis time in each experiment was 7.0 min at 11 kV (current 2.8 μA) using a 47-cm (inlet to detector), 50-μm I.D. open, uncoated quartz capillary. MO (open square) and HHM (open circle) were used as internal standards. The asterisk (*) and cross (+) are discussed in the text. (Reprinted with permission from Reference 51.)

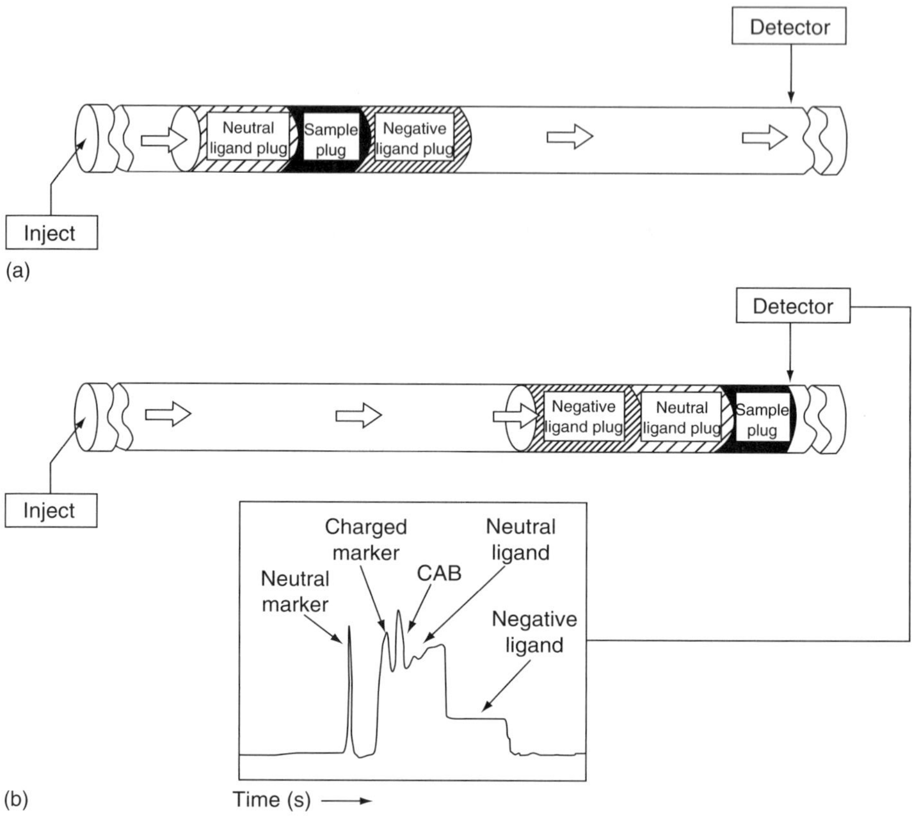

FIGURE 3.4. Schematic of a competitive binding FTPFACE experiment. The sample plug is enlarged to best pictorially represent the technique. (Reprinted with permission from Reference 52.)

TABLE 3.3. Experimental factors and levels used in the Box–Behnken design (reprinted with permission from Reference 52)

Factor	Level (–)	Level (0)	Level (+)
Injection time (min)	1	2.5	4
Voltage (kV)	5.0	12.5	20
$[L_0]$ (μM)	1	5	9

neutral arylsulfonamide and CAB relative to two noninteracting standards (Eq. 3.1):

$$RMTR = (t_r - t_{s'})/(t_{s'} - t_s) \quad \text{(Eq. 3.1)}$$

Here, $t_{s'}$ and t_s are the measured migration times for the noninteracting markers (MO and HHM), and t_r is the migration time for CAB. Equation 3.2 is used to

TABLE 3.4 Box–Behnken design matrix including mean actual (experimental) and model predicted responses (reprinted with permission from Reference 52)

Experiment	Injection Time (min)	Voltage (kV)	$[L_o]$ (μM)	Mean Actual Response ($\Delta RMTR$) (n = 3)	Model Predicted Response ($\Delta RMTR$) (n = 3)
1	1	5	5	1.83842	1.84415
2	4	5	5	1.80813	1.81802
3	1	20	5	1.83929	1.83974
4	4	20	5	1.87217	1.87679
5	2.5	12.5	1	2.01261	2.02874
6	2.5	12.5	9	1.75861	1.73236
7	2.5	12.5	5	1.83419	1.82945
8	2.5	12.5	5	1.82916	1.82945
9	2.5	12.5	5	1.82501	1.82945
10	2.5	5	5	1.84946	1.83383
11	2.5	20	5	1.86609	1.86101
12	1	12.5	1	2.02052	2.01454
13	4	12.5	1	2.0476	2.03746
14	1	12.5	9	1.76667	1.76646
15	4	12.5	9	1.75883	1.75445

obtain the value for K_b of L_0, from the change in RMTR as a function of the $[L_0]$ ($\Delta RMTR/[L_0]$) for the interaction of L_-, and CAB on a relative time scale with noninteracting markers (21). Equation 3.2 represents the general equation used to obtain a linear plot and hence the K_b of the competitive binding system we are investigating.

$$\Delta RMTR_{R,L}/[L_0] = (1 - \Delta RMTR_{R,L})/(K_{bo}^{-1})(1 + ([L](K_b)) \quad \text{(Eq. 3.2)}$$

Here, K_{b-} and L_- are the known values for the binding constant and concentration of ligand **1** in the running buffer, respectively. In this experiment, K_{b-} and L_- are $4.16 \times 10^6\,M^{-1}$ and 20 μM, respectively. The values of $\Delta RMTR_{R,L}/[L_0]$ over a range of concentrations of neutral ligand (0, 2.5, 5.0, 7.5, 9.0, 15, 20, 30, 40, and 50 μM) were then used for analysis. A maximum $\Delta RMTR_{R,L}/[L_0]$ response value is expected to yield a target K_{bo} of $2.50 \times 10^6\,M^{-1}$ that is based on previous work on FTPFACE and with the charged arylsulfonamide system.

We found that voltage and $[L_o]$ had significant single effects on the response ($\Delta RMTR$) with Prob > F values of 0.0135 and <0.0001, respectfully. Interestingly, injection time did not have a significant single effect (Prob > F = 0.5505) on $\Delta RMTR$ but was significant (Prob > F = 0.0186) as an interactive effect when combined with voltage. The shift in the migration time of the peak for CAB is predicated on both the amount of time the zone of sample exists within the plug of neutral ligand and the concentration of that neutral ligand, the former being greatly dependent on the voltage. The values of injection time chosen for this study are sufficient to cause equilibrium to be achieved between ligand and receptor unbeknownst to voltage. This is not to say that any voltage will be sufficient to create an equilibrium but that typical voltage values (most ACE studies are run in excess of 20 kV) will yield accurate values for the binding constant. In the present experiment, the extreme values for the voltage outweighed the injection time and had an effect on the experimental values for $\Delta RMTR$. Such an interaction would not have been detectable by use of classical univariate optimization methods.

A graphical plot display of all single effects and the significant interactive effect in relation to $\Delta RMTR$ leverage residuals can be obtained. Such a plot allows closer examination and maximum insight into how the fit carries the data and shows for each point what the residual would be both with and without that effect in the model.

The quadratic model (Eq.3.3) allowed the generation of the 3-D response surface image (Fig. 3.5) for the main interaction between injection time and voltage. The quadratic terms in this equation models the curvature in the true response function. The shape and orientation of the curvature results from the eigenvalue decomposition of the matrix of second-order parameter estimates. After the parameters are estimated, critical values for the factors in the estimated surface can be found. For this study, a post hoc review of our model

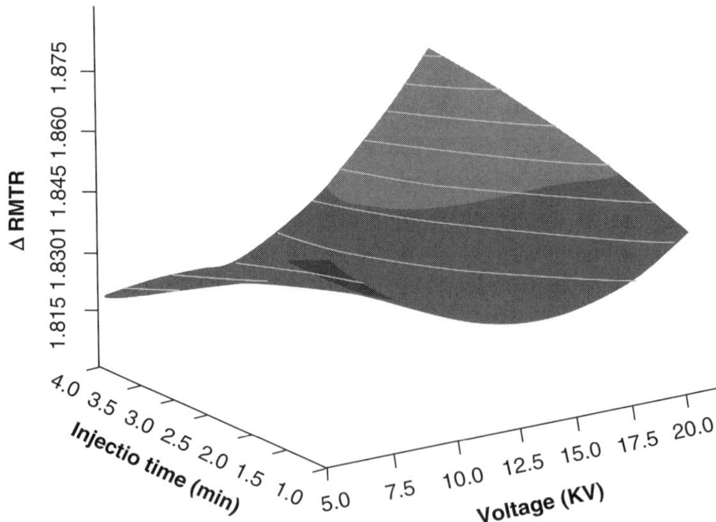

FIGURE 3.5. Response surface generated plot showing the main interactive effect injection time × voltage. (Reprinted with permission from Reference 52.)

revealed optimum critical values of injection time = 2.3 min, voltage = 11.6 kV, and $[L_0]$ = 1.4 μM.

$$Y = \beta_0 + \beta_1 X_1 + \beta_2 X_2 + \beta_3 X_3 + \beta_{12} X_1 X_2 + \beta_{13} X_1 X_3 + \beta_{23} X_2 X_3 + \beta_{11} X_1^2 + \beta_{22} X_2^2 + \beta_{33} X_3^2 \quad \text{(Eq. 3.3)}$$

The generated optimized model was then validated experimentally by a representative series of replicate ($n = 6$) electropherograms (Fig. 3.6) of CAB and markers (HHM and MO) in capillaries partially filled with increasing $[L_0]$ = 1.4 μM run at the conditions of injection time = 2.3 min, voltage = 11.6 kV. Under these conditions, the experimental and model predicted had a percent discrepancy difference of only 10.1%. The generated model predicted optimal conditions were further validated experimentally by an ACE experiment (results not shown).

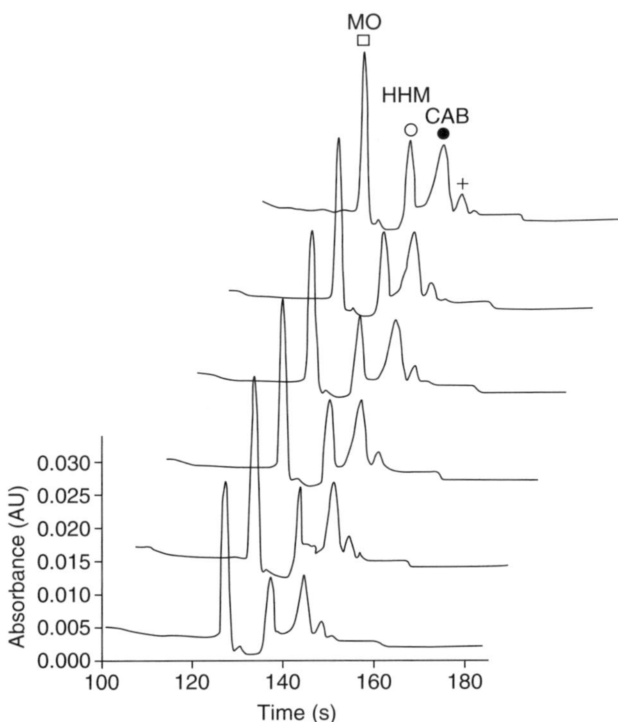

FIGURE 3.6 A representative set of stacked electropherograms of CAB in 0.192 M glycine-0.025 M Tris buffer (pH 8.3) containing **2** (1.4 μM) using the competitive binding FTPFACE technique. The total analysis time in each experiment was 3.0 min at 11.6 kV (current, 6.0–13.6 μA) using a 30.5-cm (inlet to detector), 50-μm ID open, uncoated quartz capillary. MO (open square) and HHM (open circle) were used as internal standards. (Reprinted with permission from Reference 52.)

3.2.3. EMMA

In a third study, we used RSM in EMMA by examining the optimization of reaction conditions for the conversion of nicotinamide adenine dinucleotide (NAD) to nicotinamide adenine dinucleotide, reduced form (NADH) by glucose-6-phosphate dehydrogenase (G6PDH, EC 1.1.1.49) in the conversion of glucose-6-phosphate (G6P) to 6-phosphogluconate (53). Experimental factors including voltage (V), enzyme concentration (E), and mixing time of reaction (M) at the applied voltage were selected at three levels and tested in a Box–Behnken response surface design. Upon migration in a capillary under CE conditions, plugs of substrate and enzyme are injected separately in buffer and allowed to react at variable conditions (Fig. 3.7). Extent of reaction and product ratios were subsequently determined by CE. The model predicted results are shown to be in good agreement (7.1% discrepancy difference) with experimental data.

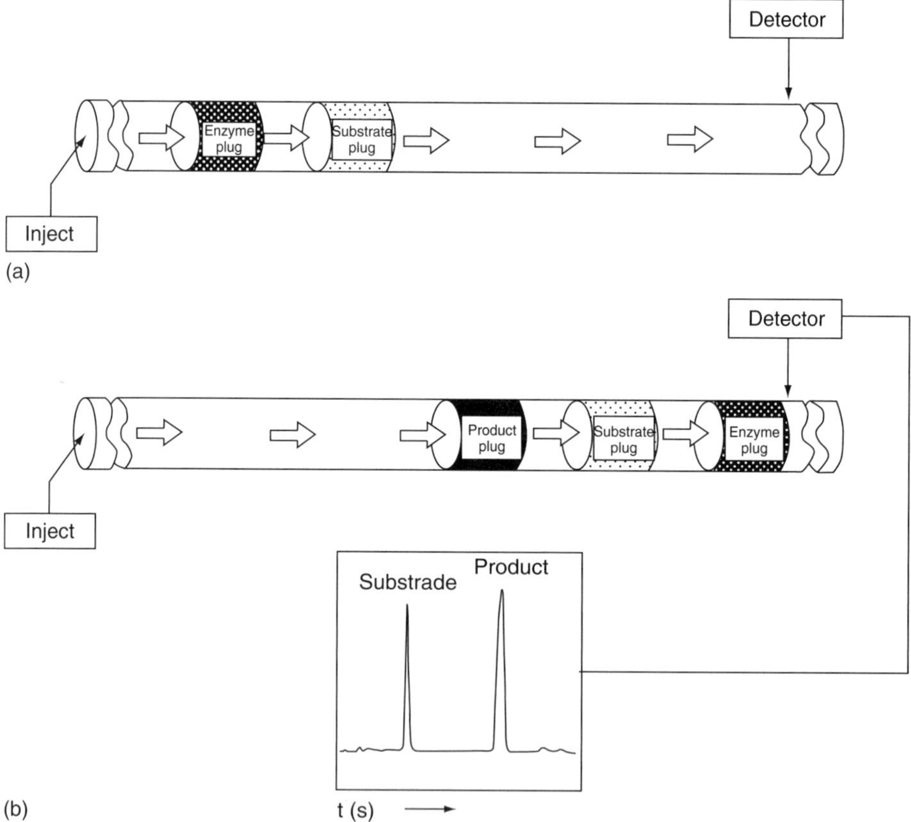

FIGURE 3.7. Schematic representation of an in-capillary enzyme-catalyzed microreactor (a) before reaction and (b) after reaction. (Reprinted with permission from Reference 53.)

Table 3.5 shows the three electrophoretic factors and levels selected in which experimental optimization, in terms of overall response (% conversion), could be performed. A design matrix was then generated for the Box–Behnken study (Table 3.6). It was found that voltage and mixing time, when combined, had a significant effect on % conversion. Here, the extent of contact between substrate and enzyme is dictated by the difference in electrophoretic mobilities, which is in turn dictated by mixing time and voltage. Such an interaction would not have been possible by use of classical univariate optimization methods.

The quadratic model from the Box–Behnken design allowed us to generate a response surface image (Fig. 3.8) for the main interaction voltage and mixing time. Here, we assessed how the predicted responses change with respect to changing these factors simultaneously, while keeping enzyme concentration constant. A post hoc review of our model revealed optimum critical values of: mixing time = 0.78 min, voltage = 13.2 kV, enzyme concentration =

TABLE 3.5. Experimental factors and levels used in the Box–Behnken design (reprinted with permission from Reference 53)

Factor	Level (–)	Level (0)	Level (+)
Mixing time (*M*) (min)	0.2	0.8	1.4
Voltage (*V*) (kV)	1.0	13	25
Enzyme concentration (*E*) (mg/L)	0.5	2.0	3.5

TABLE 3.6. Box–Behnken design matrix with mean predicted and experimental responses (reprinted with permission from Reference 53)

Experiment	Mixing Time (min)	Voltage (kV)	Enzyme Concentration (mg/mL)	Mean Experimental Response (% Conversion) (n = 3)	Mean Model Predicted Response (% Conversion) (n = 3)
1	0.2	1.0	2.0	24.1	21.2
2	1.4	1.0	2.0	24.3	22.4
3	0.2	25	2.0	26.3	22.8
4	1.4	25	2.0	24.9	22.5
5	0.8	13	0.5	8.10	4.30
6	0.8	13	3.5	38.8	32.0
7	0.8	13	2.0	30.6	30.8
8	0.8	13	2.0	32.4	30.8
9	0.8	13	2.0	29.3	30.8
10	0.8	1.0	2.0	15.0	19.7
11	0.8	25	2.0	14.7	20.5
12	0.2	13	0.5	3.60	5.90
13	1.4	13	0.5	5.40	6.80
14	0.2	13	3.5	30.1	33.9
15	1.4	13	3.5	31.0	33.9

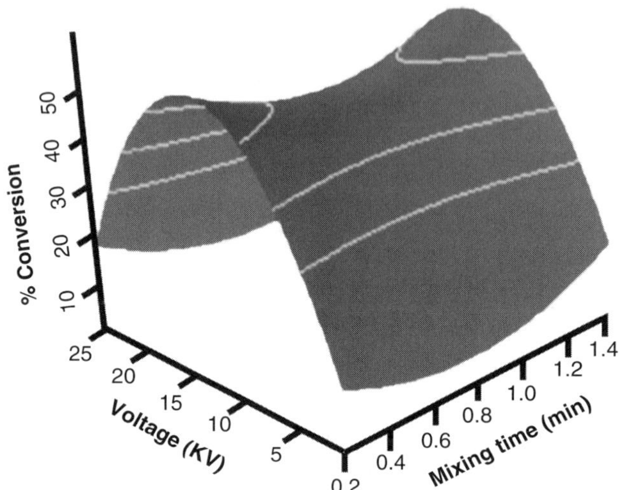

FIGURE 3.8. Response surface image for the main interactive effect of voltage/mixing time at predicted critical values with enzyme concentration kept constant. (Reprinted with permission from Reference 53.)

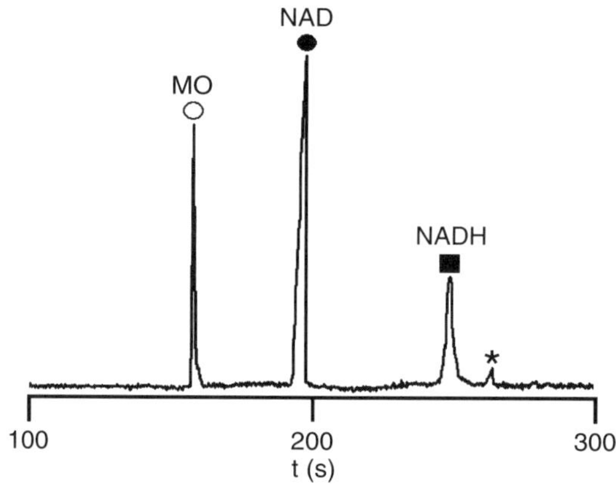

FIGURE 3.9. Representative electropherogram showing the separation of NAD and NADH after reaction with G6PDH in 30 mM Tris buffer (pH 7.85). The total analysis time in this experiment was 8.0 min at 13.2 kV (current 22.8 µA) using a 40.5-cm (inlet to detector), 50-µm I.D. open, uncoated capillary. Mesityl oxide (MO) was used as an internal standard. The peak marked * is an impurity. (Reprinted with permission from Reference 53.)

2.82 mg/mL, and a predicted conversion of 31.2%. A series of five validation experiments using the optimum critical values were performed. A mean experimental conversion of 29.0% was obtained with a 7.1% discrepancy difference from the model predicted. The generated model was validated experimentally by a representative electropherogram (Fig. 3.9) showing the separation of NAD and NADH after reaction with G6PDH.

3.3. CONCLUSIONS

There is both a great need to assess many compounds expeditiously and accurately and to optimize their experimental protocols via high-throughput techniques and those involving CE than at any time in history. Chemometrical experimental design and optimization techniques in CE have been instrumental in separating multicomponent environmental samples, DNA fragments, soluble organic acids, and chiral molecules that otherwise proved troublesome. We have described three applications (FTPFACE, CBFTPFACE, and EMMA) in CE that have benefited from chemometrics. It can be concluded that this approach yielded a large amount of information while minimizing the number of experimental runs. Such an approach is having significant impacts in separation science and will no doubt be a major area of study for years to come. This work provides further basis for integrating chemometrics in CE and especially in applications where optimizing experimental conditions are time-consuming, require large amounts of expensive reagents, and/or where a univariate approach to optimization yields results of marginal confidence and accuracy.

ACKNOWLEDGMENT

The authors gratefully acknowledge financial support for this research by grants from the National Science Foundation (CHE-0515363 and DMR-0351848).

REFERENCES

1. Clohs, L. and McErlane, K.M. (2001) *J Pharm Biomed Anal*, **24**, 545–554.
2. Guzman, N.A. (2004) *Anal Bioanal Chem*, **378**, 37–39.
3. Flurer, C.L. (2001) *Electrophoresis*, **22**, 4249–4261.
4. Thormann, W., Theurillat, R., Wind, M., and Kuldvee, R. (2001) *J Chromatogr A*, **924**, 429–437.
5. Amundsen, L.K. and Siren, H. (2007) *Electrophoresis*, **28**, 99–113.
6. Villareal, V., Zhang, Y., Zurita, C., Moran, J., Silva, I., and Gomez, F.A. (2003) *Anal Lett*, **36**, 451–463.
7. Novotny, M.V., Hong, M., Cassely, A., and Mechref, A. (2001) *J Chromatogr A*, **752**, 207–213.
8. Busby, B.M. and Vigh, G. (2005) *Electrophoresis*, **26**, 3849–3860.
9. Simal-Gándara, J. (2004) *Crit Rev Anal Chem*, **34**, 85–94.
10. Landers, J.P. (1997) *Handbook of Capillary Electrophoresis*, CRC Press LLC, Boca Raton, FL.
11. Villareal, V., Kaddis, J., Azad, M., Zurita, C., Silva, I., Hernandez, L., Rudolph, M., Moran, J., and Gomez, F.A. (2003) *Anal Bioanal Chem*, **376**, 822–831.

12. Kraak, J.C., Bush, S., and Poppe, H. (1992) *J Chromatogr*, **608**, 257–264.
13. Chu, Y.-H. and Whitesides, G.M. (1992) *J Org Chem*, **57**, 3524–3525.
14. Heegaard, N.H.H. and Robey, F.A. (1992) *Anal Chem*, **64**, 2479–2482.
15. Chu, Y.-H., Avila, L.Z., Biebuyck, H.A., and Whitesides, G.M. (1992) *J Med Chem*, **35**, 2915–2917.
16. Gomez, F.A., Mirkovich, J.N., Dominguez, V.M., Liu, K.W., and Macias, D.M. (1996) *J Chromatogr A*, **727**, 291–299.
17. Rundlett, K.L. and Armstrong, D.W. (1997) *Electrophoresis*, **18**, 2194–2202.
18. Qian, X.-H. and Tomer, K.B. (1998) *Electrophoresis*, **19**, 415–419.
19. Colton, J.J., Carbeck, J.D., Rao, J., and Whitesides, G.M. (1998) *Electrophoresis*, **19**, 367–382.
20. Heintz, J., Hernandez, M., and Gomez, F.A. (1999) *J Chromatogr A*, **840**, 261–268.
21. Mito, E., Zhang, Y., Esquivel, S., and Gomez, F.A. (2000) *Anal Biochem*, **280**, 209–215.
22. Varenne, A., Gareil, P., Colliec-Jouault, S., and Daniel, R. (2003) *Anal Biochem*, **315**, 152–159.
23. Buchanan, D.D., Jameson, E.E., Perlette, J., Malik, A., and Kennedy, R.T. (2004) *Electrophoresis*, **24**, 1375–1382.
24. Taga, A., Yamamoto, Y., Maruyama, R., and Honda, S. (2004) *Electrophoresis*, **25**, 876–881.
25. Castagnola, M., Rossetti, D.V., Inzitari, R., Lupi, A., Zuppi, C., Cabras, T., Fadda, M.B., Onnis, G., Petruzzelli, R., Giardina, B., and Messana, I. (2004) *Electrophoresis*, **25**, 846–852.
26. Azad, M., Brown, A., Silva, I., and Gomez, F.A. (2004) *Anal Bioanal Chem*, **379**, 149–155.
27. Zhang, Y., Kodama, C., Zurita, C., and Gomez, F.A. (2001) *J Chromatogr A*, **928**, 233–241.
28. Mito, E. and Gomez, F.A. (1999) *Chromatographia*, **50**, 689–694.
29. Azad, M., Hernandez, L., Plazas, A., Rudolph, M., and Gomez, F.A. (2003) *Chromatographia*, **57**, 339–347.
30. Zhang, Y. and Gomez, F.A. (2000) *J Chromatogr A*, **897**, 339–347.
31. Harmon, B.J., Patterson, D.H., and Regnier, F.E. (1993) *Anal Chem*, **65**, 2655–2662.
32. Patterson, D.H., Harmon, B.J., and Regnier, F.E. (1994) *J Chromatogr A*, **662**, 389–394.
33. Patterson, D.H., Harmon, B.J., and Regnier, F.E. (1996) *J Chromatogr A*, **732**, 119–132.
34. Zhao, D.S. and Gomez, F.A. (1998) *Electrophoresis*, **19**, 420–426.
35. Zhao, D.S. and Gomez, F.A. (1997) *Chromatographia*, **44**, 514–520.
36. Kwak, E.-S., Esquivel, S., and Gomez, F.A. (1999) *Anal Chim Acta*, **397**, 183–190.
37. Zhang, Y., El-Maghrabi, R., and Gomez, F.A. (2000) *Analyst*, **125**, 685–689.
38. Avila, L.Z. and Whitesides, G.M. (1993) *J Org Chem*, **58**, 5508–5512.

39. Van Dyck, S., Van Schepdael, A., and Hoogmartens, J. (2002) *Electrophoresis*, **23**, 2854–2859.
40. Whisnant, A.R., Johnston, S.E., and Gilman, S.D. (2000) *Electrophoresis*, **21**, 1341–1348.
41. Xue, Q. and Yeung, E. (1995) *Nature*, **373**, 681–683.
42. Burke, B.J. and Reginer, F.E. (2003) *Anal Chem*, **75**, 1786–1791.
43. Glatz, Z. (2006) *J Chromatogr A*, **841**, 23–28.
44. Lewis, L.M., Engle, L.J., Pierceall, W.E., Hughes, D.E., and Shaw, K.J. (2004) *J Biomol Screen*, **9**, 303–308.
45. Brown, A., Desharnais, R., Roy, B.C., Mallik, S., and Gomez, F.A. (2005) *Anal Chim Acta*, **540**, 403–409.
46. Li, G., Zhou, X., Wang, Y., El-Shafey, A., Chiu, N.H., Krull, I.S. (2004) *J Chromatogr A*, **1053,** 253–263.
47. Dinc, E., Ozdemir, A., Aksoy, H., Ustundag, O., and Baleanu, D. (2006) *Chem Pharm Bull*, **54**, 415–421.
48. Damiani, P.C., Orraccetti, M.D.B., and Olivieri, A.C. (2002) *Anal Chim Acta*, **471**, 87–96.
49. Lonni, A.A.S.G., Scarminio, I.S., Silva, L.M.C., and Ferreira, D.T. (2003) *Anal Sci*, **19**, 1013–1017.
50. Duarte, A. and Capelo, S. (2006) *J Liq Chromatogr Rel Technol*, **29**, 1143–1176.
51. Xu, F., Gong, F., Dixon, S.J., Brereton, R.G., Soini, H.A., Novotny, M.V., Oberzaucher, E., Grammer, K., and Penn, D.J. (2007) *Anal Chem*, **79**, 5633–5641.
52. Hanrahan, G., Montes, R.E., Pao, A., Johnson, A., and Gomez, F.A. (2007) *Electrophoresis*, **28**, 2853–2860.
53. Montes, R.E., Gomez, F.A., and Hanrahan, G. (2008) *Electrophoresis*, **29**, 375–380.

CHAPTER 4

APPLICATION OF CHEMOMETRIC METHODS IN DRUG PURITY DETERMINATION BY CAPILLARY ELECTROPHORESIS

GERHARD K.E. SCRIBA
Department of Pharmaceutical Chemistry, Friedrich Schiller University of Jena, Jena, Germany

CONTENTS

4.1. INTRODUCTION

Within the last 25 years, capillary electrophoresis (CE) has developed as a high-resolution analytical technique that has been applied to all analytical fields including chemical, pharmaceutical, biomedical, forensic, environmental analysis, and food sciences. Based on the number of publications, drugs are actually the preferred analytes in CE. While they served as model compounds for the investigation of specific aspects in some studies, CE has been used to solve "real" pharmaceutical problems in the majority of applications.

CE can be operated at a similar performance and level of automation as high performance liquid chromatography (HPLC), and it has many

Chemometric Methods in Capillary Electrophoresis. Edited by Grady Hanrahan and Frank A. Gomez

advantages compared to HPLC in terms of rapid method development and lower operating costs due to reduced consumption of chemicals and samples. However, the major strength of CE is the fact that the separation principle is different from chromatographic techniques so that CE and HPLC are in fact a powerful combination for the analysis of complex molecules. Generally, the scope of applications of CE in pharmaceutical analysis is identical to that of HPLC. Therefore, often a choice between the two techniques has to be made. In recent years, an increasing number of pharmaceutical companies have included CE methods in early drug discovery testing and routine quality control as well as in documents for regulatory submission. CE methods are accepted by the regulatory authorities such as the U.S. Food and Drug Administration and the European Agency for the Evaluation of Medicinal Products, and the technique has been implemented as an analytical method by the United States Pharmacopeia and the European Pharmacopoeia. Numerous validated and robust CE methods for pharmaceutical analysis have been published as summarized in review papers (1, 2), book chapters (3), and monographs (4, 5).

In CE, factors such as buffer pH, concentration and type of the background electrolyte, applied voltage, and temperature of the capillary, as well as buffer additives such as surfactants, organic solvents, ion-pairing reagents, complexing agents, influence a separation. Therefore, the effects of many of these factors on the separation of the analytes are investigated and subsequently optimized during the method development process in order to obtain a reproducible and robust method. In the classical univariate approach, a given experimental parameter is varied within a specified range while the other experimental variables are held constant. Upon determination of the optimal value, the next parameter is subsequently investigated. This approach may lead to reasonable analytical conditions but requires a large number of experiments. Moreover, as many experimental variables in CE affect each other, the univariate approach is not rational and may not result in the best available experimental conditions. In contrast, chemometric methods for experimental design allow the simultaneous investigation of the interdependent experimental variables using a limited number of experiments. This represents a rational approach finding optimized and robust CE methods. Moreover, besides method development, chemometric methods can also be applied to the determination of the robustness of the analytical assay. Depending on the intended purpose, simplex, factorial, and response surface designs may be applied.

To date, the use of chemometrics for method development and robustness testing has been published for all areas of CE, including capillary zone electrophoresis (CZE), capillary electrokinetic chromatography (EKC) using chiral selectors for enantioseparations, micellar electrokinetic chromatography (MEKC), and microemulsion electrokinetic chromatography (MEEKC). A comprehensive description can be found in Chapters 5 and 13 as well as in recent reviews (6–11). Several monographs on chemometrics in analytical chemistry have been published such as References 12–14. This chapter will highlight the use of experimental design in the development of methods for the analysis of related compounds in drug substances for the determination of the impurity profile.

4.2. EXPERIMENTAL DESIGN IN METHOD DEVELOPMENT

The aim of method development in any analytical separation technique is to obtain an assay that allows the successful separation of the analytes of interest in a short analysis time, with high reproducibility and ruggedness. In recent years, chemometrics have been applied to screening for the identification of significant variables, method optimization, and robustness testing in order to minimize the number of overall experiments. The objective of screening is to explore many factors in order to reveal whether they have an influence on the responses and to identify their appropriate ranges. The purpose of optimization is to predict the response values for all possible combinations of factors within a given experimental design region and to identify the optimal experimental parameters. Robustness testing is performed to ascertain that the method is robust to small changes in the factor levels and (if nonrobustness is detected) to understand how to alter the bounds of the factors so that robustness may still be claimed.

Experimental factors to be examined include buffer pH, concentration and type of the background electrolyte, applied voltage, temperature of the capillary, as well as buffer additives such as organic solvents, ion-pairing reagents, complexing agents, surfactants, or cosurfactants. Typical dependent responses include peak resolution, analysis time, electric current, etc. Optimization can be based on a single response, but often, multiple criteria decisions utilizing two or more responses are applied.

Depending on the objective, that is, screening, optimization, or robustness testing, different experimental designs have been employed in CE. Some designs often used by analytical chemists in method development and the general information obtained from the designs are summarized in Table 4.1. In screening experiments for the detection of the most influential factors, two-level factorial, fractional factorial, or Plackett–Burman designs are frequently used. Due to their simplicity, two-level factorial designs are very useful for preliminary studies or in initial steps of an optimization while fractional factorial designs are preferred to investigate a higher number of variables as the number of experiments is decreased compared to (full) factorial designs. Highly fractional designs such as Plackett–Burman allow the screening of the effect of a large number of variables with a limited number of experiments. However, as only a low number of experiments at just two levels of the factors are investigated, the models fitted to these designs are somewhat restricted. Consequently, if more sophisticated models are required to study interrelated factors, the use of response surface models, which employ more than two factor levels to allow fitting to quadratic polynominals, is appropriate. Factorial

TABLE 4.1. Experimental designs used in method development in capillary electrophoresis

Design	Utilization	Response Surface	Interaction Studies
Factorial	Screening (two levels)	No	All
	Optimization (three levels)	Yes	All
Fractional factorial	Screening	No	Selected
Plackett–Burman	Screening	No	Selected
Central composite	Optimization	Yes	All
Box–Behnken	Optimization	Yes	All
Doehlert	Optimization	Yes	All
D-optimal	Optimization	Yes	All

designs utilizing more than two levels can be used, but the number of experiments increases exponentially with the number of factors studied. Two often-applied designs used in response surface modeling in CE methods are central composite and Box–Behnken designs. Central composite designs combine factorial or fractional factorial designs with additional points (star points) to allow estimation of curvature and at least one point at the center. Typically, three replicates of the center point are included to estimate the validity of the model. For fitting quadratic response models, central composite designs are a better alternative than (full) factorial three-level designs because the performance is comparable using a lower number of experiments.

Box–Behnken designs are based on incomplete three-level factorial designs. The special arrangement of the levels allows the number of points to increase at the same rate as the number of polynomial coefficients. Only three or four factors are typically studied, but experimenting in the corners of the range of the variables is avoided. Thus, a Box–Behnken design is especially appropriate when predicting the response at the extremes is not required. Less frequently applied models include Doehlert and *D*-optimal designs, which apply fewer experiments so that they are especially attractive when a large number of factors have to be studied. A two-level Doehlert design consists of a hexagon, a three-level design consist of a dodecahedron so that these can be extended in any direction by adding new experiments. *D*-optimal designs create an "irregular" experimental matrix and appear attractive for several situations, for example, when a large number of factors (six or more) have to be studied or when a certain experimental section cannot be investigated. For a detailed discussion of the various experimental designs, see Chapter 5 and monographs such as References 12–14.

The selection of the appropriate design largely depends on the requirements of the study. For example, an initial screening approach using a fractional factorial or a Plackett–Burman design can be employed to identify the significant variables, which are subsequently studied in more detail by response surface methodology such as a central composite design. With regard to robustness testing, fractional factorial and Plackett–Burman designs are often applied. In addition, robustness can be estimated from the curvature of the response surfaces of central composite, Box–Behnken designs, etc. (15). Several commercial software packages are available, that is, Design Ease, Design Expert, MODDE, StatGraphics, etc., which can assist in design selection and statistical evaluation of the generated data.

4.3. APPLICATIONS OF CHEMOMETRIC METHODS IN DRUG PURITY DETERMINATION

CE has been employed in pharmaceutical analysis for the determination of drugs, including small organic molecules, peptides and proteins, or oligonucleotide pharmaceuticals as well as inorganic ions. CE methods have been developed for main component analysis, the determination of drug-related impurities and inorganic counter ions, chiral analysis as well as for the bioanalysis of drugs and metabolites in biological fluids. The analysis of pharmaceuticals by CE has been summarized in reviews (1, 2), book chapters (3), and books (4, 5).

In pharmaceutical analysis, the demonstration of the purity of a drug as a substance or in a formulation is essential. Besides known impurities that can be explained as reaction by-products or degradation products, often, unknown impurities may be present. As a high-resolution technique, CE is suitable for analyzing closely related substances in drugs as demonstrated by a large number of sensitive, validated methods published in the literature. CZE as well as MEKC assays have been elaborated and CE methods were included in regulatory submission files. Often, identical operational parameters suitable for main component analysis can be applied to the determination of the impurities. In addition to the analysis of the purity of pharmaceuticals, CE may also be used for the profiling of illicit drugs in forensic sciences.

Currently, regulatory agencies demand the identification and quantitation of impurities at the 0.1% level. The International Conference on Harmonization (ICH) guideline Q3A(R2) (16) as well as the United States Pharmacopeia and the European Pharmacopoeia state that impurities have to be reported if they are present above 0.05% (reporting threshold), identified if above 0.1% (identification threshold), and qualified if above 0.15% (qualification threshold). These limits apply to drugs with a maximum daily dose of 2 g per day or below; lower limits apply for drugs with a higher daily intake. Substances isolated from natural sources or produced by fermentation and by DNA recombinant technology are explicitly excluded.

In drug purity analysis when several (closely related) compounds have to be separated, the methods have to be optimized with regard to multiple criteria, including the resolution between analytes that react sensitively to changes of the experimental conditions (so-called critical pairs) and/or analysis time. Sometimes, multiple critical pairs exist. Thus, experimental design appears to be well suited for the development and optimization of such methods. Despite the fact that the majority of drug purity determinations were developed by the univariate approach, an increasing number of experimental design applications have been reported in the recent literature. Examples for the analysis of related compounds of drugs by CE are summarized in Table 4.2. These include methods by CZE, EKC, which employs a chiral selector as pseudostationary phase, MEKC, as well as MEEKC.

4.3.1. Analysis of Related Substances in Drugs

Several studies have employed chemometric designs in CZE method development. In most cases, central composite designs were selected with background electrolyte pH and concentration as well as buffer additives such as methanol as experimental factors and separation selectivity or peak resolution of one or more critical analyte pairs as responses. For example, method development and optimization employing a three-factor central composite design was performed for the analysis of related compounds of the tetracycline antibiotics doxycycline (17) and metacycline (18). The separation selectivity between three critical pairs of analytes were selected as responses in the case of doxycycline while four critical pairs served as responses in the case of metacycline. In both studies, the data were fitted to a partial least square (PLS) model. The factors buffer pH and methanol concentration proved to affect the separation selectivity of the respective critical pairs differently so that the overall optimized methods represented a compromise for each individual response. Both methods were subsequently validated and applied to commercial samples.

In most studies, the related impurities were available as reference compounds for method development. This may not be the case in the analysis of drugs from natural sources as illustrated by the analysis of kanamycin (20). Kanamycin is an aminoglycoside antibiotic produced by fermentation of a streptomyces strain. The antibiotic is a mixture of five closely related triglycosides, kanamycin A–D and 1-N-(1-hydroxymethyl-2-hydroxyethyl)-kanamycin B, and three diglcyosides, paromamine, 4-O-(6-amino-6-deoxy-α-D-glucopyranosyl)-deoxystreptamine, and 6-O-(3-amino-3-deoxy-α-D-glucopyranosyl)-deoxystreptamine (Fig. 4.1a) with kanamycin A as the major component. Further derivatives may be present. Although the impurity criteria stated in ICH guideline Q3A(R2) (16) do not apply to such compounds, their analytical characterization is necessary to ensure their safe use. Initially, a borate buffer, pH 10, containing 7.5 mM β-cyclodextrin and 12.5% methanol, was investigated because this background electrolyte successfully separated the components of the related aminoglycoside antibiotic gentamycin. However, only an unsatisfactory separation of the components of kanamycin was achieved. Upon investigation of MEKC conditions and further buffer additives, the authors settled for a borate buffer containing methanol. This background electrolyte was optimized initially by a two-level factorial design with four variables (pH, borate concentration, temperature, and methanol content) and

TABLE 4.2. Examples of CE methods for the determination of related compound optimized by chemometric design

Drug	CE Mode	Chemometric Design	Studied Factors	Optimized Responses	Reference
Doxycycline	CZE	Central composite	pH, BGE concentration, % MeOH	Separation selectivity between three critical pairs	(17)
Metacycline	CZE	Central composite	pH, BGE concentration, % MeOH	Separation selectivity between four critical pairs	(18)
Mirtazapine	CZE	Central composite	pH, BGE concentration, % MeOH	Separation selectivity critical pair, analysis time	(19)
Kanamycin sulfate	CZE	Fractional factorial (screening), Central composite (optimization)	pH, % MeOH, temperature	Number of peaks separated	(20)
Mizolastine	EKC	Doehlert	CD concentration, BGE concentration, temperature, voltage	Peak resolution between three critical pairs, analysis time	(21)
Ibuprofen and codeine	MEKC	Fractional factorial	pH, BGE concentration, % ACN, SDS concentration, temperature, voltage	Peak resolution between two critical pairs, analysis time	(22)
Ketorolac	MEKC	*D*-optimal	pH, BGE concentration, SDS concentration, temperature, voltage	Peak resolution for three critical pairs, analysis time	(23)
Ketorolac	MEEKC	Mixture (Scheffé)	% buffer, % n-heptane, % SDS/n-butanol	Peak resolution for four critical pairs, analysis time	(24)
		Central composite	Voltage, temperature	Peak resolution, analysis time	

TABLE 4.2. ***Continued***

Drug	CE Mode	Chemometric Design	Studied Factors	Optimized Responses	Reference
Tyr-D-Arg-Phe-PheNH$_2$	EKC	Plackett–Burman (screening)	CD concentration, TEA concentration, BGE concentration, % MeOH, % ACN, ionic strength, temperature, voltage	Peak resolution, analysis time	(25)
		Factorial (optimization)	CD concentration, TEA concentration, % MeOH, % ACN	Peak resolution, analysis time	
		Central composite (optimization)	CD concentration, % ACN	Peak resolution, analysis time	
Calcium levofolinate	EKC	Central composite	pH, BGE concentration, temperature, voltage	Peak resolution between two critical pairs, analysis time	(26)
Escitalopram	EKC	Central composite	CD concentration, BGE concentration, temperature, voltage	Peak resolution between two critical pairs, analysis time, current	(27)
R209130	EKC	Box–Behnken	BGE concentration, concentration of two CDs, voltage	Peak resolution between seven pairs of analytes, analysis time	(28)

BGE = background electrolyte; MeOH = methanol; ACN = acetonitrile; TEA = triethanolamine; CD = cyclodextrin; SDS = sodium dodecyl sulfate.

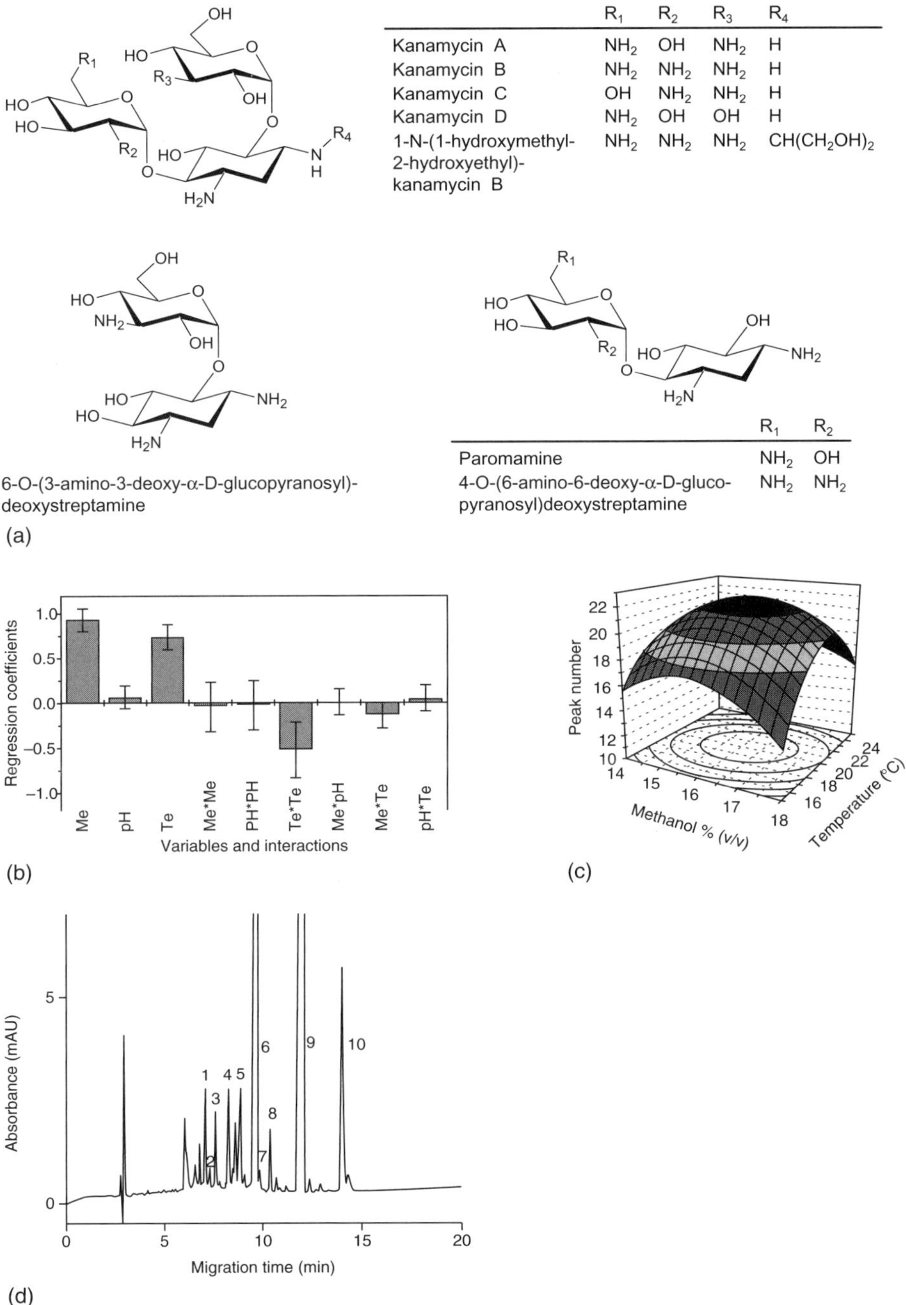

	R_1	R_2	R_3	R_4
Kanamycin A	NH_2	OH	NH_2	H
Kanamycin B	NH_2	NH_2	NH_2	H
Kanamycin C	OH	NH_2	NH_2	H
Kanamycin D	NH_2	OH	OH	H
1-N-(1-hydroxymethyl-2-hydroxyethyl)-kanamycin B	NH_2	NH_2	NH_2	$CH(CH_2OH)_2$

	R_1	R_2
Paromamine	NH_2	OH
4-O-(6-amino-6-deoxy-α-D-gluco-pyranosyl)deoxystreptamine	NH_2	NH_2

FIGURE 4.1. (a) Structures of kanamycin and related aminoglycosides. (b) Regression coefficients of variables. (c) Response surface plots of peak number as a function of the significant factors methanol content and capillary temperature. (d) Resulting electropherogram utilizing the optimized conditions. 1: reagent, 2: 2-deoxystreptamine, 3: kanamycin D, 4: 6-O-(3-amino-3-deoxy-α-D-glucopyranosyl)deoxystreptamine, 5: 4-O-(6-amino-6-deoxy-α-D-glucopyranosyl)deoxystreptamine, 6: kanamycin A, 7: 1-N-(1-hydroxymethyl-2-hydroxyethyl)kanamycin B, 8: kanamycin C, 9: picric acid (internal standard), 10: kanamycin B. (Adapted with permission from Reference 20.)

three center points. The influence of buffer concentration was insignificant so that further optimization was achieved by a central composite design considering only pH, temperature, and methanol concentration as variables in a narrow range. The response was the number of peaks separated. Figure 4.1b illustrates the regression coefficients and Figure 4.1c the response surface for the significant factors methanol concentration and temperature. The resulting electropherogram using the optimized experimental conditions is shown in Figure 4.1d. As the compounds have no chromophore, derivatization by o-phthalaldehyde and mercaptoacetic acid was performed prior to the CE analysis. This derivatization reaction was also optimized using chemometrics. Upon identification of o-phthalaldeyhde concentration and mercaptoacetic acid concentration as significant factors out of four parameters investigated by a two-level factorial design, the method was further optimized by a central composite design. The response was the ratio of the corrected peak areas of kanamycin A and the internal standard picric acid (20).

Furlanetto and coworkers optimized MEKC and MEEKC methods for the analysis of the nonsteroidal anti-inflammarory drug ketorolac trometamine and three known impurities by chemometric design. In MEKC, a borate/phosphate buffer was selected as background electrolyte and sodium dodecyl sulfate (SDS) as surfactant. Subsequent optimization employed a *D*-optimal design with buffer pH, buffer concentration, SDS concentration, voltage, and capillary temperature as variables resulting in a 25-run matrix with three replicates at the center (21). As responses, the resolutions between the respective pairs of compounds and migration time of the last migrating analyte were selected. The optimized conditions were derived from a Derringer desirability function analysis resulting in a run time of about 6 min. The method was subsequently validated according to the ICH guideline Q2(R1) (29) and applied to the analysis of the drug in tablets. In a follow-up study, an MEEKC assay was developed for ketorolac (24). The Scheffé mixture design was used to optimize the microemulsion as mixture designs appeared especially suitable for blending problems. The percentage of the aqueous phase (10 mM borate buffer, pH 9.2), n-heptane as oil phase, and the surfactant/cosurfactant ratio (SDS/n-butanol) was investigated by a 13-run matrix with peak resolution between analytes and analysis time as responses. Two microemulsion systems were derived from a Derringer desirability function. Using the two optimized microemulsions, applied voltage and column temperature were studied as further factors in a five-level central composite design in order to shorten the overall analysis time while retaining peak resolution. One system proved to result in better overall performance resulting in an analysis time of less than 3.5 min when using short-end injection. Robustness was checked by a six-factor 11-run *D*-optimal design. The method was validated and compared to the previously developed MEKC assay as well as a capillary electrochromatography method.

A complex sequence of experimental designs was applied in the separation of the tetrapeptide Tyr-(D)Arg-Phe-PheNH_2 from related di-, tri-, and tetrapeptides resulting from hydrolysis and/or side reactions during synthesis of the peptide (25). The CE method was developed using three experimental designs in a four-step procedure in which eight variables were investigated in a total of 47 experiments. The aim of the initial experiments (step 1) was the selection of the type of the run buffer and the pH range. Based on the pK_a values of the peptides, an acidic pH was selected. A malonic acid/malonate buffer, pH 2.5, separated all 10 analytes using a polyvinyl alcohol-coated capillary to suppress adsorption of the basic peptides to the capillary wall. In step 2, a Plackett–Burman design investigating eight variables was applied that reduced the number of experiments required to test eight variables in a two-level factorial design from 2^8 to 12. Three replicates for estimation of reproducibility were included so that overall 15 experiments were conducted. The concentration of the additives 2,6-dimethyl-β-cyclodextrin, triethanolamine, methanol, and acetonitrile proved to significantly affect peak resolution and migration time. These were subsequently investigated in a two-level factorial design with a total of 19 experiments again including three replicates at the center point for reproducibility estimation (step 3). The cyclodextrin concentration influenced both peak resolution and migration time, while methanol increased analysis time and acetonitrile reduced the migration times. Thus, only cyclodextrin and acetonitrile concentration were further optimized in an 11-run circumscribed central composite design including axial points and three replicates at the center point. The response surfaces or resolution and migration time are shown in Figure 4.2. No distinguishable optimum could be found. Furthermore, the results indicated that acetonitrile did not improve analysis

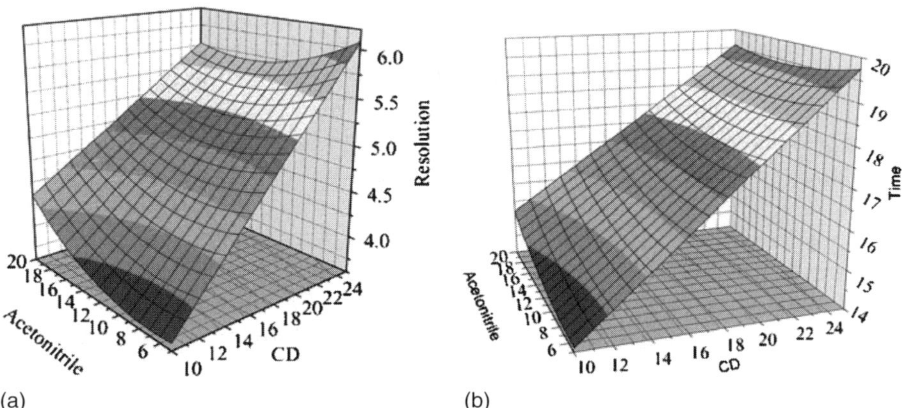

FIGURE 4.2. Response surface plots of the (a) resolution between (D)Arg-Phe-PheOH and Tyr-(D)Arg-Phe-PheNH$_2$ and (b) migration time of Phe-PheOH as the last migrating impurity as a function of the significant variables acetonitrile content and concentration of 2,6-dimethyl-β-cyclodextrin. (Reproduced with permission from Reference 25.)

time, and a low cyclodextrin concentration results in short analysis time while retaining good resolution. The final conditions were validated.

4.3.2. Simultaneous Determination of Chiral and Achiral Impurities

For chiral drugs, the desired pharmacological activity may reside in one stereoisomer while the other(s) may be less active, inactive, posses a different pharmacological activity, or may even be toxic. In such cases, the regulatory authorities demand the development of the stereochemically pure drug. Despite the fact that stereochemical impurities are excluded from the requirements of ICH guideline Q3A(R2) (16), there is general consent that they should be treated in the same manner as the related substances (2, 30, 31). Consequently, analytical methods for the determination of the stereochemical purity have to be implemented. Traditionally, the stereochemical composition of a drug is determined by optical rotation. While still being frequently applied by the pharmacopeias in the case of pure drug substances, the method is too inaccurate for regulatory purposes where HPLC and, more recently, CE methods are included. In CE, chiral separation is achieved by complex formation between the analyte stereoisomers and a chiral selector added to the background electrolyte. The resulting transient diastereomeric complexes differ in complexation constants and/or complex mobilities, resulting in a chiral separation. As complex formation is a chromatographic principle, while transport of analytes through the capillary is an electrophoretic principle, such CE methods are also termed EKC. Cyclodextrins are by far the most frequently used chiral selectors (32, 33).

CE has developed into the premier technique for enantioseparations, and chemometric designs for method development and optimization have been applied in many cases which will not be discussed here. However, as a high-resolution technique, CE offers the possibility for the simultaneous determination of the related substances as well as stereochemical impurities in drug substances. This is rarely achieved by chromatographic techniques. An example is the stereospecific CE assay for the simultaneous analysis of related substances and the enantiomeric purity of the antidepressant drug escitalopram (27). The compounds are shown in Figure 4.3. Based on published enantioseparations of the racemate citalopram, several neutral and charged cyclodextrin derivatives were screened as chiral selectors for citalopram and the precursor citadiol in the pH range 2.3–6.2. Baseline separation of the analyte enantiomers was observed in a phosphate buffer, pH 2.5, using 5 mg/mL sulfated β-cyclodextrin, but strong peak tailing was observed. Reversing the polarity of the applied voltage and exploiting the carrier ability of the charged cyclodextrin, good resolution of the enantiomers but considerable peak fronting was observed at a concentration of 15 mg/mL sulfated β-cyclodextrin. Peak shape improved using a dual cyclodextrin system by adding 0.5–1.0 mg/mL native β-cyclodextrin. As escitalopram is marketed as the oxalate or the bromide, the anions were included in the study. Employing a

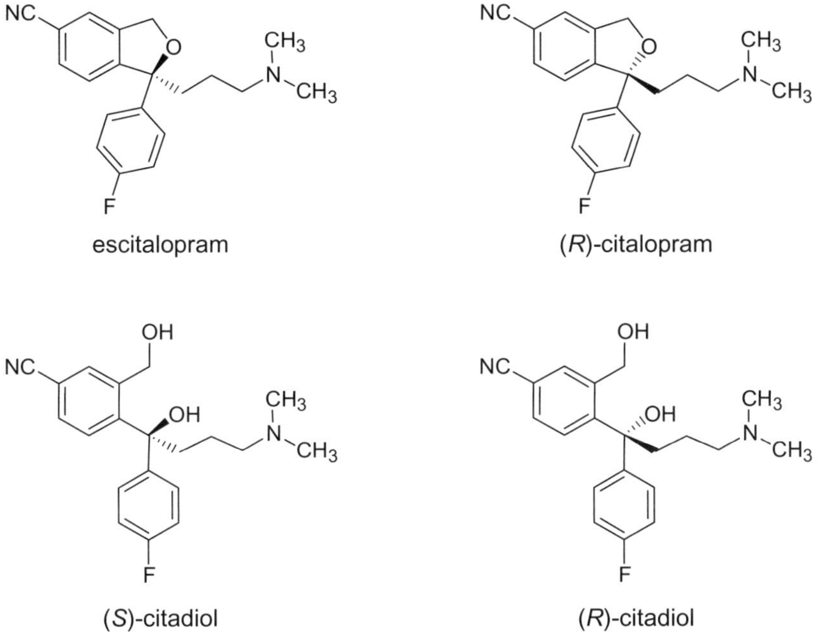

escitalopram (*R*)-citalopram

(*S*)-citadiol (*R*)-citadiol

FIGURE 4.3. Structures of citalopram and citadiol enantiomers.

35-mM sodium phosphate buffer, pH 2.5, containing 15 mg/mL sulfated β-cyclodextrin and 0.5 mg/mL β-cyclodextrin as background electrolyte under reversed polarity led to the migration order bromide > oxalate > (*R*)-citalopram > escitalopram > (*S*)-citadiol > (*R*)-citadiol. Four factors, concentration of sulfated β-cyclodextrin, buffer concentration, applied voltage, and column temperature, were subsequently studied in a central composite face-centered design including three center points. The concentration of β-cyclodextrin was kept constant at 0.5 mg/mL as preliminary experiments revealed no significant effect in the range of 0.5–2.0 mg/mL. Moreover, pH was set at 2.5 because the drug and the related substances are all basic and always protonated in the acidic pH range. The resolution between the citalopram enantiomers as well as the resolution between oxalate and (*R*)-citalopram, the migration time of the last migrating compound, and the electrical current were selected as responses. The current was included because it increases with increasing concentrations of sulfated β-cyclodextrin. High currents will lead to loss in resolution and unstable run conditions due to extensive Joule heating. The individual experiments carried out in random order and the respective results are summarized in Table 4.3.

The resolution between the citalopram enantiomers always exceeded 4.0 and was therefore excluded from further considerations. The resolution between oxalate and (*R*)-citalopram exceeded 2 except for runs with low concentrations of sulfated β-cyclodextrin. Thus, essentially, only the

TABLE 4.3. Central composite face-centered design matrix for method optimization for escitalopram and related substances showing the factors sulfated β-cyclodextrin (S-β-CD) concentration, buffer concentration, voltage and temperature, and the results for the responses resolution, R_S, between the citalopram enantiomers and between oxalate and (R)-citalopram as well as migration time, MT, and electric current (Modified from Reference 27 with permission)

Experiment Number	S-β-CD Conc [mg/mL]	Buffer Conc [mM]	Voltage [kV]	Temp [°C]	R_S (*S*)-cit/(*R*)-cit	R_S ox/(*R*)-cit	MT (min)	Current (μA)
1	10	20	15	20	5.82	2.73	15.5	27.5
2	30	20	15	20	6.89	3.98	13.8	52.5
3	10	20	15	30	5.35	2.39	14.9	33
4	30	20	15	30	5.88	3.61	10.9	66
5	10	20	25	20	5.06	4.13	9.8	48.5
6	30	20	25	20	8.49	4.06	7.0	102.5
7	10	20	25	30	4.62	2.26	9.7	60
8	30	20	25	30	6.48	3.79	6.6	130
9	10	50	15	20	11.45	0.93	28.6	39
10	30	50	15	20	4.56	5.36	14.6	79
11	10	50	15	30	7.00	0.74	22.0	48
12	30	50	15	30	4.07	5.00	12.0	98
13	10	50	25	20	8.52	0.96	14.6	72
14	30	50	25	20	4.33	5.74	7.9	164
15	10	50	25	30	6.56	0.39	17.2	88
16	30	50	25	30	5.43	4.00	7.5	195
17	10	35	20	25	6.38	1.71	17.3	49.5
18	30	35	20	25	6.95	3.07	9.7	92
19	20	35	20	20	8.70	7.64	11.8	71.5
20	20	35	20	30	7.26	3.09	10.4	88.4
21	20	35	15	25	7.60	3.06	14.8	56.5
22	20	35	25	25	6.30	2.69	7.9	110.5
23	20	20	20	25	5.43	3.99	9.2	63.5
24	20	50	20	25	7.37	2.78	12.4	90
25	20	35	20	25	6.87	3.01	9.9	81
26	20	35	20	25	6.91	3.14	10.5	80
27	20	35	20	25	6.87	3.16	10.5	82

concentration of the cyclodextrin is affecting this response as was also concluded from inspection of the respective coefficients. The scaled and centered coefficients of migration time and current are displayed in Figure 4.4a. Cyclodextrin concentration and applied voltage had a positive effect on the current while a negative effect on migration time was found. Increasing the buffer concentration led to an increase of migration time and current.

As peak resolution between the citalopram enantiomers and between oxalate and (*R*)-citalopram was not really an issue in this assay, only migration time of the last migrating compound and current were minimized by the software used in setting the respective values to maxima of 10 min and 80 μA,

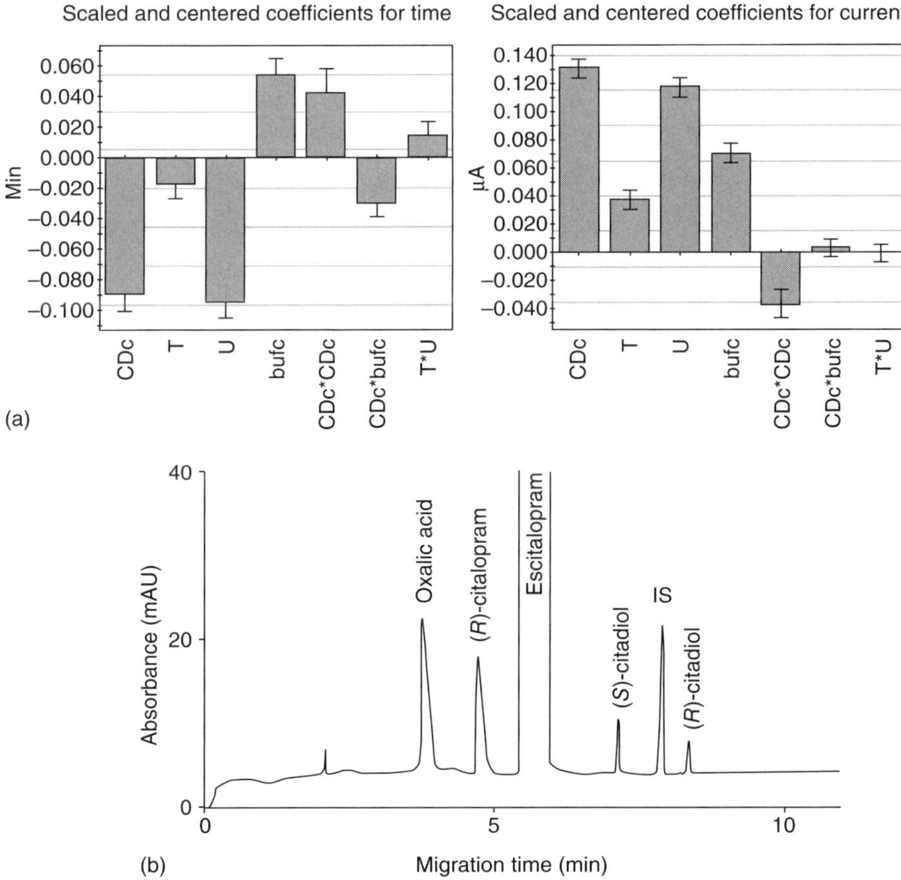

FIGURE 4.4. (a) Scaled and centered coefficients of the logarithmic of migration time and electrophoretic current. (b) Electropherogram of 5 mg/mL escitalopram oxalate containing approximately 2.4% (*R*)-citalopram spiked with 0.1% of citadiol enantiomers under optimized CE conditions; IS: internal standard salicylic acid. (Adapted with permission from Reference 27.)

respectively. This optimization is based on the Nelder–Mead simplex method (34). Optimized conditions 20 mM sodium phosphate buffer, pH 2.5, containing 22 mg/mL sulfated β-cyclodextrin and 0.5 mg/mL β-cyclodextrin at a capillary temperature of 22 °C using an applied voltage of –20 kV resulted in the electropherogram shown in Figure 4.4b. The predicted versus observed values for migration time were 8.7 min and 8.4 min, respectively. For the electric current –73 μA were predicted and –68 μA were found. Salicylic acid was used as internal standard to correct for minor fluctuations in migration time and injection errors. The optimized conditions were validated according to the ICH guideline Q2(R1) (29), and the final method proved to be suited for the impurity profiling of escitalopram in drug substance as well as commercial tablets.

A similar approach using a central composite face-centered design led to a sensitive and robust method for the impurity profiling of calcium levofolinate including the (6*R*,2′S)-diastereomer (26). Following initial screening, buffer pH, buffer concentration, applied voltage, and column temperature were further investigated using peak resolutions between levofolinic acid and the (6*R*,2′S)-diastereomer and between the two impurities migrating last, that is, N-(4-aminobenzoyl)-L-glutamic acid and 10-formylfolic acid, as well as analysis time as responses. A 20 mg/mL of 2,6-dimethyl-β-cyclodextrin were added to enhance the resolution between the folinic acid diastereomers. The concentration of the cyclodextrin was not included in the design but kept constant. The response surface plots of the dependence of the peak resolution on the main significant factors, buffer pH and column temperature, are shown in Figure 4.5. In both cases, the resolution increased with pH. In contrast, the resolution between the folinic acid diastereomers increased when the column

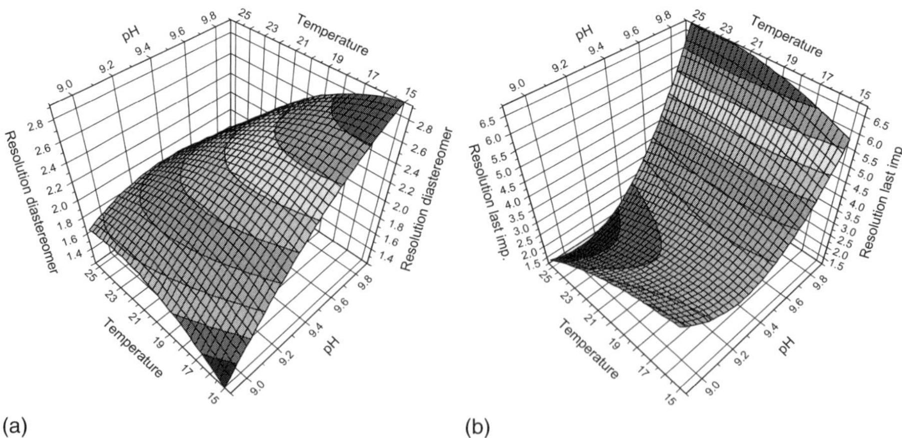

FIGURE 4.5. Response surface plots of the resolution between (a) the folinic acid diastereomers and (b) N-(4-aminobenzoyl)-L-glutamic acid and 10-formylfolic acid as the last migrating impurities. (Reproduced with permission from Reference 26.)

temperature was lowered while the R_S values of the last two migrating impurities increased when column temperature was raised. Method optimization using the optimization function of the software package that is based on the Nelder–Mead simplex method (34) maximized the peak resolution, and minimized migration times resulted in a background electrolyte consisting of 40 mM sodium tetraborate, pH 9.9, containing 20 mg/mL 2,6-dimethyl-β-cyclodextrin, using an applied voltage of 16 kV and a column temperature of 16 °C. The predicted resolution values exactly matched the experimental data ($R_S = 2.8$ for the folinic acid diastereomers and $R_S = 6.1$ for the last migrating impurities), and only minor deviation of the migration time of the last migrating compound was observed (predicted 20.7 min, observed 20.2 min). This final method was validated and applied to the analysis of commercial samples.

4.3.3. Determination of Stereoisomeric Impurities in Compounds with Multiple Chiral Centers

The determination of chiral impurities in drugs with multiple chiral centers is a challenging task in analytical chemistry as the number or stereoisomers increases exponentially with the number of the stereocenters. The analysis of the propriety compound R209130 containing three chiral carbon atoms (Fig. 4.6) was studied by Jimidar et al. (28). Initial screening conditions indicated the necessity of α-cyclodextrin and a negatively charged derivative, sulfated β-cyclodextrin, in a phosphate buffer, pH 3.0, containing 10% methanol for the separation of all eight stereoisomers. The final conditions for the four experimental factors, α-cyclodextrin concentration, sulfated β-cyclodextrin concentration, buffer molarity, and applied voltage, were optimized by a three-level Box–Behnken design including three center points resulting in a matrix of 27 experiments. The resolution between the individual pairs of stereoisomers, that is, a total of seven pairs, and the migration time of the last analyte were selected as responses. Figure 4.6 illustrates prediction of the influence of the individual factors on the responses derived from the Box–Behnken design. The optimized conditions were then generated by defining a target minimum resolution expressed as a desirability function shown as the dashed horizontal lines. The final conditions were validated, including rinsing procedures resulting in a method that was able to determine the stereochemical impurities at the 0.1% level (Fig. 4.6).

4.4. CONCLUSIONS AND OUTLOOK

As illustrated by several examples, experimental design methods proved to be very useful in the development of reproducible and robust CE methods for the analysis of related substances in drugs. This includes the analysis of complex mixtures of substances isolated from natural sources and the simultaneous separation of chiral and achiral impurities as well as compounds with multiple

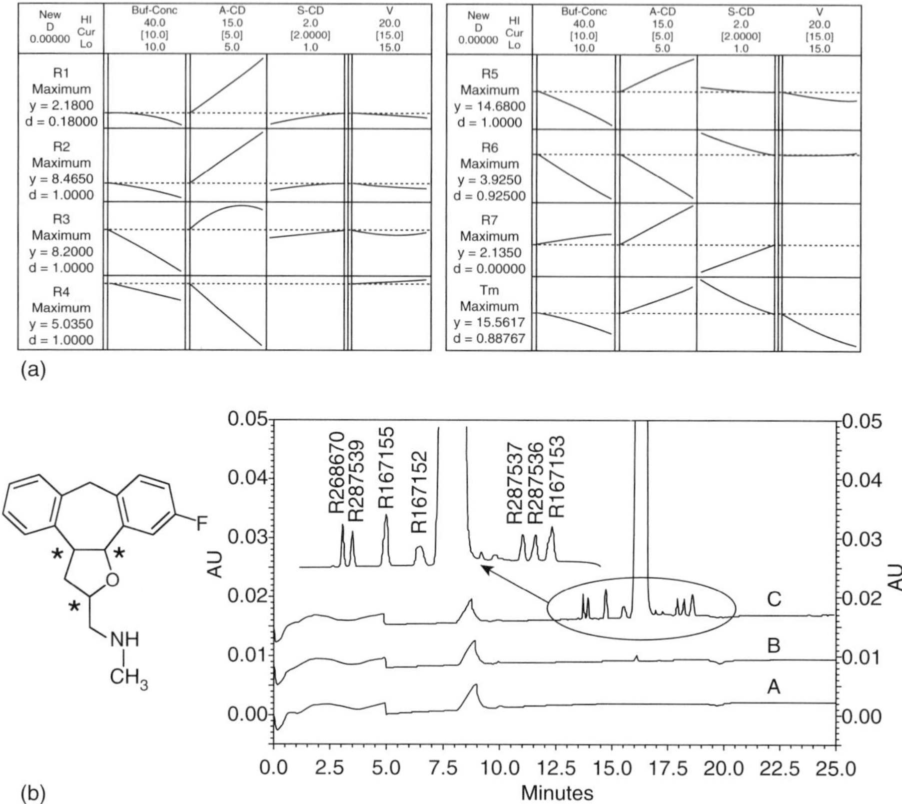

FIGURE 4.6. (a) Influence of factors on each response based on the results of the Box–Behnken design. The separation of each peak is predicted as a function of the investigated factors. The closeness of a response that is targeted for each response is presented by a desirability valued. The horizontal dotted lines predict the expected resolution value. (b) Structure of R209130 and electropherograms of the compound spiked with 1.0% of the stereoisomeric impurities (line C) under optimized separation conditions. Lines A and B represent a blank solution and the reporting threshold of 0.1%, respectively. (Adapted with permission from Reference 28.)

chiral centers. Because several parameters have to be optimized in CE and because of the interdependence of the experimental variables, the application of chemometrics is highly recommended for rational and economical method development. The user-friendly commercial software packages allow even the relatively inexperienced researcher to use chemometric design in his experiments so that design methods will be increasingly used in analytical chemistry including CE.

Striving to further miniaturization, analytical chemists have been also developing analytical methods for various analytes utilizing CE chips. However, although feasible for chip CE, experimental design has hardly been used during the development of such assays but can be expected to be used in the near future.

REFERENCES

1. Altria, K.D., Chen, A.B., and Clohs, L. (2001) *LCGC Europe*, **19**, 972–985.
2. Altria, K.D., Marsh, A., and Sänger-van de Griend, C.E. (2006) *Electrophoresis*, **27**, 2263–2282.
3. Scriba, G.K.E. (2005) *Encyclopedia of Analytical Sciences*, 2nd ed. (eds. P. Worsfold, A. Townshend, and C. Poole), Elsevier, Amsterdam, pp. 343–354.
4. Altria, K.D. (1997) *The Analysis of Pharmaceuticals by Capillary Electrophoresis*, Vieweg, Wiesbaden.
5. Ahuja, S. and Jimidar, M.I. (2008) *Capillary Electrophoresis Methods for Pharmaceutical Analysis*, Academic Press, Amsterdam.
6. Hanrahan, G., Montes, R., and Gomez, F.A. (2008) *Anal Bioanal Chem*, **390**, 169–179.
7. Sentellas, S. and Saurina, J. (2003) *J Sep Sci*, **26**, 875–885.
8. Altria, K.D., Clark, B.J., Filbey, S.D., Kelly, M.A., and Rudd, D.R. (1995) *Electrophoresis*, **16**, 2143–2148.
9. Siouffi, A.M. and Phan-Tan-Luu, R. (2000) *J Chromatogr A*, **892**, 75–106.
10. Bianchi, F. and Careri, M. (2008) *Curr Anal Chem*, **4**, 55–74.
11. Hanrahan, G. and Lu, K. (2006) *Crit Rev Anal Chem*, **36**, 141–151.
12. Brereton, R.G. (2007) *Applied Chemometrics for Scientists*, John Wiley & Sons, Chichester.
13. Cox, D.R. and Reid, N. (2000) *Theory of Design of Experiments*, CRC Press, Boca Raton, FL.
14. Massart, D.L., Vandeginste, B.G.M., Buydens, L.M.C., De Jong, S., Lewi, P.J., and Smeyers-Verbeke, J. (1997) *Handbook of Chemometrics and Qualimetrics*, Elsevier, Amsterdam.
15. Goupy, J. (2005) *Anal Chim Acta*, **544**, 184–190.
16. ICH Guideline Q3A(R2) (2006) Impurities in new drug substances, http://www.ich.org (accessed July 7, 2009).
17. Gil E.C., Van Schepdael, A., Roets, E., and Hoogmartens, J. (2000) *J Chromatogr A*, **985**, 43–49.
18. Gil E.C., Dehouck, P., Van Schepdael, A., Roets, E., and Hoogmartens, J. (2001) *Electrophoresis*, **22**, 497–502.
19. Wynia, G.S., Windhorst, G., Post, P.C., and Maris, F.A. (1997) *J Chromatogr A*, **773**, 339–350.
20. Kaale, E., Van Schepdael, A., Roests, E., and Hoogmartens, J. (2001) *J Chromatogr A*, **924**, 451–458.
21. Orlandini, S., Gioanni, I., Gotti, R., Pinzauti, S., La Porta, E., and Furlanetto, S. (2007) *Electrophoresis*, **28**, 395–405.
22. Persson-Stubberud, K. and Aström, O. (1998) *J Chromatogr A*, **798**, 307–314.

23. Orlandini, S., Fanali, S., Furlanetto, S., Marras, A.M., and Pinzauti, S. (2004) *J Chromatogr A*, **1032**, 253–263.
24. Furlanetto, S., Orlandini, S., Marras, A.M., Mura, P., and Pinzauti, S. (2006) *Electrophoresis*, **27**, 805–818.
25. Brunnkvist, H., Karlberg, B., Astervik, A., and Granelli, I. (2004) *J Chromatogr B*, **807**, 293–300.
26. Süß, F., Harang, V., Sänger-van de Griend, C.E., and Scriba, G.K.E. (2004) *Electrophoresis*, **25**, 766–777.
27. Sungthong, B., Jac, P., and Scriba, G.K.E. (2008) *J Pharm Biomed Anal*, **46**, 959–965.
28. Jimidar, M.I., Vennekens, T., Van Ael, W., Redlich, D., and De Smet, M. (2004) *Electrophoresis*, **25**, 2876–2884.
29. ICH Guideline Q2(R1) (2005) Validation of analytical procedures: Text and methodology, http://www.ich.org (accessed July 7, 2009).
30. Scriba, G.K.E. (2002) *J Pharm Biomed Anal*, **27**, 373–399.
31. Scriba, G.K.E. (2003) *Electrophoresis*, **24**, 2409–2421.
32. Fanali, S. (2000) *J Chromatogr A*, **875**, 89–122.
33. Scriba, G.K.E. (2008) *J Sep Sci*, **31**, 1991–2011.
34. Nelder, J.A. and Mead, R. (1965) *Computer J*, **7**, 308–313.

CHAPTER 5

OPTIMIZATION OF MICELLAR ELECTROKINETIC CHROMATOGRAPHY SEPARATION CONDITIONS BY CHEMOMETRIC METHODS

JESSICA L. FELHOFER and CARLOS D. GARCIA
Department of Chemistry, The University of Texas at San Antonio, San Antonio, TX

CONTENTS

5.1. MICELLAR ELECTROKINETIC CHROMATOGRAPHY (MEKC)

The versatility of capillary electrophoresis (CE) arises from the different modes of separation available. Of these modes, MEKC can be used to separate neutral species in addition to charged species (1, 2). Since its development in 1984 by Terabe et al. (3), MEKC has been used to separate and quantify a wide variety of analytes including amino acids, biomarkers, antiretroviral agents, drugs, pharmaceutical preparations, dyes, flavonoids, antioxidants, and pesticides (4–7). To perform MEKC, a surfactant (a molecule with a hydrophobic tail and a polar head group) must be included in the running buffer at a concentration higher than its critical micellar concentration (CMC). Above the CMC, monomer surfactant molecules are entropically driven to aggregate into spherical structures, called micelles, in which the hydrophobic tails are

Chemometric Methods in Capillary Electrophoresis. Edited by Grady Hanrahan and Frank A. Gomez

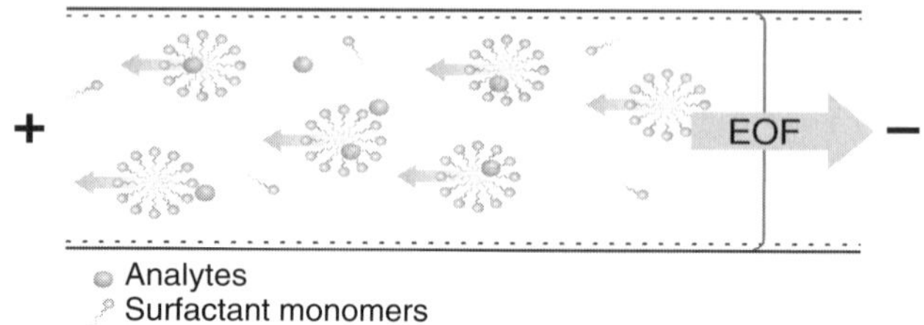

FIGURE 5.1. The separation principle of MEKC.

oriented within to avoid contact with the aqueous environment, and the polar head groups are oriented toward the surface of the aggregation, in contact with the aqueous environment. The micelles form a pseudostationary phase inside of the capillary, creating a hybrid system of electrophoresis and chromatography (8).

Due to the polar head groups, the micelles are charged and move with or against the electroosmotic flow (EOF), depending on the charge of the head group. During migration, neutral species in solution can interact with the micelles through hydrophobic and electrostatic interactions, resulting in partitioning in and out of the micelles (Fig. 5.1). The "retention" time of each analyte is proportional to the strength of the interaction with the micelles, and the differential analyte/micelle interactions are the key factors in determining the resolution of MEKC (9). Although more hydrophobic analytes typically show more affinity for the micelles with respect to analytes of a more hydrophilic character, other factors such as molecular weight, shape, and dipolar moment may play a fundamental role in the separation. An additional advantage of MEKC is that in some cases, the surfactant used to create the micelles can be used to control the EOF (10–12), minimize analyte–wall interactions (and therefore improve peak shape), and improve the performance of the detector (13, 14).

5.2. APPLYING CHEMOMETRICS TO MEKC

Chemometrics has played two major roles in MEKC: for analysis of the data collected from the separation and detection of analytes, and for efficient optimization of the separation conditions. Regarding data analysis, chemometrics can allow deconvolution of poorly resolved peaks (15, 16) and quantification of the corresponding analytes. Chemometrics can also be employed for multivariate calibration (17), characterization of complex samples, and to study peak purity. Sentellas and Saurina have recently reviewed the role of chemometrics applied to data analysis in CE (18). For MEKC in particular, chemometrics has been used more widely as a tool for optimization of separation conditions. The focus of this chapter is to exemplify the utility of chemometric methods for the optimization of separation conditions in MEKC.

5.2.1. The Utility of Chemometrics for Optimization

5.2.1.1. Separation Conditions. The outcome of a separation of a series of compounds by MEKC depends heavily on a number of factors defining the conditions inside of the capillary. Some factors that can be adjusted to optimize a separation are surfactant type and concentration, pH of the running buffer, buffer type and concentration, organic modifier type and concentration, and type and concentration of any additives, such as cyclodextrins (CDs), if used. The temperature of the capillary and the electric field applied across the capillary can also be adjusted (19, 20). The selectivity of MEKC can be manipulated by varying the concentration or by using different types of surfactants: anionic, cationic, nonionic, and zwitterionic, or even mixtures (9). Additionally, the chain length of the surfactants can be varied to change the physical nature of the micelles. Increasing the amount of surfactant in the running buffer increases the concentration of micelles and therefore can improve resolution. However, the analysis time may be prolonged due to the higher ionic strength (which decreases the EOF) (10, 21, 22). High ionic strength can increase generated current and may lead to Joule heating, so care must be taken in choosing the optimal concentration of surfactant. The charge of a silica capillary wall depends on the deprotonation of the silanol groups. The extent of deprotonation affects the zeta potential of the double layer and, consequently, determines the EOF. Therefore, the pH of the running buffer should be controlled in order to obtain the optimal rate of EOF. The pH of the running buffer also determines the charge of the analytes, so the pH also has an effect on selectivity. The type of buffer chosen is based on the buffer capacity at the relevant pH range.

Like the concentration of the surfactant, the concentration of the buffer must be selected with regard to the ionic strength of the running buffer and the EOF desired. Including an organic modifier to the running buffer is useful in solubilizing hydrophobic compounds and in controlling the EOF (23–25). More importantly, organic modifiers can alter the selectivity of MEKC in two ways. One, they can reduce the hydrophobic interactions between the solutes and the micelles to change the partition coefficients of the analytes, affecting resolution and retention (19, 26–28). Also, the addition of an organic solvent can decrease the hydrophobic interactions among the tails of the surfactants within the micelles, resulting in more rapid chromatographic kinetics (8, 9, 19, 20). The temperature of the capillary must be carefully controlled to avoid fluctuations, as increases or decreases in temperature alter the viscosity of the running buffer and, consequently, can alter the rate of EOF. The temperature can also affect the pK_a values of the analytes and the capillary walls, the pH of the running buffer, and the CMC of the surfactants (29). Finally, the separation potential must be chosen so that the analysis time is efficient, yet not so high as to cause Joule heating. Other factors that affect separations include capillary length and diameter, method of sample injection, and the injection plug size/time. Evidently, there are numerous factors that affect the quality of

MEKC separations. Traditionally, separation conditions have been optimized by simple univariate techniques, in which each factor is optimized individually and sequentially until the desired result is obtained (30–33). This method is generally time-consuming and labor-intensive. Relatively recently, chemometric applications that have been used for optimizing chromatographic separations and standard CE separations have become more frequently used in MEKC.

Multivariate chemometric optimization is superior to univariate optimization of the many factors that affect an MEKC separation. Because factors can be varied simultaneously, fewer experiments are required, which improves cost- and time-efficiency of the optimization. More information can be gathered, as the factors most influential on the separation are elucidated along with systematic relationships among factors. These interactive effects among factors are assumed absent in univariate approaches. As discussed below, the quality of a separation can be quantified by a response function tailored to the specific characteristics desired by the analyst. In contrast to sequential optimization methods such as simplex, multivariate optimization elucidates the global, rather than a local, optimum of the response (34). Table 5.1 shows how several groups have used chemometrics for optimization of MEKC separations. Examples from this table will be highlighted throughout the chapter as the process of using chemometrics for MEKC optimization is detailed.

5.2.1.2. Response Functions. The goal of optimizing the conditions of an MEKC separation is the output of a quality electropherogram. The criterion defining a "quality" electropherogram varies from analyst to analyst. As evident in Table 5.1, this commonly means good resolution. However, some researchers may be interested in minimizing band broadening as Thorsteinsdóttir et al. were for separations of enkephalin-related peptides (35, 36). Other analysts value maximizing signal intensities in order to decrease detection limits. For instance, Gotti et al. focused on maximizing the peak area for an MEKC separation of polyphenols and methylxanthine theobromine (37). Other criteria include short total analysis time, detection of a maximum number of peaks, symmetrical peak shape, and maximum separation efficiency and selectivity. In order to get the most information possible from a separation of human urine components, Alfazema et al. looked for the highest number of peaks detected (38).

It has also proven advantageous to use a combination of individual responses in order to optimize as many parameters as possible (39–43). The two performance goals for a separation of bisphenols by MEKC were good resolution among five peaks and short total analysis time (42). Thus, a chromatographic response function (*CRF*) was employed that was a product of two types of desirability functions, as used by Divjak et al. (30–33, 44). Resolution (*R*) between two adjacent peaks in an electropherogram was calculated using

TABLE 5.1. A summary of factors, responses, chemometric designs and methods, and validation criteria used by different groups for the optimization of the separation of various samples by MEKC

Sample	Factors	Response	Chemometric Designs and Methods	Validation	Reference
Enkephalin-related peptides	[Surfactant], [organic modifier], *T*, ionic strength	t_m, R between peptide pairs	Four-factor central composite design, RSM, PLS	N, t_m window, retention factor	(36)
Anionic metal complexes	[Surfactant], [organic modifier]	μ_{eff}	Nonlinear regression	Agreement of predicted and experimental μ_{eff}	(69)
Tropane alkaloids	pH, [SDS], [organic modifier]	R, T, generated power and current	Doehlert design, MLR, RSM	t_m and peak area precision, linearity, sensitivity, accuracy	(63)
Isoniazid, pyridoxine HCl	pH, buffer type, [buffer], [SDS], *T*, *V*, injection time	R, peak symmetry, T	Two-level full factorial design, RSM, MLR	Stability, linearity, LOD, LOQ, precision, accuracy, specificity, robustness	(64)
Shuangdan Chinese medicine components	[Borate], [phosphate], [SDS], [ACN]	Modified chromatographic exponential function (R_{tot} and T)	Genetic algorithm experimental designs, RSM, MLR	Precision, linearity, recovery	(39)
Neurotransmitter amino acids	[Buffer], [SDS], *V*	R, N, t_m	Central composite design, MLR	Repeatability, peak symmetry, sensitivity, impurity determination	(62)
Enkephalin-related peptides	[Surfactant], injection plug length, *V*, *T*, ionic strength	Peak width	Fractional factorial design, RSM, PLS, central composite face design	Band broadening	(35)

TABLE 5.1. ***Continued***

Sample	Factors	Response	Chemometric Designs and Methods	Validation	Reference
Pesticides	Type CD, [CD], [buffer], pH, [micelles], [organic modifier]	Response function based on information theory (R_{tot})	ORM, Plackett–Burman design	Unambiguous identification, accurate quantification, acceptable T	(40)
Phenols and amino acids	pH, [primary surfactant], [secondary surfactant], [buffer] *T*, [ACN], [urea] (as an organic modifier), *V*	Arc tangens resolution (R_{tot})	Plackett–Burman design, full factorial design, circumscribed central composite design	R	(41)
Ibuprofen, codeine phosphate, and their main degradation products and impurities	[SDS], pH, [ACN], [borate], *V*, *T*	t_m, R	Two-level fractional factorial design with replicating center point	Selectivity, linearity, accuracy, precision, LOD, LOQ, robustness and range	(61, 70)
Fungal metabolites	[Phosphate], [borate], ionic strength, pH, [SDS], [sodium deoxycholate], [ACN], [methanol], *V*	R, n	Two-level fractional factorial design with three center points, RSM, full factorial design, MLR	Optimum of RSM verified	(66)
Organic solvents	[SDS], [veronal buffer], [barbitone buffer]	R	Central composite design, RSM	Precision, linearity, LOD	(71)

Cefalexin and related substances	pH, [SDS], [buffer]	Selectivity	Full-fraction factorial design, central point combination, MLR	Linearity, precision, LOD, LOQ, repeatability	(72)
Human urine components	[Methanol], [SDS], [CD], *V*, pH, *T*, [electrolyte additives] (urea, Brij 35)	Maximum n in the shortest T	Full factorial design, simplex optimization, RSM, ANOVA	R, reproducibility, accuracy	(38)
Epoxy fatty acids	[SDS], [organic modifier]	R, N	Two-factor full factorial design with three center points, RSM	Repeatability	(73)
Inhibitors of angiotensin-converting enzyme	pH, [alkylsulfonates]	R	Three-level full factorial design, RSM	Method transferability, peak shape, T	(74)
Bisphenols	[Borate], [ACN], pH, [SDS]	Chromatographic response function (R_{tot}, T)	Univariate optimization, Box–Behnken design, RSM, ANOVA	Repeatability	(42)
γ-amino butyric acid and amino acids	[Borate], [CD], [SDS], pH, *V*	R, T	Factorial design with three center points, ANOVA, MLR	Validation of model prediction, LOD, LOQ, N, interday precision, recovery	(46)
Polyphenols and methylxanthine theobromine	Extraction conditions prior to separation: time, *T*, type of solvent, [solvent]	Peak area	Fractional factorial design, central composite design	Robustness, selectivity, sensitivity, linearity, range, accuracy, precision	(37)
Steroids	[Ethanol], [ACN], [THF]	Retention, selectivity	LSER	Real sample analysis, coefficient of variation for t_m and peak area, interday precision	(28)

TABLE 5.1. ***Continued***

Sample	Factors	Response	Chemometric Designs and Methods	Validation	Reference
Isoflavones in soy germ pharmaceutical capsules	[Methanol], [SDS]	Response function based on the productory of the μ_{EOF} differences, μ_{EOF} of the first and last eluting peaks, and the electrolyte conductance	3^2 factorial design, RSM	Repeatability, intermediate precision, recoveries, linearity, LOQ	(43)
Bisbenzylisoquinoline alkaloids	[Surfactant], [organic modifier], pH	R	ORM	R, T, N, LOD	(60)
Rhubarb anthraquinones and bianthrones	[Surfactant], [organic modifier], [buffer], pH, *V*	Geometric mean of overall resolution (R_{tot}), T, and peak asymmetry	Fractional factorial design, central composite face-centered design	Repeatability, reproducibility, precision of peak area ratios, linearity, recovery	(75)
Ketorolac tromethamine and related impurities	*V*, [buffer], pH, [SDS], *T*	R, T	*D*-optimal design, Plackett–Burman design, RSM	Selectivity, robustness, linearity and range, precision, accuracy, LOD, LOQ, and system suitability	(76)

Pesticides	Type of surfactant, [surfactant], [buffer], pH, [urea]	Response function based on information theory (R_{tot})	Orthogonal array design	None reported	(77)
Arbutin, kojic acid, and hydroquinone	[SDS], pH, [buffer]	R_{tot}	Full factorial design	Linearity, recoveries	(78)
Letrozole, citalopram and their metabolites	11 factors including [buffer], [SDS], [organic modifiers], *V*, *T*, injection time	R	Plackett–Burman fractional factorial model	Specificity, linearity, recovery, precision, LOD, LOQ	(79)
β-lactams antibiotics	pH, [buffer], [SDS], *V*	R	Face-centered Draper–Lin small composite design with four central points	Linearity, LOD, LOQ precision	(80)
Angiotensin-II-receptor antagonists	pH, [buffer], [SDS]	Relative t_m, T	Face-centered central composite design	Linearity, precision, and accuracy	(81)
Food-related seleno amino acids	pH, [buffer], [SDS]	N, R, and T combined into one response function	Central composite design, RSM	Repeatability, precision	(82)
Glucosinolates	[SDS], [tetramethyl-ammo-nium hydrox-ide], [methanol]	R	Doehlert design	Linearity, repeatability, reproducibility	(83)

T = temperature; *V* = applied separation potential; SDS = sodium dodecyl sulfate (surfactant); ACN = acetonitrile (organic modifier); CD = cyclodextrin (additive); THF = tetrahydrofuran (organic modifier); t_m = migration time; T = total analysis time; R = resolution between two peaks; R_{tot} = total resolution among all peaks; μ_{EOF} = electrophoretic mobility; μ_{eff} = effective mobility; N = theoretical plates; n = number of peaks; RSM = response surface methodology; PLS = partial least squares; MLR = multiple linear regression; ORM = overlapping resolution mapping; ANOVA = analysis of variance; LSER = linear solvation energy relationship; LOD = limit of detection; LOQ = limit of quantitation.

Equation 5.1 where t_m^k, w^k, t_m^{k+1}, and w^{k+1} are the migration time and the peak width at base for each of the two consecutive peaks named k and $k + 1$, respectively:

$$R^{k,k+1} = \frac{t_m^{k+1} - t_m^k}{\left(w^{k+1} + w^k\right)/2} \qquad \text{(Eq. 5.1)}$$

Two analytes that give only one peak (comigration) result in the resolution of zero. Two Gaussian-shaped peaks are theoretically resolved at $R^{k,k+1} > 0.6$, and therefore, resolution values lower than 0.5 were considered not acceptable. A resolution of 1.5 implies a slight overlap of two equal width peaks and is generally considered sufficient for baseline resolution of equal height peaks. The value $R^{k,k+1}$ can be transformed by Equation 5.2 to give $S^{k,k+1}$, a dimensionless value between 0 and 1 corresponding to poor resolution ($R^{k,k+1} < 0.5$) and maximum resolution ($R^{k,k+1}$ approaching 2.5), respectively.

$$S^{k,k+1} = \frac{1}{1 + e^{-2.20R^{k,k+1}} + 3.30} \qquad \text{(Eq. 5.2)}$$

To account for the resolution between adjacent peaks in a separation of N analytes, the final form of the desirability function for resolution (f) was calculated in Equation 5.3 as the geometrical average of all individual desirability values $S^{k,k+1}$:

$$f = \left(\prod_{k=1}^{N-1} S^{k,k+1}\right)^{1/(N-1)} \qquad \text{(Eq. 5.3)}$$

The desirability function (g) that scaled the total analysis time (T) was also a sigmoidal transformation that gave values close to zero for analysis times greater than 45 min and values approaching one for total analysis times close to 6 min. Preliminary experiments, mostly performed by a univariate approach, were used to set these limits.

$$g = \frac{1}{1 + e^{0.09T - 1.94}} \qquad \text{(Eq. 5.4)}$$

The final *CRF* was the product of the desirability function for resolution (f) and the desirability function for analysis time (g):

$$CRF(f, g) = f \cdot g \qquad \text{(Eq. 5.5)}$$

Equation 5.5 enabled evaluating an entire electropherogram on a dimensionless scale from 0 (poor) to 1 (desirable) corresponding to poor separation and/or long total analysis time to good resolution with short analysis time, respectively.

Figure 5.2 shows representative electropherograms from three separate experiments. The quantitative measure of optimization, the *CRF*, scales with the qualitative assessment one can make by inspection. For example, the top electropherogram had the best average resolution (with a relatively short analysis time) among the three electropherograms, and consequently, the highest *CRF* value (0.7). Further, inspection of the lower electropherograms shows an obvious decrease in quality of resolution and accordingly, a decrease in *CRF* (42).

Similarly, Yu et al. required a separation of the active components in Shuangdan, a Chinese medicine, to have good resolution and minimum total analysis time (39). The researchers utilized a modified chromatographic exponential function (*MCEF*):

$$MCEF = \left(\sum_{i=1}^{n-1} e^{\alpha(R_{opt} - R_i)} + 1 \right) \left(1 + \frac{t_f}{t_{\max}} \right) \qquad \text{(Eq. 5.6)}$$

in which α is used to weight the resolution term, R_i is the resolution of the ith peak, R_{opt} the optimum resolution (set to 2.5), n the number of peaks, t_f the migration time of the final peak, and $t_{\max}$ the maximum acceptable migration time of the final peak (set to 30 min). Thus, the response of the *MCEF* becomes lower as the resolution becomes better and the analysis time shorter.

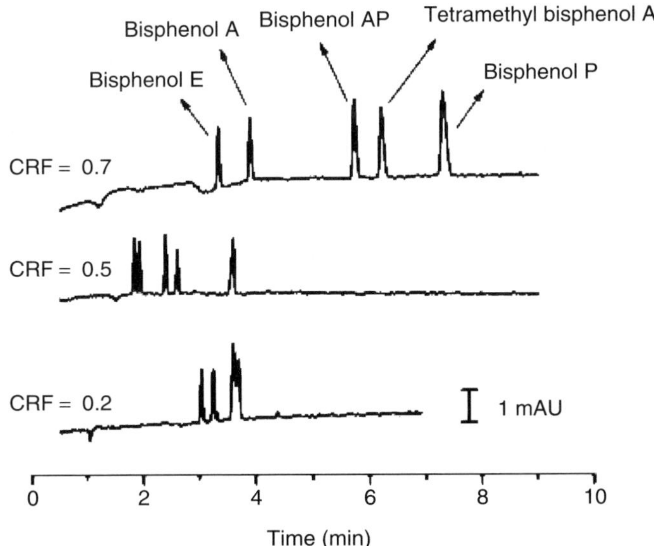

FIGURE 5.2. Representative electropherograms for three of the experiments of a Box–Behnken design and corresponding *CRF* (chromatographic response function) values. Used to optimize a separation of Bisphenols E, A, AP, and P, and Tetramethyl bisphenol A. Extracted from Reference 42.

To optimize the separation buffer conditions of CD-modified MEKC for the separation of pesticides, He and Kee Lee adapted a response function based on information theory which summed the maximum individual contribution of each factor to the resolution of the electropherograms (40). Mikaeli et al. optimized the conditions for the separation of phenols (and then validated their optimization technique for a separation of amino acids) by using the arc tangent resolution response function, which allows the researcher to define the acceptable values of resolution between peaks (41). More examples of response functions used in CE are discussed in Reference 45. Overall, the flexibility of user-defined response functions and the ability to weight and combine individual responses allows for optimizing conditions to the specific goals of the given separation. Once the desired responses are established, the optimization experiments can then be statistically designed to vary each experimental factor.

5.2.1.3. Experimental Designs. As mentioned, multivariate optimization techniques are superior to univariate ones which assume the absence of interactions between experimental variables. In order to discover the interactive effects, simultaneous variations in the levels of these factors should be considered in order to optimize the separation to the greatest extent (46). In this regard, statistically designed experiments, traditionally used to optimize chromatographic separations (47–52), have been applied to MEKC (see Table 5.1). These experimental designs include overlapping resolution mapping (ORM), full factorial, fractional factorial, Plackett–Burman, central composite, and Box–Behnken designs as well as response surface methods. The principles, advantages, and limitations of each design are detailed in References 34 and 53–55. The application of optimization experimental designs to MEKC has been reviewed (29). More recent reviews of chemometric designs applied to MEKC and to CE in general are References 45, 56, and 57. The selection of factors and the reasonable ranges of each variable studied affect the success of the experimental design, and is typically chosen based on the experience of the analyst. If necessary, initial experiments could be defined by the general guidelines of References 19, 41, 58, and 59.

ORM requires plotting resolution versus the separation conditions of each factor and overlaying plots for each factor. ORM is limited to optimization of a few factors (40). Sun and Wu used this method to optimize the pH and the concentration of surfactant and organic modifier of the buffer for the separation of bisbenzylisoquinoline alkaloids (60). Factorial designs allow the elucidation of the factors and interactions that have the most significant impact on the response. However, as the number of factors considered for optimization increases, so does the number of experiments required. To limit the number of experiments, fractional factorial designs can be used, including Plackett–Burman designs (34). An illustration of the reduced number of experiments required by a fractional factorial design is Persson-Stubberud and Åström's optimization of six parameters for a separation of ibuprofen, codeine phosphate, and their main degradation products and impurities in only 16 experiments (61). Frequently, fractional factorial designs are used to quickly screen many parameters to identify the most significant ones. Then, studies by more exhaustive designs like full factorial designs, central composite designs, and Box–Behnken designs can be used for further optimization (56). Table 5.2 shows an example of a central composite design that was used by Wan et al. for the optimization of a separation of neurotransmitter amino acids (62). The optimization of sodium dodecyl sulfate (SDS) and acetonitrile (ACN) concentrations as well as buffer pH was performed using a Doehlert experimental design for the separation of tropane alkaloids in belladonna extract (63). Replicating center points can be included in chemometric designs as a check for curvature and interactions (61), to provide an estimation of the experimental variance, and to examine the loss of linearity between the levels chosen for each variable.

Ehlen et al. and Gotti et al. show the efficiency of screening factors for significance before delving into a more exhaustive optimization in separations of microdialysates and catechins in *Theobroma cacao* beans, respectively (37, 46). Mikaeli et al. used a Plackett–Burman design to screen eight factors

TABLE 5.2. A central composite design used for the optimization of a separation of neurotransmitter amino acids. Adapted from Reference 62

Name	Buffer (mM)	SDS (mM)	HV (kV)
Low	20	10	15
High	60	40	25
Exp01	20	10	15
Exp02	60	10	15
Exp03	20	40	15
Exp04	60	40	15
Exp05	20	10	25
Exp06	60	10	25
Exp07	20	40	25
Exp08	60	40	25
Exp09	6.36	25	20
Exp10	73.64	25	20
Exp11	40	10	20
Exp12	40	50.23	20
Exp13	40	25	11.59
Exp14	40	25	28.41
Exp15	40	25	20
Exp16	40	25	20
Exp17	40	25	20
Exp18	40	25	20
Exp19	40	25	20

for significance (41). After determining that the pH and the concentrations of ACN, SDS, and sodium deoxycholate (secondary surfactant) would have the most influence on the resolution of the separation of phenols, they investigated these factors using a full factorial design to elucidate any interactive effects. Finally, the concentrations of ACN and SDS were optimized using a full factorial design, including center points. Optimization of the eight factors took only 48 experiments. Similarly, Nemutlu et al. employed an initial screening followed by optimization by two full factorial designs for a separation of a pharmaceutical formulation (64).

5.2.1.4. Modeling the Experimental Data. The data collected from each experiment in a given experimental design can be mathematically modeled so that the response, such as migration time, resolution, and so on, can be correlated with the experimental conditions that produced it. This way, by using the model, the desired output can be maximized and the corresponding experimental conditions defined in a predictive manner. Frequently, data are fitted to quadratic polynomial functions similar to Equation 5.7,

$$y = b_0 + b_1x_1 + b_2x_2 + b_3x_3 + b_{12}x_1x_2 + b_{13}x_1x_3 + b_{23}x_2x_3 + b_{11}x_1^2 + b_{22}x_2^2 + b_{33}x_3^2 \qquad \text{(Eq. 5.7)}$$

which contains linear terms for all factors (x), squared terms for all factors, and products of all pairs of factors. The regression coefficient, b, gives a measure of the rate of change in response (y) per unit change in each of the factors. In other words, the regression coefficients show the influence of each factor on the response. Data can be fit to such a model by statistical treatments, such as multiple linear regression (MLR), which fits the data by minimizing the sum of the squared y-residuals (62). If a model has two independent variables, they can each be plotted against the response to form a three-dimensional response surface, such as in Figure 5.3. This figure displays a response surface generated for the optimization of a separation of isoflavones in soy germ pharmaceutical capsules conducted by Micke et al. (43). Used in response surface methodology (RSM), the maximum (or minimum) of the response surface is located, and the corresponding optimal factor levels are determined. For a model with more than two independent variables, all but two factors can be held at a constant value to produce a response surface that can be visualized. Factor significance can also be determined by analysis of variance (ANOVA) statistics. ANOVA for a linear regression partitions the total variation of a sample into components, which are used to compute an F-ratio that evaluates the effectiveness of the model. Prob > F is the significance probability for the F-ratio, which states that if the null hypothesis is true, a larger F-statistic would only occur due to random error. It is the probability of obtaining a greater F-value by chance alone if the variation due to lack of fit variance and the pure error are the same. Significance probabilities of 0.05 or less are often considered evidence that there is at least one significant

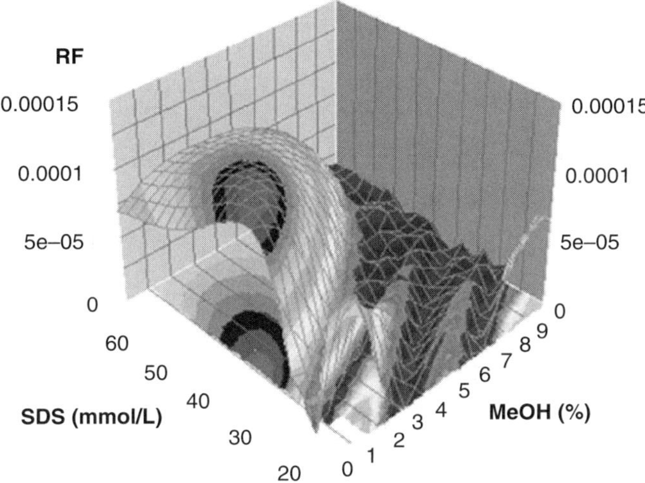

FIGURE 5.3. Response surface from a plot of the response function (RF) versus SDS and methanol (MeOH) concentration. Extracted from Reference 43.

regression factor in the model. Silva et al. showed the utility of linear solvation energy relationships (LSERs) to study solvent effects of ethanol, ACN, and tetrahydrofuran on the separation of natural and synthetic steroids (28). Other multivariate modeling methods including partial least squares (PLS) regression, nonlinear methods, and artificial neural networks are described in References 34, 55, 56, and 65.

5.2.1.5. Validation of Optimized Conditions. Once the relationship between the experimental parameters and the response has been modeled and the optimum conditions predicted, experiments should be performed to verify that the response is in fact the desired one. Most commonly, the resolution among the peaks should meet a quantitative requirement. Another method of verification is to compare the predicted response (defined by the model-predicted optimal conditions) to the actual experimental response. In the case of Nielsen et al., the experimental response fell within the confidence intervals of the predicted response, and therefore, the model used to optimize the separation of fungal metabolites was a success (66). In the case of the MEKC separation of anionic metal complexes by Breadmore et al., in which the model predicted the electrophoretic mobility of each complex, the model-predicted separation was overlaid with an actual separation, shown in Figure 5.4. Inspection of the coinciding peaks shows that the prediction was, in fact, accurate. Once the separation is deemed optimized, validation of criteria by figures of merit such as precision, dynamic range, selectivity, limit of detection, limit of quantitation, and robustness (see Table 5.1) are typically performed to ensure reproducible and secure results (34).

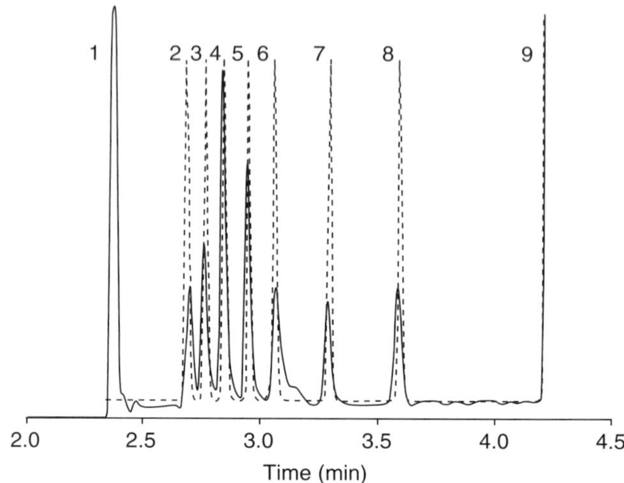

FIGURE 5.4. Experimental (solid line) and predicted (broken line) separations of metal HEDTC (bis[2-hydroxyethyl]dithiocarbamate) complexes at 9% methanol and 12 mM SDS. Peaks: 1, EOF; 2, Cd(II); 3, Pb(II); 4, Ni(II); 5, Co(II); 6, Bi(III); 7, Cu(II); 8, Hg(II); 9, HEDTC. Extracted from Reference 69.

5.3. CONCLUDING REMARKS

Multivariate chemometric modeling techniques have proven to have great utility in optimizing the many factors involved in a separation. MEKC is a relatively new technique and adapting it to existing analytical separation and quantitation problems as well as performing new separations will always require an optimization step first. Additionally, new modifications are being developed to enhance the sensitivity and resolution of MEKC separations (reviewed in Reference 67), such as online sample preconcentration, in-capillary derivatization, and coupling MEKC with flow-injection systems. New additives, such as ionic liquids, vesicles, carbon nanostructures, and ionic polymers are being added to the separation buffer as well (67, 68). In addition to providing better separations, these additional parameters of MEKC separations will all require additional time spent on optimization. In this regard, chemometrics applied to MEKC can be a time-efficient, information-rich option for the optimization of separation conditions.

REFERENCES

1. Terabe, S. (2004) *Anal Chem*, **76**, 240A–246A.
2. Watanabe, T. and Terabe, S. (2000) *J Chromatogr A*, **880**, 295–301.
3. Terabe, S., Otsuka, K., and Ando, T. (1985) *Anal Chem*, **57**, 834–841.
4. Molina, M. and Silva, T. (2002) *Electrophoresis*, **23**, 3907–3921.

5. Iadarola, P., Cetta, G., Luisetti, M., Annovazzi, L., Casado, B., Baraniuk, J., Zanone, C., and Viglio, D. (2006) *Electrophoresis*, **26**, 752–766.
6. Manuel, S. (2007) *Electrophoresis*, **28**, 174–192.
7. Kostal, V., Katzenmeyer, J., and Arriaga, E.A. (2008) *Anal Chem*, **80**, 4533–4550.
8. Pappas, T.J., Gayton-Ely, M., and Holland, L.A. (2005) *Electrophoresis*, **26**, 719–734.
9. Heiger, D.N. (1992) *High Performance Capillary Electrophoresis—An Introduction*, 2nd ed., Hewlett-Packard Company, France.
10. Garcia, C.D., Dressen, B.M., Henderson, A., and Henry, C.S. (2005) *Electrophoresis*, **26**, 703–709.
11. Mora, M.F., Giacomelli, C.E., and Garcia, C.D. (2007) *Anal Chem*, **79**, 6675–6681.
12. Mora, M.F., Felhofer, J., Ayon, A., and Garcia, C.D. (2008) *Anal Lett*, **41**, 312–334.
13. Ding, Y. and García, C.D. (2006) *Electroanalysis*, **22**, 2202–2209.
14. Ding, Y., Mora, M.F., Merrill, G.N., and Garcia, C.D. (2007) *Analyst*, **132**, 997–1004.
15. Zhang, F. and Li, H. (2006) *Chemom Intell Lab Syst*, **82**, 184–192.
16. Vera-Candiotti, L., Olivieri, A.C., and Goicoechea, H.C. (2008) *Electrophoresis*, **29**, 4527–4537.
17. Nepote, A.J., Vera-Candiotti, L., Williner, M.R., Damiani, P.C., and Olivieri, A.C. (2003) *Anal Chim Acta*, **489**, 77–84.
18. Sònia Sentellas, J.S. (2003) *J Sep Sci*, **26**, 1395–1402.
19. Tonin, F.G., Jager, A.V., Micke, G.A., Farah, J.P., and Tavares, M.F. (2005) *Electrophoresis*, **26**, 3387–3396.
20. Jager, A.V., Tonin, F.G., and Tavares, M.F. (2005) *J Sep Sci*, **28**, 957–965.
21. Garcia, C.D. and Henry, C.S. (2004) *Anal Chim Acta*, **24**, 1–9.
22. Garcia, C.D. and Henry, C.S. (2005) *Electroanalysis*, **17**, 1125–1131.
23. James, S.F. (2003) *Electrophoresis*, **24**, 1530–1536.
24. Berzas Nevado, J.J., Castaneda Penalvo, G., and Pinilla Calderon, M.J. (2002) J Chromatogr B, **773**, 151–158.
25. Wang, M., Wu, D., Yao, Q., and Shen, X. (2004) *Anal Chim Acta*, **519**, 73–78.
26. Nuñez, O., Kim, J.-B., Moyano, E., Galceran, M.T., and Terabe, S. 2002. *J Chromatogr A*, **961**, 65–75.
27. Roman, G.T., McDaniel, K., and Culbertson, C.T. (2006) *Analyst*, **131**, 194–201.
28. Silva, C.A., Pereira, E.A., Micke, G.A., Farah, J.P.S., and Tavares, M.F.M. (2007) *Electrophoresis*, **28**, 3722–3730.
29. Corstjens, H., Billiet, H.A.H., Frank, J., and Luyben, K.C.A.M. (1995) *J Chromatogr A*, **715**, 1–11.
30. Hompesch, R.W., Garcia, C.D., Weiss, D.J., Vivanco, J.M., and Henry, C.S. (2005) *Analyst*, **130**, 694–700.
31. Mejia, E., Ding, Y., Mora, M.F., and Garcia, C.D. (2007) *Food Chem*, **102**, 1027–1033.

32. Zhao, X., Wang, Y., and Sun, Y. (2007) *J Pharm Biomed Anal*, **44**, 1183–1188.
33. Liu, X., Zhang, J., and Chen, X. (2007) *J Chromatogr*, **B852**, 325–330.
34. Otto, M. (2007) *Chemometrics, Statistics and Computer Application in Analytical Chemistry*, 2nd ed., Wiley-VCH, Weinheim.
35. Thorsteinsdóttir, M., Westerlund, D., Andersson, G., and Kaufmann, P. (1998) *J Chromatogr A*, **809**, 191–201.
36. Thorsteinsdóttir, M., Ringbom, C., Westerlund, D., Andersson, G., and Kaufmann, P. (1999) *J Chromatogr A*, **831**, 293–309.
37. Gotti, R., Furlanetto, S., Pinzauti, S., and Cavrini, V. (2006) *J Chromatogr A*, **1112**, 345–352.
38. Alfazema, L.N., Hows, M.E.P., Howells, S., and Perrett, D. (1997) *Electrophoresis*, **18**, 1847–1856.
39. Yu, K., Lin, Z., and Cheng, Y. (2006) *Anal Chim Acta*, **562**, 66–72.
40. He, Y. and Kee Lee, H. (1998) *J Chromatogr A*, **793**, 331–340.
41. Mikaeli, S., Thorsén, G., and Karlberg, B. (2001) *J Chromatogr A*, **907**, 267–277.
42. Felhofer, J., Hanrahan, G., and García, C.D. (2009) *Talanta*, **77**, 1172–1178.
43. Micke, G.A., Fujiya, N.M., Tonin, F.G., de Oliveira Costa, A.C., and Tavares, M.F.M. (2006) *J Pharm Biomed Anal*, **41**, 1625–1632.
44. Divjak, B., Moder, M., and Zupan, J. (1998) *Anal Chim Acta*, **358**, 305–315.
45. Siouffi, A.M., and Phan-Tan-Luu, R. (2000) *J Chromatogr A*, **892**, 75–106.
46. Ehlen, J.C., Albers, H.E., and Breyer, E.D. (2005) *J Neurosci Methods*, **147**, 36–47.
47. Havel, J., Peña, E.M., Rojas-Hernández, A., Doucet, J.P., and Panaye, A. (1998) *J Chromatogr A*, **793**, 317–329.
48. Farková, M., Peña-Méndez, E.M., and Havel, J. (1999) *J Chromatogr A*, **848**, 365–374.
49. Mutihac, L. and Mutihac, R. (2008) *Anal Chim Acta*, **612**, 1–18.
50. Pierce, K.M., Hoggard, J.C., Mohler, R.E., and Synovec, R.E. (2008) *J Chromatogr A*, **1184**, 341–352.
51. Ferreira, S.L.C., Bruns, R.E., da Silva, E.G.P., dos Santos, W.N.L., Quintella, C., David, J.M., de Andrade, J.B., Breitkreitz, M.C., Jardim, I.C.S.F., and Neto, B.B. (2007) *J Chromatogr A*, **1158**, 2–14.
52. Marini, F., Bucci, R., Magrì, A.L., and Magrì, A.D. (2008) *Microchem J*, **88**, 178–185.
53. Deming, S.N. and Morgan, S.L. (1993) *Experimental Design: A Chemometric Approach*, 2nd ed., Elsevier Science Publishers, Amsterdam.
54. Brereton, R.G. (2007) *Applied Chemometrics for Scientists*, John Wiley & Sons, Ltd., West Sussex.
55. Bezerra, M.A., Santelli, R.E., Oliveira, E.P., Villar, L.S., and Escaleira, L.A. (2008) *Talanta*, **76**, 965–977.
56. Sònia Sentellas, J.S. (2003) *J Sep Sci*, **26**, 875–885.
57. Hanrahan, G., Montes, R., and Gomez, F. (2008) *Anal Bioanal Chem*, **390**, 169–179.
58. Rodriguez Delgado, M.A., Pérez, M.L., Corbella, R., González, G., and García Montelongo, F.J. (2000) *J Chromatogr A*, **871**, 427–438.

59. Baher, E., Fatemi, M.H., Konoz, E., and Golmohammadi, H. (2007) Microchim Acta, **158**, 117–122.
60. Sun, S.-W. and Wu, A.-C. (1998) *J Chromatogr A*, **814**, 223–231.
61. Persson-Stubberud, K. and Åström, O. (1998) *J Chromatogr A*, **798**, 307–314.
62. Wan, H., Öhman, M., and Blomberg, L.G. (2001) *J Chromatogr A*, **916**, 255–263.
63. Mateus, L., Cherkaoui, S., Christen, P., and Veuthey, J.-L. (1998) Use of a Doehlert design in optimizing the analysis of selected tropane alkaloids by micellar electrokinetic capillary chromatography. *J Chromatogr A*, **829**, 317–325.
64. Nemutlu, E., Çelebier, M., Uyar, B., and Altinöz, S. (2007) *J Chromatogr B*, **854**, 35–42.
65. Havel, J., Breadmore, M., Macka, M., and Haddad, P.R. (1999) *J Chromatogr A*, **850**, 345–353.
66. Nielsen, M., Nielsen, P.V., and Frisvad, J.C. (1996) *J Chromatogr A*, **721**, 337–344.
67. Silva, M. (2008) *Electrophoresis*, **30**, 1–15.
68. Palmer, C.P. (2008) *Electrophoresis*, **30**, 1–6.
69. Breadmore, M.C., Macka, M., and Haddad, P.R. (1999) *Anal Chem*, **71**, 1826–1833.
70. Persson-Stubberud, K. and Åström, O. (1998) *J Chromatogr A*, **826**, 95–102.
71. Altria, K.D. and Howells, J.S. (1995) *J Chromatogr A*, **696**, 341–348.
72. Yong-Min Li, Y.Z., Vanderghinste, D., Van Schepdael, A., Roets, E., and Hoogmartens, J. (1999) *Electrophoresis*, **20**, 127–131.
73. Hong Wan, L.G.B. and Hamberg, M. (1999) *Electrophoresis*, **20**, 132–137.
74. Hillaert, S., Vander Heyden, Y., and Van den Bossche, W. (2002) *J Chromatogr A*, **978**, 231–242.
75. Kuo, C.-H. and Sun, S.-W. (2003) *Anal Chim Acta*, **482**, 47–58.
76. Orlandini, S., Fanali, S., Furlanetto, S., Marras, A.M., and Pinzauti, S. (2004) *J Chromatogr*, **1032**, 253–263.
77. Zhang, Y., Li, X., Yuan, Z., and Lu, Y. (2002) *Microchem J*, **73**, 307–315.
78. Lin, Y.-H., Yang, Y.-H., and Wu, S.-M. (2007) *J Pharm Biomed Anal*, **44**, 279–282.
79. Rodríguez Flores, J., Salcedo, A.M., Llerena, M.J.V., and Fernández, L.M. (2008) *J Chromatogr A*, **1185**, 281–290.
80. Pérez, M.I.B., Rodríguez, L.C., and Cruces-Blanco, C. (2007) *J Pharm Biomed Anal*, **43**, 746–752.
81. Hillaert, S., De Beer, T.R.M., De Beer, J.O., and Van den Bossche, W. (2003) *J Chromatogr A*, **984**, 135–146.
82. Wang, J., Mannino, S., Camera, C., Chatrathi, M.P., Scampicchio, M., and Zima, J. (2005) *J Chromatogr A*, **1091**, 177–182.
83. Paugam, L., Ménard, R., Larue, J.-P., and Thouvenot, D. (1999) *J Chromatogr A*, **864**, 155–162.

CHAPTER 6

CHEMOMETRIC METHODS FOR THE OPTIMIZATION OF CE AND CE–MS IN PHARMACEUTICAL, ENVIRONMENTAL, AND FOOD ANALYSIS

JAVIER HERNÁNDEZ-BORGES,[1]
MIGUEL ÁNGEL RODRÍGUEZ-DELGADO,[1] and
ALEJANDRO CIFUENTES[2]
[1]Department of Analytical Chemistry, Nutrition and Food Science, University of La Laguna (ULL), Canary Islands, Spain
[2]Department of Food Analysis, Institute of Industrial Fermentations (CSIC), Madrid, Spain

CONTENTS

6.1. INTRODUCTION

The emergence of capillary electrophoresis (CE) in the early 1980s (1,2) introduced a new separation technique with several advantages over more common liquid chromatographic approaches, including high speed of analysis, high efficiencies, low sample and reagent requirements, and a wide number of applications. Likewise, during the development of a CE method, many factors can

Chemometric Methods in Capillary Electrophoresis. Edited by Grady Hanrahan and Frank A. Gomez

have influence on the separation. These parameters comprise composition of the background electrolyte (BGE), including the addition of organic modifiers, surfactants, polymers or chiral selectors (depending on the CE separation mode), its pH, and ionic strength. Apart from the BGE-related parameters, other factors including length, internal diameter (i.d.) and temperature of the capillary, sample injection mode and volume, and separation voltage can also have an important impact on the figures of merit of the final separation. Additionally, some detection systems used on-line with CE also require specific optimization. For instance, if mass spectrometry (MS) detection is involved, several parameters should also be optimized in order to achieve the best sensitivity without losing separation efficiency: that is, dry gas flow and temperature, the composition and flow rate of the sheath liquid (if a sheath-flow interface is used), etc. As a result, a large number of parameters are involved during the development of a new CE (or CE–MS) method. Therefore, the use of suitable optimization strategies can be a helpful procedure for this task.

Whenever a new CE method is being developed, optimization strategies are usually applied to improve analysis speed, sensitivity, and resolution, using these three parameters or a combination of them as the monitored output (also called response or performance criteria). Very frequently, a step-by-step approach in which each factor is varied sequentially is followed. In this case, all parameters are kept constant, while the parameter of interest is varied and the response is measured. Depending on the problem (especially when the number of factors to optimize is very low) and on the performance criteria, univariate optimization can be useful, that is, the analysis of a single compound with only one component of the BGE. However, in most cases, a step-by-step optimization is laborious and tedious because it typically requires a high number of experiments. Furthermore, and more important, it does not consider possible interactions between factors.

It should also be taken into account that optimum response is not a universal concept. These conditions may be optimum for some authors and not for others, depending on the priority or the purpose of the work. In some cases, the analysis time and separation efficiency (resolution) might be the main responses to handle, while for others, only the peak efficiency or sensitivity might be of importance (especially if short analysis times are already involved).

In general, a very useful approach for CE optimization is to take advantage of the use of chemometrics (3–6). The use of chemometrics brings about the possibility to vary each factor at the same time in a more programmed and coherent way, in which the results obtained can be interpreted following a more rational and fruitful approach, and optimal analytical conditions can be reached faster, and ultimately provide a considerable reduction of the number of experiments. For this purpose, nowadays, there are different types of statistical software available (Statgraphics Plus, SPSS, etc) that allow the use of various chemometrics-based techniques including multivariate experimental design (ED), response surface methodology (RSM), artificial neural networks (ANNs).

An appropriate use of ED ensures that experimental data contain maximum information and provide answers to real problems, such as the case in analytical chemistry. Whenever an ED is used, it is necessary first to develop a univariate approach to explore the experimental domain and check the pertinence of its limits before undertaking a multivariate study, since ED approaches are planned in order to homogenously cover the experimental space. Afterward, a screening ED is developed to clearly establish the factors to be considered in the following optimization experiments, in the so-called response surface designs. These designs permit one to define an empirical model (usually quadratic polynomials) that accurately describes the behavior of the responses at all values of the experimental factors. The most popular methodology applied to multiple response optimization is the desirability function approach, as proposed by Derringer and Suich (7). In order to calculate quadratic regression model coefficients, each factor must be studied in terms of three levels. For this purpose, a central composite design (CCD) is often used to provide estimation of a second-order equation. The CCD is very effective with respect to the number of runs required and therefore, it is one of the most commonly used ED, especially in regard to the optimization of CE and CE–MS parameters.

An additional chemometrical approach for this optimization is the use of ANNs (8, 9), which consist of a large number of simple, highly interconnected processing elements in an architecture inspired on the brain's structure. A relatively large number of experimental data is frequently necessary to train the network so that it "learns" the behavior of data and can develop further predictions. In this sense, the use of EDs provides an appropriate source of experimental data that can greatly help to train the ANN. The combination of both approaches, ED and ANNs, constitutes an effective strategy toward optimization of CE separation that has not been fully studied.

In view of the importance of the application of these techniques in CE analysis, the chapter presents an overview on the most recent applications of chemometrics to optimize CE and CE–MS parameters, focusing on pharmaceutical, environmental, and food analysis applications mainly in the last 5 years. The chapter has been divided into six main sections corresponding to an introduction, three main applications (pharmaceutical, environmental, and foods), an additional section summarizing other recent studies in differing fields, and a final section including concluding remarks and future perspectives.

6.2. PHARMACEUTICAL APPLICATIONS

The development of new separation methods for the analysis of drugs is continuously requested during drug design, development, production, and use. In this regard, some analytical issues are of particular importance, including purity assessment, analysis of isomers, detection of impurities, and/or related compounds. Although most of the methods used for pharmaceutical analysis involve the use of high performance liquid chromatography (HPLC), CE has also found its room in this important field of applications, while chemometrics has also demonstrated to be a helpful tool in the optimization of CE method development. Table 6.1 shows several examples of the most recent applications.

In one of the above studies, Bailón Pérez et al. (15) explored the use of ED for the optimization of micellar electrokinetic chromatography (MEKC) separation of nine β-lactam antibiotics (cloxacillin, dicloxacillin, oxacillin, penicillin G, penicillin V, ampicillin, nafcillin, piperacillin, and amoxicillin) and *p*-aminobenzoic acid (used as an internal standard). BGE nature, pH, and concentration, as well as the concentration of the micellar medium, separation voltage, and temperature were the factors considered in a face-centered Draper–Lin design (scarcely applied in chemistry). The optimization criteria were to obtain a maximum for a response function that considers either the peak efficiencies (the highest possible), peak resolution (the highest possible), or analysis time (the lowest possible). A good separation that met all the requirements was achieved using a BGE containing 26 mM sodium tetraborate at pH 8.5 with 100 mM sodium dodecyl sulfate (SDS) (25 kV and 30 °C) (see Fig. 6.1). After appropriate validation of the method, it was applied to the analysis of the above-mentioned compounds in Orbenin capsules (GlaxoSmithKline, S.A., Madrid, Spain), Britapen tablets (Reig Jofre, S.A., Madrid, Spain), and Veterin-Micipen injectables (Intervet, S.A., Madrid, Spain).

Capella-Peiró et al. (28) used a 3^2 full factorial design to optimize the capillary zone electrophoresis (CZE) separation of a group of seven antihistamines (brompheniramine, chlorpheniramine, cyproheptadine, diphenhydramine, doxylamine, hydroxyzine, and loratadine). In this case, critical parameters such as pH (a concentration of 20 mM phosphate was kept constant in all the experiments) and the applied voltage were studied to evaluate their effect on the resolution and efficiency. Maximum response was achieved at pH 2.0 and an applied voltage of 5 kV. After a repeatability study to check the precision of the electrophoretic method, as well as a suitable calibration, the usefulness of this optimized method was demonstrated through the determination of the listed histamines in pharmaceuticals, urine, and serum samples (recoveries were in agreement with the stated contents). Urine samples were diluted and directly injected in the CE system, while serum samples were previously extracted by means of a solid-phase extraction (SPE) procedure.

In recent years, special attention has been given to the use of ionic liquids in analytical chemistry. Room temperature ionic liquids are salts with melting points at or close to room temperature, and are currently considered as green solvents with use in a wide variety of applications, including synthesis, catalysis, and electrochemistry (33). Moreover, they have also attracted some attention regarding their use in separation chemistry, including their application as additives in CE (12, 34–36). ED can also be used to provide a deeper insight into analyte interactions with components of the BGE (including ionic liquids) and, therefore, its optimization. This is the case of the work developed by François et al. (12) who used a four-factor *D*-optimal ED to evaluate

TABLE 6.1. Applications of chemometrics to CE and CE–MS optimization in pharmaceutical analysis

Analytes	Matrix	CE Mode	Buffer	Chemometric Approach	Comments	Reference
Norfloxacin, tinidazole	Pharmaceuticals	CZE–DAD (301 nm)	32.5 mM phosphate, pH 2.5	ED (BGE optimization). Two-level full factorial design. Factors: BGE concentration and pH. Response: resolution, migration time, peak area.	—	(10)
Tetracycline, chlortetracycline, oxytetracycline, doxycycline	Pharmaceuticals	CZE–DAD (270 nm)	50 mM sodium carbonate, 1 mM EDTA pH 10	ED (BGE optimization). Two-level full factorial design and CCD. Factors: BGE concentration, pH, temperature, voltage. Response: number of separated compounds.	CCD used for significant factors: BGE concentration, pH, temperature	(11)
Carprofen, ketoprofen, naproxen, suprofen	—	NACE–UV (200, 230, 240, 254, 300 nm)	Different BGEs	ED (BGE). *D*-optimal design. Factors: ionic liquid concentration, salt concentration, alcohol percentage, and nature. Response: electrophoretic mobility.	Evaluation of the interactions between ionic liquid contra-ion and the anionic analytes.	(12)
Streptomycin, oxytetracycline	Bactericidal products to be used in agriculture	CZE–DAD (195 nm)	0.10 M sodium phosphate, pH 2.6	ED (BGE optimization). CCD. Factors: pH, temperature, voltage. Response: peak area.	A second CCD was used for significant factors: voltage, temperature.	(13)

TABLE 6.1. *Continued*

Analytes	Matrix	CE Mode	Buffer	Chemometric Approach	Comments	Reference
Arbutin, kojic acid, hydroquinone	Cosmetics	MEKC–UV (200 nm)	20 mM phosphate, 100 mM SDS, pH 6.5	ED (BGE optimization). Three-level full factorial design. Factors: SDS concentration, pH, phosphate concentration. Response: resolution and migration time.	—	(14)
Cloxacillin, dicloxacillin, oxacillin, penicillin G, penicillin V, ampicillin, nafcillin, piperacillin, amoxicillin	Pharmaceuticals	MEKC–DAD (220 nm)	26 mM sodium tetraborate, pH 8.5, 100 mM SDS	ED (BGE optimization). Face-centered Draper–Lin design. Factors: pH, buffer concentration, micelle concentration, voltage. Response: multiple response function (efficiencies, resolution, analysis time).	—	(15)
Naphazoline, diphenhydramine, phenylephrine	Nasal solutions	CZE–DAD (210 nm)	63 mM phosphate, pH 3.72	ED (BGE optimization). CCD. Factors: pH, buffer concentration, voltage. Response: resolution, migration time.	—	(16)

2-[(4′-benzoyloxy-2′ -hydroxy)phenyl-propionic acid]	—	CZE–UV (195 nm)	Britton–Robinson buffer at pH 6.4, 7 mM vancomycin	ED (BGE optimization). CCD. Factors: pH, chiral selector, temperature. Response: resolution, migration time.	Enantiomeric separation. Partial-filling-countercurrent method. Derringer's desirability function.	(17)
Rufloxacin	Coated tablets	CZE–UV (240 nm)	0.10 M boric acid, pH 8.8	ED (BGE optimization). Doehlert design. Factors: pH, buffer concentration, temperature, voltage. Response: efficiency, peak area/migration time ratio.	Use of pefloxacin mesylate as internal standard.	(18)
Ethambutol, 2-amino-1-butanol, phenylephrine (internal standard)	Pharmaceuticals	CZE–DAD	58 mM borate, pH 9.50	ED (BGE optimization). Box–Behnken. Factors: pH, buffer concentration, voltage. Response: efficiency, migration time, resolution.	—	(19)
Norfloxacin and its carboxylated degradant	Pharmaceuticals	CZE–DAD (285, 301 nm)	10 mM phosphate, pH 2.5	ED (BGE optimization). Two-level full factorial design. Factors: pH, buffer concentration. Response: resolution, peak area, migration time, RSD migration time, RSD peak area.	—	(20)

TABLE 6.1. ***Continued***

Analytes	Matrix	CE Mode	Buffer	Chemometric Approach	Comments	Reference
Chloramphenicol, danofloxacin, ciprofloxacin, enrofloxacin, sulfamethazine, sulfaquinoxaline, sulfamethoxazole	Pharmaceuticals	CZE–DAD (203, 270 nm)	60 mM phosphate, 20 mM tetraborate, pH 8.5	ED (BGE optimization). Two-level full factorial design. Factors: temperature, voltage. Response: resolution.	Buffer concentration and pH were optimized following an univariate approach.	(21)
Ephedrine, pseudoephedrine, norephedrine, norpseudoephedrine	Urine	CZE–DAD (195 nm)	260 mM Tris-phosphate, pH 3.5, 13.3 mM dimethyl-β -cyclodextrin (CD)	ED (BGE optimization). CCD. Factors: buffer concentration, pH, CD concentration. Response: resolution, separation time, and current.	—	(22)
Omeprazole, 5-hydroxyomeprazole	—	NACE–DAD (301 nm)	30 mM ammonium acetate 1 mM formic acid in methanol, 30 mM HDMS-β-CD	ED (BGE optimization). Factors: buffer concentration, CD concentration, voltage.	Enantiomeric separation.	(23)
Clenbuterol, salbutamol, terbutaline	Pharmaceuticals	NACE–DAD (220 nm), NACE–MS	18 mM ammonium acetate in MeOH:ACN:acetic acid (66:33:1%, v/v/v)	ED (BGE optimization). Two-level full factorial design. Factors: buffer concentration, organic solvent, injection time, voltage, temperature. Response: resolution, efficiency, tailing factor, migration time.	Preliminary experiments with NACE–MS.	(24)

Atenolol, celiprolol, propanolol. Bupivacaine, mepivacaine, prilocaine	—	NACE–DAD (230 nm)	Different BGE containing: HDMS-β-CD, MeOH, ammonium formate, potassium camphor SO_3^-,	ED (BGE optimization). *D*-optimal design. Factors: cationic BGE component, anionic BGE component, CD concentration. Response: resolution.	Enantiomeric separation. Study of the effect of salts on the enantioseparation.	(25)
Ofloxacin	Physiological solutions	CZE-UV (300 nm)	50 mM phosphate, pH 2.8, 4.0% methyl-β-CD	ED (BGE optimization). CCD. Two-level full factorial design. Factors: BGE concentration, CD concentration, pH, temperature. Response: resolution, peak area, migration time, current.	Enantiomeric separation. Investigation of the absorption of ofloxacin enantiomers *in vitro*.	(26)
Enalapril, lisinopril, quinapril, fosinopril, perindopril, ramipril, benazepril, cilazapril	—	MEKC–UV (214 nm)	100 mM sodium phosphate, pH 2.0, 65 mM sodium octanesulfonate	ED (BGE optimization). Three-level full factorial design. Factors: pH, sodium octanesulfonate concentration. Response: migration time, peak width, resolution.	Inhibitors of angiotensin-converting enzyme.	(27)

TABLE 6.1. *Continued*

Analytes	Matrix	CE Mode	Buffer	Chemometric Approach	Comments	Reference
Bromphenitramine, chlorphenitramine, cyproheptadine, diphenyldramine, doxylamine, hydroxyzine, loratadine	Pharmaceuticals, urine, serum	CZE–DAD (214 nm)	20 mM phosphate, pH 2.0	ED (BGE optimization). Three-level full factorial design. Factors: pH, voltage. Response: resolution.	—	(28)
Thiazinamium, promazine, promethazine	Pharmaceuticals	CZE–DAD (254 nm)	100 mM Tris, pH 8.0, 15% ACN	ED (BGE optimization). Face centered Draper–Lin small composite design. Factors: BGE concentration, pH, acetonitrile percentage, temperature, voltage. Response: efficiency.	—	(29)
Epinastine, lidocaine	Human serum	CZE–DAD (200 nm)	160 mM phosphate, pH 8.5	ED (BGE optimization). Plackett–Burman design, CCD. Factors: BGE concentration, pH, injection voltage, injection time, separation voltage. Response: resolution, migration time.	CCD used for significant factor: BGE concentration, pH, injection voltage, separation voltage.	(30)

Artemisinin	Medicinal plant (*Artemisia annua L.*)	CZE–DAD (292 nm)	10 mM phosphate, pH 10.5	ED (BGE and flow injection conditions optimization). Five-level full factorial design. Factors: BGE concentration, pH, voltage, alkali concentration, heating time, reaction temperature. Response: peak area.	Simultaneous optimization of flow injection conditions.	(31)
Tamsulosin	—	CZE–DAD (nm)	100 mM Tris buffered with phosphoric acid to pH = 2.5, sulfated-β-CD, 0.15% (w/v),	ED (BGE optimization). Box–Behnken design, central composite face-centered design, central composite circumscribed design. Factors: CD concentration, voltage, temperature. Response: resolution, migration time.	Enantioseparation.	(32)

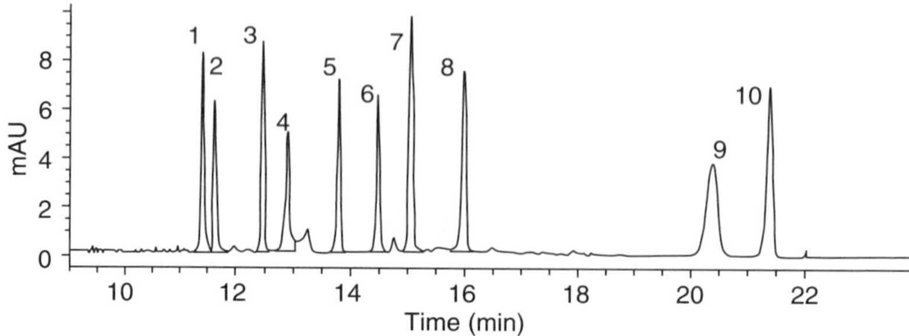

FIGURE 6.1. MEKC separation of nine β-lactams antibiotics and the internal standard *p*-aminobenzoic acid under optimized conditions: pH 8.5 using 26 mM sodium tetraborate buffer containing 100 mM SDS. Separation voltage: 25 kV; capillary temperature: 30 °C. (1) ampicillin; (2) amoxicillin; (3) penicillin G; (4) piperacillin; (5) oxacillin; (6) penicillin V; (7) *p*-aminobenzoic acid; (8) cloxacillin; (9) nafcillin; (10) dicloxacillin. Reprinted from Reference 15 with permission from Elsevier.

the interactions between an achiral ionic liquid (1-butyl-3-methylimidazolium bis[tribluoromethanesulfonyl]imide, BMIm-NTf_2) used in ACN/alcohol (methanol or ethanol)-based BGE and four arylpropionic acids (carprofen, ketoprofen, naproxen, and suprofen). In the initial step, factors that affected the electrophoretic mobilities of the profens were studied by a univariate approach to establish the experimental domain as well as its limits. Next, the ED was applied (25 experiments) taking into account the factors: BMIm-NTf_2 concentration, buffer salt concentration, alcohol proportion, and nature of the BGE. The relationship between the response (profen electrophoretic mobility) and the factors was defined as a quadratic multilinear regression model. From the obtained response surface plots, competitive interactions of ion-pair type interactions between the anionic profens and the BGE contra-ion (i.e., the ionic liquid cation, either adsorbed onto the capillary wall or in free solution) were proposed.

Chiral separation of enantiomeric isomers is one of the most challenging tasks for any analytical technique including CE. Since the first report in 1985 showing the great possibilities of CE for the separation of chiral compounds, the number of publications concerning this topic has quickly increased, especially for the enantioseparation of drugs and pharmaceuticals. Compared to empirical methods, chemometrics can greatly simplify the optimization of chiral CE analysis (important in pharmaceutical applications) allowing the rapid determination of appropriate experimental conditions (17, 23, 25, 26, 32, 37). One example of this type of application is the work of Siouffi and Phan-Tan-Luu (37), who employed ED methodology for the enantioseparation of a nonsteroidal anti-inflammatory drug (an arylpropionic acid) using CZE. Chiral selector concentration, pH, and temperature were the factors selected in a CCD approach in which resolution and migration times were selected as experimental responses. The partial-filling (PF) technique was used to avoid the presence of the chiral selector (vancomycin) in the detection window and to thus improve the sensitivity (vancomycin has a strong UV absorption). A buffer at pH 6.4 containing 7 mM of vancomycin at 22 °C was the optimal experimental condition providing suitable enantioresolution in a short analysis time (8.5 min). Servais et al. (25) studied the influence of the nature of the electrolyte on the chiral separation of basic compounds like three β-blockers (atenolol, celiprolol, and propanolol) and three local anesthetics (bupivacaine, mepivacaine, and prilocaine) in non-aqueous capillary electrophoresis (NACE) using heptakis(2,3-di-*O*-methyl-6-*O*-sulfo)-β-cyclodextrin (CD) as a chiral selector. For this purpose, two *D*-optimal designs with 33 and 26 experimental points were applied. The influence of the type of cation (sodium, ammonium, and potassium) and anion (chloride, formate, methanesulfonate, and camphorsulfonate) of the BGE was studied. The results obtained for the six compounds were examined individually. It was found that enantiomeric resolution was highly influenced by both cationic and anionic components of the BGE, with the cationic component exerting the highest influence. Two BGEs were recommended (i.e., ammonium formate and potassium camphorsulfonate in methanol) to achieve an efficient enantioresolution of the compounds. Olsson et al. (23) also optimized BGE with the same CD for the NACE enantioseparation of omeprazole and its metabolite 5-hydroxyomeprazole, ultimately making use of an ED.

As stated previously, various papers (38–40) have described the importance of optimizing the different factors that affect the electrospray interface (ESI)–MS signal. However, as described above, it has generally been done using a step-by-step procedure. Rudaz et al. (41), however, employed an ED to optimize the CE–MS analysis of enantiomers of methadone using PF techniques (to avoid the chiral selector entering in the MS ion source). Separation was studied using a BGE composed of 20 mM ammonium acetate at pH 4.0 and either sulfobutyl ether-β-CD, carboxymethylated-β-CD, and hydroxypropyl-β-CD, and a polyvinyl alcohol coated capillary. Chiral selector concentration, percentage of the capillary filled with the chiral selector, and drying gas nebulization pressure were the relevant factors taken into account. A full factorial design was used to examine the effects and significance of the factors, while a central composite face-centered design was used to establish the mathematical model of the selected responses in function of the experimental factors. Enantiomeric resolution, migration time, and efficiency were used as responses. Under different conditions, each CD was able to separate the two enantiomers. However, since this work involved the enantioseparation of standards (i.e., R- and S-methadone), no sensitivity problems were addressed and, as a consequence of the many ESI–MS parameters implicated, only the drying gas nebulization pressure was included in the ED.

6.3. ENVIRONMENTAL APPLICATIONS

Currently, special attention is being drawn toward the state of the environment and the level at which human activities are affecting it. Of particular importance is the presence of contaminants or residues in water systems as well as soils, and the ways they are entering the food chain or the way they are affecting animal species. Due to the ultra low levels at which these compounds are typically present in the environment, most of the methods used for the identification and determination of contaminants or residues in representative matrices include an appropriate preconcentration step followed by a separation process. Capillary electromigration approaches have also found their place in this field of research, especially when one of the main problems of CE (its low sensitivity when compared with gas chromatography [GC] or HPLC) can be overcome by using online and/or offline preconcentration strategies (see section 6.2 for applications in pharmaceutical analysis). Table 6.2 compiles the most recent applications of chemometrics to the optimization of CE and CE–MS methods applied to environmental analyses.

Drover and Bottaro (45) developed a CD-modified MEKC-UV method for the analysis of 12 widely used pharmaceutical compounds (ibuprofen, diclofenac, naproxen, bezafibrate, gemfibrozil, ofloxacin, norfloxacin, carbamazepine, primidone, sulfamethazine, sulfadimethoxine, sulfamethoxazole) commonly found in environmental waters. The separation of the 12 compounds was first optimized by means of a univariate approach that resulted in a full separation of the analytes in approximately 24 min (with the BGE composed of 10 mM ammonium hydrogen phosphate at pH 11.5, 60 mM SDS, 6 mg/mL sulfated-β-CD and 10% [v/v] isopropanol at 25 °C). Ammonium acetate was employed since the authors planned to use the optimized method in a future CE–MS analysis (ammonium acetate is a suitable volatile BGE compatible with CE–MS). However, when a multivariate optimization approach was applied, the analysis time was reduced to 6.7 min, with good resolution between the peaks (resolution and analysis time were selected as response). In this case, the optimum BGE was 10 mM ammonium hydrogen phosphate at pH 11.5, 69 mM SDS, 6 mg/mL sulfated-β-CD, 8.5% (v/v) isopropanol at 30 °C. In the ED, only SDS concentration, percentage of isopropanol, and capillary temperature were selected as factors to be optimized, while the rest were kept constant (based on the previous univariate approach experiments). After developing the calibration and repeatability study, the method was applied to the analysis of water samples. For this purpose, SPE Strata-X cartridges were used to extract the analytes from water samples (a preconcentration factor of 100-fold was achieved). Limits of detection (LODs) of the method ranged from 4 to 30 μg/L.

Felhofer et al. (46) reported an application describing the separation of five bisphenols (bisphenol E, bisphenol A, bisphenol AP, tetramethyl bisphenol A, and bisphenol P) by MEKC. It has been well established that bisphenols can reach the environment, and also the human body (47). Bisphenols are widely employed in the manufacture of plastics, especially those used in food and beverage packages, baby bottles, and water supply pipes. In this study, a univariate approach was first developed using a BGE composed of borate, SDS, and acetonitrile. The goal was to achieve the best separation of the

TABLE 6.2. Applications of chemometrics to CE and CE–MS optimization in environmental analysis

Analytes	Matrix	CE Mode	Buffer	Chemometric Approach	Comments	Reference
Zinc (II), sodium (I), calcium (II), magnesium (II)	Water	CZE–UV (214 nm)	2 mM 1, 10-phenanthroline, 3 mM 4-methylbenzylamine, pH 3.7	ED (BGE optimization). Factorial design. Factors: concentration of complexing and visualization agent, pH, injection time, voltage, capillary length. Response: peak area, peak height, migration time.	—	(42)
38 carboxylic acids	Atmospheric particles and cloud water	CZE–MS (IT, μTOF)	20 μM ammonium acetate, 10% (v/v) MeOH, pH 9.1	ED (ESI optimization). Factorial design and CCD. Factors: isopropanol percentage in the sheath liquid, flow rate, nebulizer gas pressure, dry gas temperature, dry gas flow rate. Response: sum of peak heights.	CCD used for significant factors: isopropanol percentage and nebulizer gas pressure.	(43)

TABLE 6.2. ***Continued***

Analytes	Matrix	CE Mode	Buffer	Chemometric Approach	Comments	Reference
Mecoprop, dichlorprop	—	CE–UV (214 nm)	45 mM Na_2HPO_4 Ph 5 (citric acid), 8.5–9.2 mM ethylcarbonate-β-CD	ED (BGE optimization). Two-level full factorial designs. Factors: CD concentration, methanol percentage. Response: differences in migration times of enantiomers and congeners.	—	(44)
Ibuprofen, diclofenac, naproxen, bezafibrate, gemfibrozil, ofloxacin, norfloxacin, carbamazepine, primidone, sulfamethazine, sulfadimethoxine, sulfamethoxazole	Water	MEKC–UV (200, 230, 274, 289 nm)	10 mM $(NH_4)_2HPO_4$, pH 11.5, 69 mM SDS, 6 mg/mL sulfated β-CD, 8.5% (v/v) isopropanol	ED (BGE optimization). Face-centered composite design. Factors: SDS concentration, percentage of isopropanol, temperature. Response: resolution, analysis time.	Comparison with a univariate approach.	(45)

Bisphenol A, bisphenol E, bishphenol AP, tetramethyl bisphenol A, bisphenol P	—	MEKC–UV (280 nm)	14.6 mM borate, 15% (v/v) ACN, pH 9.25, 28.5 mM SDS	ED (BGE optimization). Box–Benhken design. Factors: BGE concentration, pH, percentage of organic acetonitrile, SDS concentration. Response: migration time, peak width, resolution, total analysis time.	Comparison with a univariate approach. Also applicable in food analysis applications.	(46)

compounds in the shortest analysis time. The applied univariate method determined that 20 mM borate, 30 mM SDS, 15% (v/v) acetonitrile, and pH 9.3 were the best conditions to separate the selected bisphenols. Analysis times of lower than 8 min were achieved. Overall, 120 experiments were required (preliminary results developed for selecting a set of conditions were not counted). For the multivariate analysis, borate concentration, pH, amount of organic solvent (acetonitrile), and concentration of surfactant were the factors selected in a Box–Behnken design. The electropherograms obtained were processed to obtain the migration time and peak width for each analyte, resolution, and the total analysis time. A chromatographic response function was calculated using a modified version of a method proposed by Divjak et al. (48). In this case, optimum conditions found were slightly different from the ones obtained with the univariate approach: 14.6 mM borate, 28.5 mM SDS, 15% (v/v) acetonitrile, and pH 9.25. A lower analysis time (approximately 5.5 min) with a lower number of experiments (twenty-seven) was achieved.

Although chiral electromigration methods have mainly been used for enantioseparation of drugs and pharmaceuticals, they have also been applied to analyze chiral pollutants (49), being chemometric methods also used for optimization purposes (44). Zerbinatti et al. (44) optimized the CE enantioseparation of two phenoxy acid herbicides (mecoprop and dichlorprop) using an ED. In general, (*R*)-isomers of phenoxy acid herbicides showed much higher herbicide activity and different metabolism than their (*S*)-isomers, which is also the case of these two herbicides. Thus, the chiral separation of pesticides is a very important challenge that will allow optimizing enantioselective production processes, assessing the enantiopurity of formulations and monitoring their presence in the environment. In the work of Zerbinatti et al. (44), the effects of three chiral selectors as additives of the BGE (i.e., an ethylcarbonate derivative of β-CD with three substituents per molecule, hydroxypropyl-β-CD, and native α-CD) were evaluated by a two-level full factorial design. CD concentration as well as methanol content were selected as experimental factors to be optimized. The differences in the migration times of the enantiomers and in the migration times of the two congeners were the experimental responses chosen for evaluation. The main effects of the factors as well as their interaction were calculated by means of the Yates algorithm. Ethylcarbonate-β-CD and α-CD were concluded to be the best chiral reagents.

6.4. FOOD ANALYSIS APPLICATIONS

One of the most important tasks of modern analytical chemistry is the analysis of foods, including fundamental aspects as food safety, quality, and authenticity. In the last decade, and especially in the last years, CE has also gained popularity in food analysis as an alternative to GC or HPLC because of the inherent characteristics/advantages of the technique in terms of analysis speed, efficiency, and low sample and reagent consumption (50–54). Table 6.3 shows

TABLE 6.3. Applications of chemometrics to CE and CE–MS optimization in food analysis

Analytes	Matrix	CE Mode	Buffer	Chemometric Approach	Comments	Reference
Resveratrol, herperidin, L-ascorbic acid, vitamin B2, caffeic acid, p-coumaric acid, ferulic acid, sinapic acid, kuromarin, narirutin, acesulfame K	Nutraceuticals (resveratrol)	CZE–DAD (280 nm)	23 mM borate, pH 10.0	ED (BGE optimization). CCD. Factors: BGE concentration, ACN percentage, voltage. Response: critical resolution and analysis time.	RSM: Derringer desirability function. Robustness testing using a multivariate approach (Plackett–Burman).	(55)
Icariin, epimedin A, epimedin B, epimedin C	Herba Epimedii	CZE–DAD (270 nm)	50 mM borate, pH 10.0, 22% (v/v) ACN	ED (BGE optimization). CCD. Factors: BGE concentration, pH, ACN percentage. Response: total resolution.	RSM	(56)
Huperzine A	Pharmaceutical products, human serum	CZE–DAD (230 nm)	50 mM acetate, pH 4.6	ED–ANN (BGE optimization). Factors: BGE concentration, voltage. Response: peak area or peak height or migration time.	ANN network: 2:3:1. Also applicable in pharmaceutical applications.	(57)

TABLE 6.3. ***Continued***

Analytes	Matrix	CE Mode	Buffer	Chemometric Approach	Comments	Reference
Cyromazine, cyprodinil, pirimicarb, pyrimethanil, pyrifenox	Water and fruit juices	CZE–ESI–MS (IT)	0.3 M HOAc, pH 4	ED (ESI optimization). CCD. Factors: isopropanol and acid percentage in the sheath liquid, nebulizer gas pressure, dry gas flow, dry gas temperature. Response: sum of MS peak intensities.	—	(58)
Cloransulam-methyl, metosulam, flumetsulam, florasulam, diclosulam	Soy milk	CZE–ESI–MS (IT)	24 mM formic acid, 16 mM ammonium carbonate, pH 6.4	ED (ESI optimization). CCD. Factors: isopropanol and acid percentage in the sheath liquid, nebulizer gas pressure, dry gas flow, dry gas temperature. Response: sum of MS peak intensities.	—	(59)

Sucralose	Sparkling beverages, yogurts, cherry candy	CZE–UV (238 nm)	3 mM dinitrobenzoic acid, 20 mM sodium hydroxide, pH 12.1	ED (BGE optimization). CCD. Factors: BGE concentration, pH, injection time, voltage, temperature. Response: resolution.	Indirect UV detection. A second CCD was used for significant factors: separation voltage, temperature.	(60)
Cadaverine, putrescine, histamine, tryptamine, tyramine, phenylethylamine, ethanolamine, agmantine, serotonin	Red wine	CZE–UV (230, 305, 360, 480 nm)	40 mM aqueous sodium tetraborate solution (pH 10.5)–2-propanol (25%, v/v)	ED (BGE optimization). Two-level full factorial design. CCD. Factors: BGE concentration, pH, voltage, percentage of 2-propanol, derivatization reagent injection time, sample injection time. Response: resolution, analysis time, peak width, number of resolved peaks.	In-capillary derivatization with 1,2-naphthoquinone-4-sulfonate (plug injection) and separation conditions optimized independently. Use of CCD used for significant factors.	(61)

TABLE 6.3. *Continued*

Analytes	Matrix	CE Mode	Buffer	Chemometric Approach	Comments	Reference
Tartaric acid, malic acid, succinic acid, acetic acid, lactic acid	Red wine	CZE–UV (214 nm)	35% (v/v) methanol, 22 mM benzoic acid at pH 6.10 adjusted with 1.0 M Tris-base buffer	ED (BGE optimization). Two-level full factorial design, CCD. Factors: temperature, separation voltage, and percentage of methanol.	Coated capillary	(62)
Fructooligosaccharides, inulin	*Bifidobacterium* cultures	CZE–UV (214 nm)	15 mM sodium benzoate, pH 6.22 (1.0 M Tris), 24% (v/v) MeOH	ED (BGE optimization). Two-level full factorial design. Factors: methanol percentage, voltage, temperature. Response: separation efficiency.	Indirect UV detection.	(63)
2,4-dichlorophenoxyacetic acid (2,4-D), dicamba, 2,4,5-trichorophenoxyacetic acid (2,4,5-T)	Tobacco	NACE–UV (200 nm)	40.0 mM ammonium acetate in 90% ACN (apparent pH 10.2)	ED (BGE optimization). Ortogonal design. Factors: BGE concentration, acetonitrile percentage, pH*, voltage. Response: resolution and analysis time.	—	(64)

L-ascorbic acid, D-isoascorbic acid	—	CZE–DAD (254 nm)	50 mM Tris-HCl, pH 8.5	ED (BGE optimization). Box–Behnken design, central composite face-centered design, full fractional design. Factors: BGE concentration, pH, voltage. Response: resolution, migration time.	Comparison of the performance of three experimental designs.	(65)
Danofloxacin, sarafloxacin, ciprofloxacin, marbofloxacin, enrofloxacin, difloxacin, oxolinic acid, flumequine	Milk	CZE–ESI–MS/MS (IT)	70 mM ammonium acetate, pH 9.1	ED (BGE and ESI optimization). Half-fraction factorial design and Doehlert design. Factors: nebulizer pressure, dry gas flow, dry gas temperature, sheath-liquid flow rate, percentage of 2-propanol in the sheath liquid, percentage of formic acid in the sheath liquid. Response: signal-to-noise ratio of danofloxacin.	Doehlert design used for significant factors: nebulizer pressure, dry gas flow, sheath-liquid flow rate, percentage of 2-propanol in the sheath liquid.	(66)

some examples of the most recent applications of chemometrics to the optimization of CE and CE–MS parameters in this field.

Although chemometrics can be helpful in the optimization of the large number of parameters usually involved in any CE–MS analysis, the number of studies in which this approach has been applied is still limited (58, 59, 66). One of these examples is from our group (58), which describes a procedure to sensitively analyze five pesticides (pyrimethanil, pyrifenox, cyprodinil, cyromazine, and pirimicarb) in grape and orange juices using CE–ESI–MS. Good overall separation of these compounds was achieved using a volatile aqueous buffer containing 0.3 M ammonium acetate/acetic acid at pH 4.0. ED methodology using a CCD was used to separately optimize the multiple parameters that can play a role either in the solid-phase microextraction (SPME) procedure used for the extraction of the analytes from the sample matrix or in the CE–MS analysis. ESI parameters selected were nebulizer pressure, dry gas flow, dry gas temperature, and percentage of organic solvent and acid in the sheath liquid. The sum of the peak intensities obtained during direct infusion experiments was selected as the response.

The combined use of chemometrics and SPME–CE–MS clearly improved the LODs that could be achieved, allowing the detection of pesticides at concentrations down to 15 ng/mL. The usefulness of this approach was demonstrated by detecting (in a single run) these pesticides in grapes and orange juice at concentrations below their maximum residue limits (MRLs) values. A similar approach was also carried out by our group for the optimization of the CE–MS separation of another group of pesticides (cloransulam-methyl, metosulam, flumetsulam, florasulam, and diclosulam) and their ultimate determination in soy milk (59).

CE–MS parameters were also optimized by Lara et al. (66). In this study, a CE–MS/MS method for the identification and simultaneous quantification of eight quinolones (danofloxacin, sarafloxacin, ciprofloxacin, marbofloxacin, enrofloxacin, difloxacin, oxolinic acid, and flumequine) of veterinary use in bovine raw milk was developed. Separation buffer composition and ESI conditions (nebulizer pressure, dry gas flow, dry gas temperature, sheath-liquid flow rate, percentage of 2-propanol in the sheath liquid, and percentage of formic acid in the sheath liquid) were optimized in order to obtain both an adequate CE separation and a high sensitivity. A half-fractional factorial screening design in two blocks plus three central points was carried out to check the significance of the factors. Signal-to-noise ratio of danofloxacin was selected as response. Neither the dry gas temperature nor the percentage of formic acid were found significant and thus, a Doehlert design was developed to optimize the remaining variables. An aqueous solution of 70 mM ammonium acetate adjusted to pH 9.1 was chosen as optimum BGE since these conditions gave the best resolution for the eight quinolones. The sheath liquid that provided the highest ESI–MS signal consisted of 2-propanol : water : formic acid (50 : 49 : 1 v/v/v). Nebulizer pressure was kept at 10 psi, dry gas flow at 6 L/min, and dry gas temperature at 150 °C. Quantification in bovine raw milk samples were also developed using SPE. LODs (<6 µg/L) were lower than the MRLs tolerated by the European Union (EU) for these compounds in milk, with recoveries ranging from 81% to 110%.

Biogenic amines are also important analytes to be monitored in foods. These analytes are mainly produced by decarboxylation of the amino acids or the transamination of aldehydes and ketones by the action of microorganisms that present amino acid-decarboxylase enzymatic activity. They appear in protein-rich food like sausages, meat, or fish, in fermented foods like cheese and salami, or in fermented beverages like wine or beer. In a recent study by García-Villar et al. (61), nine biogenic amines (histamine, tryptamine, phenylethylamine, tyramine, agmatine, ethanolamine, serotonin, cadaverine, and putrescine) were determined in red wine by CE–diode array detector (DAD) using field-amplified sample stacking and in-capillary derivatization with 1,2-naphthoquinone-4-sulfonate to improve the sensitivity of the method. BGE composition, as well as in-capillary derivatization conditions, were optimized by means of an ED. In a first step, a screening design (two-level full factorial design) was used to evaluate the effect of the reagent, buffer, and sample injection time on the in-capillary derivatization. Selectivity toward interfering impurities, sensitivity, and resolution were considered the responses of interest.

The three variables were found to be statistically significant, but the sample injection time was not related to any other experimental variable and thus, it was independently optimized. The other two variables were optimized using a CCD. Separation conditions (BGE concentration—sodium tetraborate, pH, voltage, and percentage of 2-propanol) were also studied using a two-level full factorial design. Only BGE pH and percentage of 2-propanol were found to be related and thus, a CCD was used to optimize these factors. Optimum separation conditions achieved were: 40 mM aqueous sodium tetraborate solution (pH 10.5) and 2-propanol (25% v/v). The method, which only included dilution and filtration of the samples, was finally applied to the analysis of this group of biogenic amines in red wine following a standard addition procedure. Figure 6.2 shows a representative electropherogram of a spiked red wine sample.

Another interesting application of chemometrics in food analysis is the work by Gong et al. (65) in which three EDs (Box–Behnken, central composite face-centered, and full fractional design) were comparatively used for the optimization of the BGE concentration, pH, and separation voltage. The optimized method was applied to the separation of L-ascorbic (natural antioxidant with important nutritional benefits) and D-isoascorbic acid (often added for nonvitamin purposes). Figure 6.3 shows a representation of the model of the three EDs used (three factors). Resolution and migration times of the last migrated analyte were selected as response using the Derringer's desirability function. In general, good agreement was found between predicted response and actual experimental data for all three approaches. The response surface

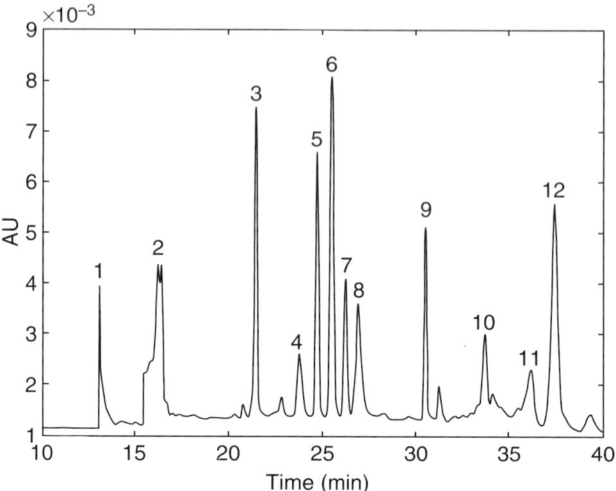

FIGURE 6.2. Representative electropherogram of a spiked red wine sample (50 μM of each biogenic amine). Peak assignment: (1) agmatine; (2) 1,2-naphthoquinone-4-sulfonate excess; (3) tryptamine; (4) cadaverine monoderivative; (5) phenylethylamine; (6) histamine; (7) serotonin; (8) putrescine; (9) ethanolamine; (10) tyramine; (11) cadaverine diderivative; (12) putrescine diderivative. Reprinted from Reference 61 with permission from Wiley-VCH Verlag.

plots revealed a separation optimum with 50 mM Tris-HCl buffer of pH 8.5 at 30 kV. Additionally, separation was achieved in less than 5 min.

Few studies have been published so far on the combined use of ED and ANNs for the optimization of CE parameters (57, 67–69). In one such study, Farkova et al. (69) compared the results obtained for the CE analysis of galanthamine under both univariate and multivariate optimization approaches. Galanthamine, used to treat Alzheimer's disease, was originally isolated from the bulbs of snowdrops of *Galanthus nivalis*. The multivariate approach used a CCD with three factors (pH, injection time, and separation voltage) in combination with ANNs using the peak height as output. An architecture of (2:7:1) was selected for the ANN. Results demonstrated that the sensitivity and efficiency were higher (as well a lower analysis time) under multivariate optimization conditions using ANNs.

More recently, Ben-Hameda et al. (57) used this combination to improve the sensitivity of the determination of Huperzine A (a natural product from *Huperzia serrata* used to treat Alzheimer's disease and incorporated as a food supplement). BGE concentration (acetate buffer pH = 4.6) and voltage were used as input parameters, while peak area, peak height, or migration time were individually studied as outputs. In each case, optimal ANN architecture was (2:3:1). To maximize the sensitivity, relatively high concentrations of the BGE and low voltages were required (optimum conditions were 50 mM sodium acetate and 10 kV).

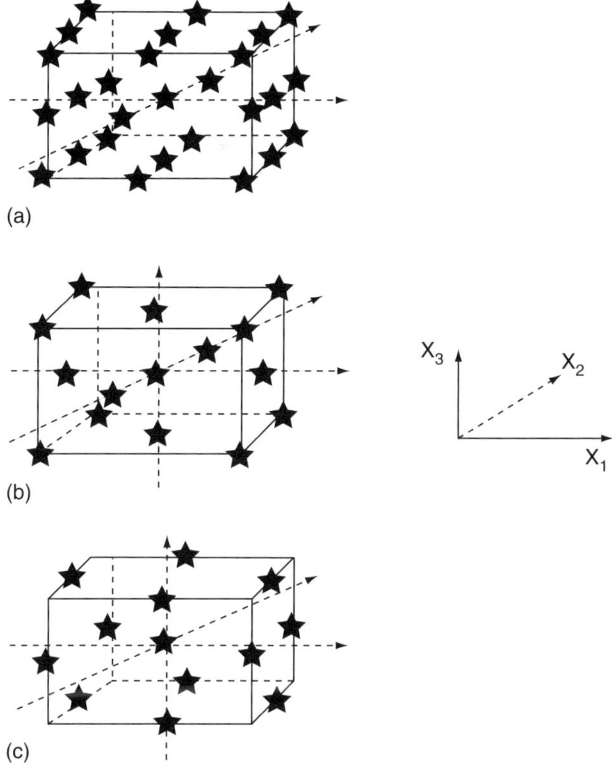

FIGURE 6.3. Representation of (a) full fractional design, (b) central composite face-centered design, and (c) Box–Behnken design models for three factors. Reprinted from Reference 65 with permission from Springer-Verlag.

6.5. RELATED APPLICATIONS

Chemometrics has also been used for optimization and application of CE in other fields of research, including clinical analysis, proteomics, DNA analysis. In Table 6.4 some examples of these applications can be found.

Affinity capillary electrophoresis (ACE) constitutes a versatile microanalytical technique that allows the estimation of affinity constants of analytes through the study of interactions such as protein–ligand, protein–antibody, and antibody–antigen. In ACE, PF techniques (whose optimization is not an easy task) can also be used to minimize the amount of sample needed. Chemometric methodology has also been applied for the optimization of the PF technique in ACE.

One example is the recent work developed by Montes et al. (70) in which a Box–Behnken design was used in flow-through PFACE. Injection time, voltage, and neutral ligand (neutral arylsulfonamides: [[[4-(aminosulfonyl) phenyl)methyl]-amino]-6-oxohexanoic acid and *p*-toluenesulfonamide) con-

TABLE 6.4 Applications of chemometrics to CE and CE–MS optimization in other different fields

Analytes	Matrix	CE Mode	Buffer	Chemometric Approach	Comments	Reference
Protein	Bovine erythrocyte, horse heart myoglobin	ACE–UV (200 nm)	192 mM glycine-25 mM Tris, pH 8.4	Partial-filling affinity conditions optimization. Box–Behnken design. Factors: injection time, voltage, neutral ligand (neutral arylsulfonamide) concentration. Response: relative migration time ratio.	Optimization of the partial-filling technique.	(70)
Protein	Bovine erythrocyte, horse heart myoglobin	ACE–UV (200 nm)	192 mM glycine-25 mM Tris, pH 8.4	Partial-filling affinity conditions optimization. Box–Behnken design. Factors: injection time, voltage, capillary length. Response: relative K_d.	Prediction of the effect of the factors on protein-ligand binding.	(71)
DNA	—	CGE–LIF (λ_{exc} = 488 nm, λ_{em} = 520 nm)	0.6 mM Tris, 0.6 mM TAPS, 0.012 mM EDTA, Ph 8.3	Simplex (injection conditions optimization). Factors: BGE concentration, injection time, injection voltage. Response: signal-to-noise ratio, resolution.	—	(72)

DNA	—	CGE–LIF (λ_{exc} = 488 nm, λ_{em} = 520 nm)	0.6 mM Tris, 0.6 mM TAPS, 0.012 mM EDTA, pH 8.3	Simplex (BGE optimization). Factors: sample buffer concentration, injection time, injection voltage, temperature, matrix concentration, separation voltage. Response: correlation coefficient of a logarithmic plot of mobility μ versus base pair.	—	(73)
Nicotinamide adenine dinucleotide, benzenesulfonamide	—	CZE–UV (220 nm)	192 mM Tris-25 mM glycine, pH 8.34	Flow injection–CE optimization. Box–Behnken design. Factors: capillary length, voltage, injection time. Response: absorbance.	In-house built flow injection–CE instrument.	(74)

TAPS = *N*-tris(hydroxymethyl)methyl-3-aminopropanesulfonic acid.

centration were the factors investigated together with their effect on protein-neutral ligand binding (carbonic anhidrase B with the neutral ligands). Predicted results were in good agreement with the experimental ones. The model was validated by experiments run under the optimal predicted conditions (2.3 min injection time, 11.6 kV, 1.4 μM ligand concentration). The achieved results clearly provided a valuable statistical tool for the study of other receptor–ligand combinations. In previous work by the same group (71), the effect of factors like capillary length, voltage, and injection time on protein–ligand binding in ACE (the ligand was 4-carboxybenzenesulfonamide) was also studied using a Box–Behnken design.

The development of coupled analytical techniques is one of the current (and future) research lines toward full automation of given analytical procedures. The hyphenation of different techniques may also increase the number of factors to be optimized, depending on their influence on analytical performance. The coupling of flow injection systems with CE (FI–CE) (74–76) is also an example showing the importance of simultaneously optimizing parameters from two separate techniques. Dahdouh et al. (74) examined FI–CE parameters (capillary length, voltage, and injection volume) as well as their interactions via an RSM in the form of a Box–Behnken design. Initial studies were developed for the assessment of the highest peak height and best peak shape of the model compound *N,N*-dimethylformamide. Figure 6.4 shows the response surface plot in which capillary length versus voltage showed a strong interactive effect that, as stated by the authors, could not have been detected by traditional univariate methods. Optimum critical values were 45 cm capillary length, 7.5 kV, and 40.13 nL injection volume. Afterward, optimum conditions were used for the injection of nicotinamide adenine dinucleotide and benzenesulfonamide mixtures.

The simplex method, first developed by Spendly et al. (77), is a relatively easy procedure for optimization that is currently not very widely used. However, its simplicity and ease of use makes this approach an interesting alternative. Thus, Catai et al. (73) reported a simplex method to optimize six separation variables simultaneously for the capillary gel electrophoresis–laser induced fluorescence detection (CGE–LIF) analysis of DNA fragments. In brief, the simplex is a geometric figure with one more vertex than the number of factors to be optimized. Each experiment is developed taking into account the results of the previous experiments. Initially, the first experiments are ranked worst to best vertices and then the next experiment is determined by reflection of the worst response through the hyperface defined by the other vertices. Afterward, the least ranked vertex is not taken into account and a new simplex is developed. The process continues until the optimum is reached. In this work, sample buffer concentration, injection time, injection voltage, temperature, matrix concentration, and separation voltage were the selected variables to be optimized. In this case, the simplex method maximized the correlation coefficient (r^2) of a logarithmic plot of mobility (μ) versus the number of base pairs (bp) for the separation of DNA fragments between 75

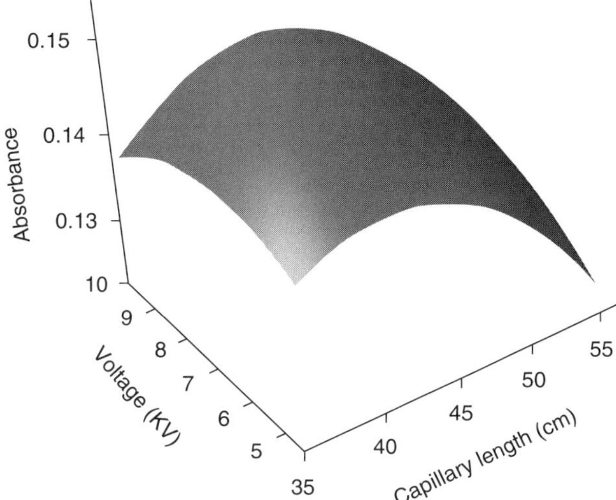

FIGURE 6.4. Response surface generated plot showing the interactive effect capillary length-voltage. Reprinted from Reference 74 with permission from Wiley-VCH Verlag.

and 4072 bp ($r^2 = 0.98$). For fragments between 201 and 2036, the r^2 increased to 0.992. Up to 38 experiments were developed, in which vertex 21 (0.6 mM buffer, 19 s injection time, 97.46 V/cm injection voltage, 25.9 °C, and 0.26% v/v separation matrix concentration) was selected as the optimum. The electrophoresis buffer stock solution was made of 100 mM Tris, 100 mM TAPS (*N*-tris[hydroxymethyl]methyl-3-aminopropanesulfonid acid), 2 mM ethylenediaminetetraacetic acid (EDTA) at pH 8.3.

In previous work of the same group (72), the electrokinetic injection of DNA fragments was optimized as well by means of a simplex method. CGE–LIF was also used. In this case, BGE concentration, sample injection voltage, and time were the factors to be optimized. The optimum conditions were reached after only nine experiments. Figure 6.5 shows the spatial evolution of the simplex method used in this work (the initial tetrahedron (vertices 1–4) and the subsequent movements of reflection and contraction). Vertex 9 was considered as the optimum for injection of the 1 kbp DNA ladder (1.0 mM TTE buffer, 20 s injection, 55 V/cm electric field injection).

6.6. CONCLUDING REMARKS AND FUTURE CONSIDERATIONS

It is clear that the current literature dealing with the application of chemometrics toward the optimization of CE and CE–MS shows that the number of studies published up to now is relatively low compared to more established chromatographic techniques. However, it is also clear that this number is growing at an important rate. The main reasons behind this growth can be found on the important advantages derived from the use of chemometrics.

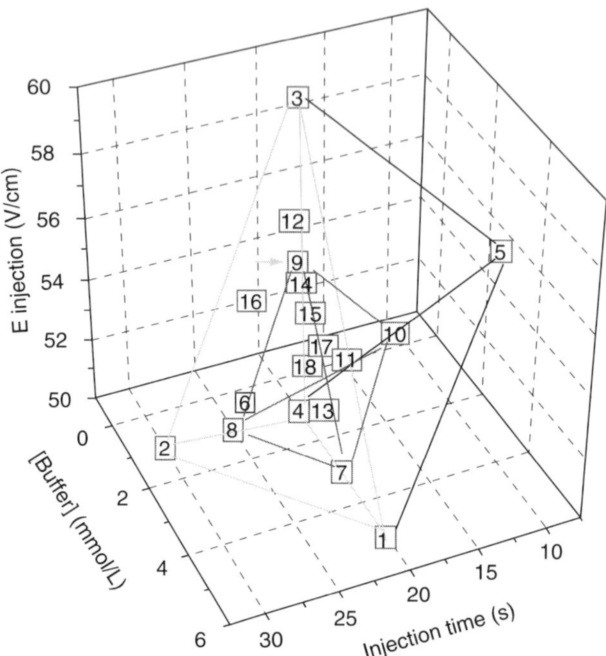

FIGURE 6.5. Spatial evolution of the simplex optimization. The solid bold lines link the initial conditions (vertices 1–4). The dashed lines show the simplex figure after the radical contraction (vertices 4, 7–9) and the first reflection after contraction (vertex 10, dotted lines). The arrow shows the best result. Reprinted from Reference 72 with permission from Wiley-VCH Verlag.

Thus, chemometrics allows one to reach, in a relatively quick and simple way, with a minimum number of experiments, optimum CE or CE–MS conditions to solve a myriad of analytical problems related to pharmaceutical, environmental, or food analysis.

In this regard, it is interesting to remark that the number of factors that may influence a CE or CE–MS separation is quite high and thus, the selection of the experimental factors to be optimized is not harmonized. Depending on the application as well as on the importance of the factors, experimental strategies followed in the literature might differ greatly. Likewise, the selection of the responses used as output can also be very different. In the majority of the applications, factorial designs as well as CCD are used for the optimization of CE and CE–MS methods. They are focused on the separation of the target analytes and/or the method sensitivity. In this sense, several studies have demonstrated the advantages of combining preconcentration strategies with stacking techniques (especially in regard to electrokinetic injection) and/or MS detection. More studies dealing with the use of ANNs are also expected, since the application of ANNs in separation is still in its infancy.

One of the current trends in separation science is the development of comprehensive or multidimensional separation systems, in which CE and CE–MS are also achieving relative importance. Chemometric approaches like the ones described in this chapter will surely be of great help for the optimization of these more complicated separation systems. Current trends toward miniaturization in separation science are also well known. Ultrafast separations, extremely low sample requirements, and automation of the arrangement are some of these goals. Chemometrics will surely provide an interesting and challenging approach for the optimization of separation conditions in these miniaturized systems, including microchips for years to come.

ACKNOWLEDGMENTS

J.H.B. wishes to thank the Spanish Ministry of Science and Innovation for the Ramón y Cajal contract at the University of La Laguna. This work was supported by projects AGL2008-00990/ALI, AGL2005-05320-C02-01, Consolider Ingenio 2010 CSD2007-00063 FUN-C-FOOD (all from Spanish Ministry of Science and Innovation) and S-505/AGR-0153 (ALIBIRD, Comunidad de Madrid).

REFERENCES

1. Jorgenson, J.W. and Luckacs, K.D. (1981) *J Chromatogr*, **218**, 209–216.
2. Jorgenson, J.W. and Luckacs, K.D. (1981) *Anal Chem*, **53**, 1298–1302.
3. Siouffi, A.M. and Phan-Tan-Luu, R. (2000) *J Chromatogr A*, **892**, 75–106.
4. Hanrahan, G., Montes, R., and Gomez, F.A. (2008) *Anal Bioanal Chem*, **390**, 169–179.
5. Hanrahan, G. and Lu, K. (2006) *Crit Rev Anal Chem*, **36**, 141–151.
6. Hanrahan, G., Zhu, J., Gibani, S., and Patil, D.G. (2005) Chemometrics: experimental design, in *Encyclopedia of Analytical Science*, 2nd ed. (eds. P.J. Worsfold, A. Townshend, and C.F. Poole), Elsevier, Oxford, pp. 8–13.
7. Derringer, G., and Suich, R. (1980) *J Quality Technol*, **12**, 214–219.
8. Gasteiger, J. and Zupan, J. (1993) *Angew Chem*, **32**, 503–527.
9. Aleksander, I. and Morton, H. (1990) *An Introduction to Neural Computing*, Chapman and Hall, London.
10. Alnajjar, A., AbuSeada, H.H., and Idris, A.M. (2007) *Talanta*, **72**, 842–846.
11. Varga Mamani, M.C., Amaya Farfán, J., Reyes Reyes, F.G., and Rath, S. (2006) *Talanta*, **26**, 236–243.
12. François, Y., Varenne, A., Juillerat, E., Servais, A.C., Chiap, P., and Gareil, P. (2007) *J Chromatogr A*, **1138**, 268–275.
13. Penido Maia, P., Amaya-Farfán, J., Rath, S., and Reyes Reyes, F.G. (2007) *J Pharm Biomed Anal*, **43**, 450–456.

14. Lin, Y.H., Yang, Y.H., and Wu, S.M. (2007) *J Pharm Biomed Anal*, **44**, 279–282.
15. Bailón Pérez, M.I., Cuadros Rodríguez, I., and Cruces Blanco, C. (2007) *J Pharm Biomed Anal*, **43**, 746–752.
16. Marchesini, A.F., Williner, M.R., Mantovani, V.E., Robles, J.C., and Goicoechea, H.C. (2003) *J Pharm Biomed Anal*, **31**, 39–46.
17. Ficarra, R., Cutroneo, P., Aturki, Z., Tommasini, S., Calabró, M.L., Phan-Tan-Luu, R., Fanali, S., and Ficarra, P.J. (2002) *J Pharm Biomed Anal*, **29**, 989–997.
18. Furlanetto, S., Orlandini, S., La Porta, E., Coran, S., and Pinzauti, S. (2002) *J Pharm Biomed Anal*, **28**, 1161–1171.
19. Ragonese, R., Macka, M., Hughe, J., and Petocz, P.J. (2002) *J Pharm Biomed Anal*, **27**, 995–1007.
20. Alnajjar, A., Idris, A.M., and AbuSeada, H.H. (2007) *Microchem J*, **87**, 35–40.
21. Vargas Mamani, M.C., Amaya-Farfan, J., Reyes Reyes, F.G., Fracassi da Silva, J.A., and Rath, S. (2008) *Talanta*, **76**, 1006–1014.
22. Mateus-Avois, L., Mangin, P., and Saugy, M. (2003) *J Chromatogr B*, **791**, 203–216.
23. Olsson, J., Stegander, F., Marlin, N., Wan, H., and Blomberg, L.G. (2006) *J Chromatogr A*, **1129**, 291–295.
24. Anurukvorakun, O., Suntornsuk, W., and Suntornsuk, L. (2006) *J Chromatogr A*, **1134**, 326–332.
25. Servais, A.C., Fillet, M., Chiap, P., Dewé, W., Hubert, P., and Crommen, J. (2005) *J Chromatogr A*, **1068**, 143–150.
26. Awadallah, B., Schmidt, P.C., and Wahl, M.A. (2003) *J Chromatogr A*, **988**, 135–143.
27. Hillaert, S., Vander Heyden, Y., and Van den Bossche, W. (2002) *J Chromatogr A*, **978**, 231–242.
28. Capella-Peiró, M.E., Bossi, A., and Esteve-Romero, J. (2006) *Anal Biochem*, **352**, 41–49.
29. Lara, F.J., García-Campaña, A.M., Alés-Barrero, F., and Bosque-Sendra, J.M. (2005) *Anal Chim Acta*, **535**, 101–108.
30. Vera-Candioti L., Olivieri, A.C., and Goicoechea, H.C. (2007) *Anal Chim Acta*, **595**, 310–318.
31. Cheng, Y.Q., Chen, H.L., Fan, L.Y., Chen, X.G., and Hu, Z.D. (2004) *Anal Chim Acta*, **525**, 239–245.
32. Zhang, Y.P., Zhang, Y.J., Gong, W.J., Wang, S.M., Xue, H.Y., and Lee, K.P. (2007) *J Liq Chromatogr Rel Technol*, **30**, 215–234.
33. Galinski, M., Lewandowski, A., and Stepniak, I. (2006) *Electrochim Acta*, **51**, 5567–5580.
34. Mwongela, S.M., Numan, A., Gill, N.L., Agbaria, R.A., and Warner, I.M. (2003) *Anal Chem*, **75**, 6089–6096.
35. Vaher, M., Koel, M., and Kaljurand, M. (2002) *Electrophoresis*, **23**, 426–430.
36. Yanes, E.G., Gratz, S.R., Baldwin, M.J., Robinson, S.E., and Stalcup, A.M. (2001) *Anal Chem*, **73**, 3838–3844.
37. Siouffi, A.M. and Phan-Tan-Luu, R. (2000) *J Chromatogr A*, **892**, 75–106.
38. Ross, G.A. (2001) *LC-GC Europe*, **1**, 2–6.

39. Huikko, K., Kotiaho, T., and Kostiainen, R. (2002) *Rapid Comm Mass Spec*, **16**, 1562–1568.
40. Moini, M. (2002) *Anal Bioanal Chem*, **373**, 466–480.
41. Rudaz, S., Cherkaoui, S., Gauvrit, J.Y., Lantéri, P., and Veuthey, J.L. (2001) *Electrophoresis*, **22**, 3316–3326.
42. Jurado-González, J.A., Galindo-Riaño, M.D., and García-Vargas, M. (2003) *Talanta*, **59**, 775–783.
43. Van Pinxteren, D. and Hermann, H.J. (2007) *J Chromatogr A*, **1171**, 112–123.
44. Zerbinatti, O., Trotta, F., and Giovannoli, C. (2000) *J Chromatogr A*, **875**, 423–430.
45. Drover, V.J. and Bottaro, C.S. (2008) *J Sep Sci*, **31**, 3740–3748.
46. Felhofer, J., Hanrahan, G., and García, C.D. (2009) *Talanta*, **77**, 1172–1178.
47. Oehlmann, J., Oetken, M., and Schulte-Oehlmann, U. (2008) *Environ Res*, **108**, 140–149.
48. Divjak, B., Moder, M., and Zupan, J. (1998) *Anal Chim Acta*, **358**, 305–312.
49. Hernández-Borges, J., Rodríguez-Delgado, M.A., García-Montelongo, F.J., and Cifuentes, A. (2005) *Electrophoresis*, **26**, 3799–3813.
50. Cifuentes, A. (2006) *Electrophoresis*, **27**, 283–303.
51. Boyce, M.C. (2007) *Electrophoresis*, **28**, 4046–4062.
52. Simó, C., Barbas, C., and Cifuentes, A. (2005) *Electrophoresis*, **26**, 1306–1318.
53. Juan-García, A., Font, G., and Picó, Y. (2005) *J Sep Sci*, **28**, 793–812.
54. García-Cañas, V. and Cifuentes, A. (2007) *Electrophoresis*, **28**, 4013–4030.
55. Orladini, S., Giannini, I., Pinzauti, S., and Furlanetto, S. (2008) *Talanta*, **74**, 570–577.
56. Liu, J.J., Li, S.P., and Wang, Y.T. (2006) *J Chromatogr A*, **1103**, 344–349.
57. Ben Hameda, A., Elosta, S., and Havel, J. (2005) *J Chromatogr A*, **1084**, 7–12.
58. Hernández-Borges, J., Rodríguez-Delgado, M.A., García-Montelongo, F.J., and Cifuentes, A. (2004) *Electrophoresis*, **25**, 2065–2076.
59. Hernández-Borges, J., Rodríguez-Delgado, M.A., García-Montelongo, F.J., and Cifuentes, A. (2005) *J Sep Sci*, **58**, 948–956.
60. McCourt, J., Stroka, J., and Anklam, E. (2005) *Anal Bioanal Chem*, **382**, 1269–1278.
61. García-Villar, N., Saurina, J., and Hernández Cassou, S. (2006) *Electrophoresis*, **27**, 474–483.
62. Bianchi, F., Careri, M., and Corradini, C. (2005) *J Sep Sci*, **28**, 898–904.
63. Corradini, C., Bianchi, F., Matteuzzi, D., Amoretti, A., Rossi, M., and Zanoni, S. (2004) *J Chromatogr A*, **1054**, 165–173.
64. Liu, H., Song, J., Han, P., Li, Y., Zhang, S., Liu, H., and Wu, Y. (2006) *J Sep Sci*, **29**, 1038–1044.
65. Gong, W.J., Zhang, Y.P., Choi, S.H., Zhang, Y.J., and Lee, K.P. (2007) *Microchim Acta*, **156**, 327–335.
66. Lara, F.J., García-Campaña, A.M., Alés-Barrero, F., Bosque-Sendra, J.M., and García-Ayuso, L.E. (2006) *Anal Chem*, **78**, 7665–7673.
67. Pokorná, L., Revilla, A., Havel, J., and Patocka, J. (1999) *Electrophoresis*, **20**, 1993–1997.

68. Elosta, S., Gajdosova, D., and Havel, J. (2006) *J Sep Sci*, **29**, 1174–1179.
69. Farkova, M., Peña-Méndez, E.M., and Havel, J. (1999) *J Chromatogr A*, **848**, 365–374.
70. Montes, R., Hanrahan G., and Gomez, F.A. (2008) *Electrophoresis*, **29**, 3325–3332.
71. Hanrahan, G., Montes, R.E., Poe, A., Johnson, A., and Gomez, F.A. (2007) *Electrophoresis*, **228**, 2853–2860.
72. Catai, J.R. and Carrilho, E. (2003) *Electrophoresis*, **24**, 648–654.
73. Catai, J.R., Formenton-Catai, A.P., and Carrilho, E. (2005) *Electrophoresis*, **26**, 1680–1686.
74. Dahdouh, F.T., Clarke, K., Salgado, M., Hanrahan, G., and Gomez, F.A. (2008) *Electrophoresis*, **29**, 3779–3785.
75. Hanrahan, G., Dahdouh, F., Clarke, K., and Gomez, F.A. (2005) *Curr Anal Chem*, **1**, 321–328.
76. Arce, L., Ríos, A., and Valcárcel, M. (1997) *J Chromatogr A*, **791**, 279–287.
77. Spendley, W., Hesat, G.R., and Himsworth, F.R. (1962) *Technometrics*, **4**, 441–461.

CHAPTER 7

OPTIMIZATION OF THE SEPARATION OF AMINO ACIDS BY CAPILLARY ELECTROPHORESIS USING ARTIFICIAL NEURAL NETWORKS

AMANDA VAN GRAMBERG, ALISON BEAVIS, LUCAS BLANES, and PHILIP DOBLE
Department of Chemistry and Forensic Science, University of Technology, Sydney, Australia

CONTENTS

Chemometric Methods in Capillary Electrophoresis. Edited by Grady Hanrahan and Frank A. Gomez

7.1. INTRODUCTION

7.1.1. Optimization Strategies for Separations by Capillary Electrophoresis

Many factors can affect the separation performance of a capillary electrophoresis (CE) electrolyte, such as the buffer, surfactant and organic modifier concentrations, pH, capillary temperature, and applied voltage (1). The efficient manipulation of these factors is critical to optimize the resolution of a given analysis in the shortest time frame.

During the method development process, an analyst will usually attempt a separation based on a previously reported method that is similar or the same as the requirements of the analysis at hand. If the separation is inadequate, a univariate approach (2) is often employed to attempt to improve the separation. This involves altering one parameter at a time in a systematic way, and viewing the results by plotting the effect of the parameter on the migration time of the analytes. In this way, suitable electrolyte compositions may be found that separate all of the analytes. If suitable conditions are not found, a second parameter is chosen and altered in a similar manner. This univariate procedure is then repeated until a suitable condition is found. This method of optimization is time-consuming, and it is unknown if the optimum is truly the global optimum. Furthermore, univariate optimization is often complex and protracted when there are a large number of factors influencing separation (3).

The alternative is to employ a multivariate optimization procedure such as Simplex. Simplex is an algorithm that seeks the vector of parameters that corresponds to the separation optimum within an n-dimensional experimental space. For example, a two-parameter CE separation optimized by Simplex would begin with three observations of the separation response at three different electrolyte conditions. These conditions are chosen by the analyst, often his or her "best guess." From the evaluation of the response of each observation, the algorithm chooses the next experimental condition for investigation (4). As with the univariate method, the experiments continue until an optimal separation condition is determined. The disadvantage of such an approach is that it is unknown how many experiments are required to achieve an optimum, or if the optimum is local or global as the entire response surface is not known.

Optimization can be simplified by employing the predictive capabilities of an artificial neural network (ANN). This multivariate approach has been shown to require minimal number of experiments that allow construction of an accurate experimental response surface (5, 6). The apposite model created from an experimental design should effectively relate the experimental parameters to the output values, which can be used to create an ANN with a strong predictive capacity for any conditions defined within the experimental space (4).

Experimental design is the process of planning a minimum sequence of experiments for altering parameters simultaneously, thus providing a mathematical framework from which the maximum amount of information can be interpolated (3). The data derived from the experimental space are used to infer a relationship between the separation conditions and electrophoretic mobilities of the analytes (2, 7). Predicted outputs from the trained network can be used to form a response surface that provides the maximum correlation between the independent variables and the optimal separation conditions (3).

As with the univariate approach, experiments are conducted until an optimum is reached. However, the predictions provide an indication as to whether the system can resolve the target compounds as the whole response surface is predicted within the experimental space. A point is eventually reached where the predictions will redirect the analyst to previously designated conditions. It is at this point that the system has reached the limits of its capability. Thus, if separation is not achieved, it is a clear sign that it cannot be achieved by altering the parameters set out in the experimental design. In this chapter we demonstrate the optimization of the separation of amino acids by CE employing an ANN.

7.2. EXPERIMENTAL

7.2.1. Standard Preparation

7.2.1.1. Amino Acids Stock Solutions. Stock solutions of 1 mg/mL of *l*-alanine, *l*-histidine, *l*-isoleucine, *l*-leucine, *l*-ornithine, *l*-phenylalanine, *l*-proline, *l*-serine, and *l*-threonine were prepared in Milli-Q water (Sartorius, Germany) and filtered with a syringe filter with a pore size of 0.25 μm. The stock solutions were stored in amber glass bottles and refrigerated at 2–4 °C and diluted weekly for derivatization with fluorescein isothiocyanate (FITC).

7.2.1.2. FITC Stock Solution. Stock solutions of FITC isomer I ~90% at a 10 mM concentration was prepared in AR grade acetone and stored in amber glass bottles wrapped in aluminium foil at −18 °C.

7.2.1.3. Derivatization. Derivatization of the amino acids was required for UV detection. The reaction is shown in Figure 7.1. The derivatization was performed by combining 100 μL of 10 mM borate, 100 μL of 1 mg/mL analyte solution, and 100 μL of 10 mM FITC in acetone. The solution was placed in a glass vial, wrapped in foil, and stored in the dark for 24 h at room temperature. After this period, derivatization was complete and the solutions were stored at −18 °C to retard the generation of hydrolysis products. The derivatized solutions were prepared freshly each week to reduce interfering signals from hydrolysis products.

7.2.2. Buffer Reagents

A 100 mM of sodium tetraborate dodecahydrate (borate, Fluka, Buchs, Switzerland), 100 mM phosphate, and 200 mM sodium dodecyl sulfate (SDS,

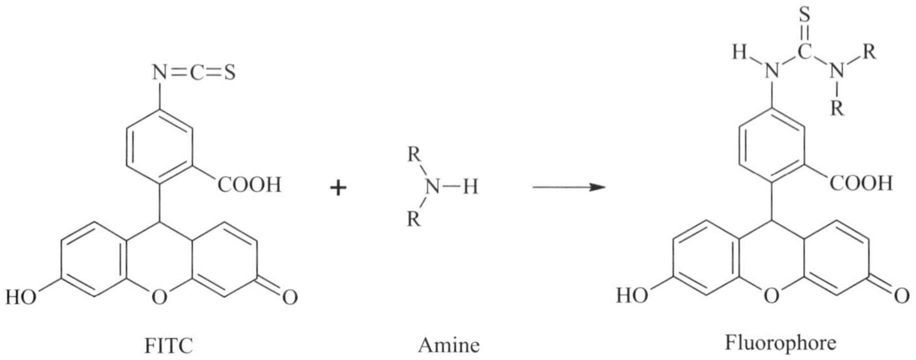

FIGURE 7.1. The thiocarbamylation reaction of FITC with an amine (where R represents an organic group of a hydrogen) to produce a fluorescent/UV-visible detectable derivative of the amine.

Sigma Aldrich, Castle Hill, NSW, Australia) stocks were made up weekly and diluted regularly for use. AR grade boric acid and sodium hydroxide at concentrations of 0.5 M were used to modify the pH.

7.2.3. Instrumentation

Experiments were conducted on an Agilent 3D Capillary Electrophoresis System (Agilent Technologies, CA, USA) with on-column photodiode array detection. All experiments were conducted on a 50 μm internal diameter fused silica capillary of 50 cm total length. A detection window was created at 8.5 cm along the capillary by burning off the polyamide coating and cleaning the capillary with acetone. As a result, the effective length of the column from the detection window to the capillary inlet was found to be 41.5 cm.

Separation voltages were 30 kV and pressure injections were at 15 mbar for 5 sec. The spectral properties of FITC derivatives allowed detection of each of the amino acids at wavelengths of 195, 200, and 488 nm (8). Specifically, UV absorption was found to be most sensitive for FITC derivatives at 488 nm differentiating their signal from other interfering compounds. All experiments were run in duplicate with the average calculated mobility used for ANN and data analysis.

7.2.4. Experimental Design

Scouting experiments were conducted to determine the boundaries of the experimental space (Table 7.1). From these experiments the following experimental design was proposed to train an appropriate ANN (see Fig. 7.2).

TABLE 7.1. Scouting experiments

Scouting Experiment No.	pH	SDS Concentration (mM)
1	8.23	50
2	9.81	50
3	10.82	50
4	9.12	25
5	9.12	50
6	9.12	75

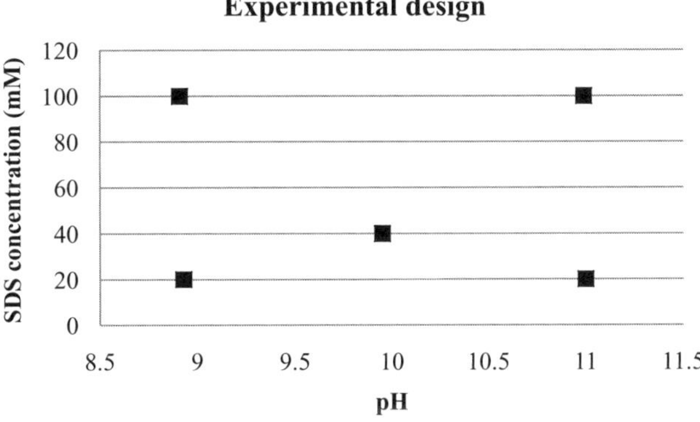

FIGURE 7.2. Experimental design for the separation of an amino acid mix in order to train an ANN and determine a response surface.

7.2.5. ANNs

The software used to construct the neural networks was Trajan Neural Networks Version 6.0 (Trajan Software Ltd., Lincs, UK). The input values for the ANN were electrolyte pH and SDS concentration. Initial networks were trained from experiments conducted at the boundaries of the experimental space. The most appropriate model was chosen that had the minimum training error after varying the number of nodes in the hidden layer.

A data grid was created for predicting the response surface. The grid described the SDS concentration from 20 to 80 mM with increments of 5 mM. The pH scale was generated by increasing the pH by units of 0.2 from 9.2 to 11.4. The network then predicted the mobilities of the analytes at each grid coordinate.

These data were then transferred to a spreadsheet in a statistical program, where the outputs were ordered from smallest to largest. From the rearranged data, the peak pair resolution was determined by calculating the difference in adjacent mobilities. From this, the minimum peak pair resolution and product peak pair resolution was determined. These values were then used to generate a response surface from which an optimum separation condition could be determined.

The optimum combination of SDS and pH was then run on a mixed amino acid standard and the resolution calculated. If baseline resolution was not achieved, the experimental condition was then reintroduced into the ANN as a verification point. Verification points were used to determine the accuracy of the model, and those with a verification error with the same magnitude as the training error were selected to predict analyte mobilities throughout the experimental space.

With the addition of a verification point, the training error and verification errors were compared. If the errors were at a minimum and were of a similar magnitude, then the corresponding ANN was selected to predict an improved response surface. This procedure is repeated until baseline resolution is achieved, or alternatively, until the optima converged.

7.2.6. Generating the Response Surface

The predicted data were processed by Minitab Release 14 Statistical Software (Minitab Inc., Pennsylvania, USA) to produce a three-dimensional response surface using the grid layout described in section 7.2.5.

7.3. RESULTS AND DISCUSSION

7.3.1. Experimental Design

The scouting experiments detailed in Table 7.1 indicated that the experimental space for the separation of amino acids was between SDS concentration of 20 mM and 80 mM, and a pH range of 9.2–11.4. The scouting experiments also indicated that at pH values below 9, the derivatized amino acids were not resolved, and had severe peak distortions (Fig. 7.3).

7.3.2. Construction of the ANNs

The designs of ANNs are based on the architecture of the cerebral networks of the brain, which learns by example, thus the structure allows for the network to mimic the mapping of multivariate data (2, 4). The basic processing units of the ANN are simulated neurons or nodes. The nodes are interconnected in groups to form a soft modeling computation tool that can be applied to a system without the need to know or establish a mathematical model (7, 9, 10). Accordingly, the fundamental structure of the data can be identified through a heuristic process (11). This heuristic process for the optimization of the CE separation is described in the following paragraphs.

Initially, networks were trained from data obtained from the experimental design conditions given in Figure 7.3. These were radial basis function (RBF) networks, multilayer perception (MLP) networks, probabilistic neural networks (PNNs), and generalized regression neural networks (GRNNs), as well

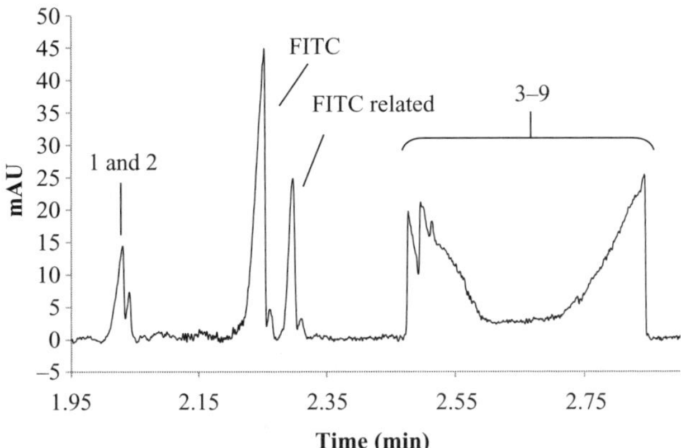

FIGURE 7.3. Separation of an amino acid mix containing 1 = ornithine-FITC; 2 = threonine-FITC; 3 = leucine-FITC; 4 = isoleucine-FITC; 5 = proline-FITC; 6 = histidine-FITC; 7 = phenylalanine-FITC; 8 = alanine-FITC; and 9 = serine-FITC. The background electrolyte consisted of 80 mM SDS, 10 mM borate, and a pH of 8.24.

as the linear (Linear) networks. The best performing networks were MLP and were therefore chosen for determinations of the response surfaces for later experiments. These MLP networks consisted of an array of nodes organized in three layers, which served to associate the nodes within one layer to those in an adjacent layer.

This structure formed a feed-forward network (Fig. 7.4) (4). Input nodes in the first layer corresponded to the independent variables characterizing each observation taken directly from the parameters of the experimental design. The input information was transmitted to layer 2 where the data were processed. Layer 2 consisted of numerous hidden nodes that connected layer 1 to layer 3. Layer 3 consisted of the output nodes, which were the mobilities of the analytes.

The root mean square (RMS) error was used to determine the suitability of the ANN (7, 12). The network was considered trained when the RMS value reached a minimum. The RMS depicted the overall error of individual errors summed and was calculated using Equation 7.1 (2):

$$RMS = \sqrt{\frac{\sum_{a=1}^{N}\sum_{b=1}^{M}(t_{ab} - y_{ab})^2}{N \times M}} \qquad \text{(Eq. 7.1)}$$

where t_{ab} are the inputs, y_{ab} are the outputs, N is the number of patterns, and M is the number of outputs for the training set derived from the experimental design.

Varying the number of nodes in the hidden layer significantly affected the network's ability to accurately define the response surface, thus the structure

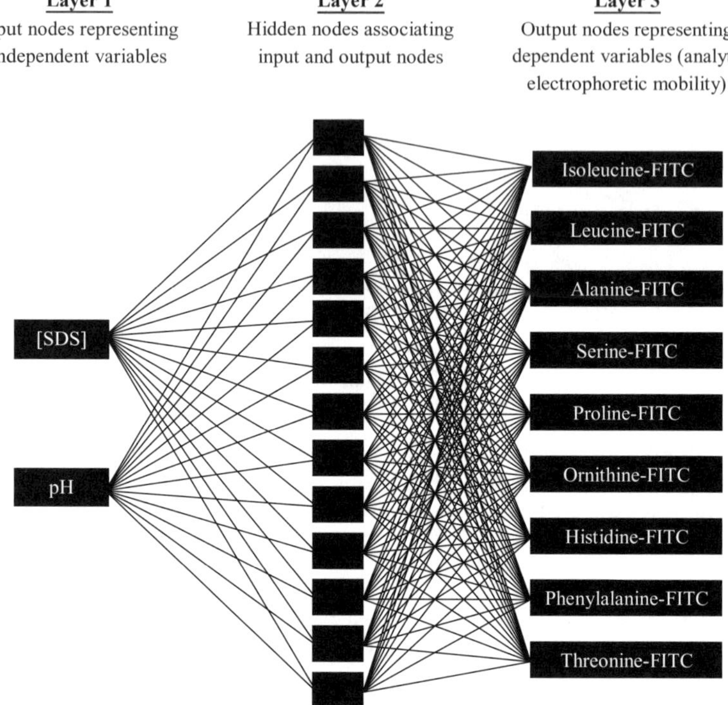

FIGURE 7.4. ANN structure for multivariate optimization of amino acids by CE.

of the network was crucial to the predictive capabilities of the network. Varying the number of nodes was a trial and error process; the number of nodes was increased until the network regressed and the error increased (13, 14). At this point, the previous network was identified as the most suitable as it had the lowest RMS error (12, 13).

The predictive ability of the best performing network was tested by construction of the response surface (see sections 7.3.3 and 7.3.4) and determination of the optimum electrolyte conditions. These conditions were then tested for agreement. Any significant difference between the predicted and experimentally determined mobilities of the analytes indicated that overlearning had occurred or insufficient data were presented to the ANN. Indeed, the first optimum prediction from the first five experiments of the experimental design was significantly different from the experimentally determined separation, indicating that the chosen ANN had overlearned (7).

Overlearning of the networks was avoided by the use of verification data, that is, the predicted optimum that was experimentally determined was reintroduced to the ANN as a check for overlearning (3, 9, 13). The RMS values for the verification points were calculated and compared against the RMS of the training data. When the verification error was smaller or was of similar

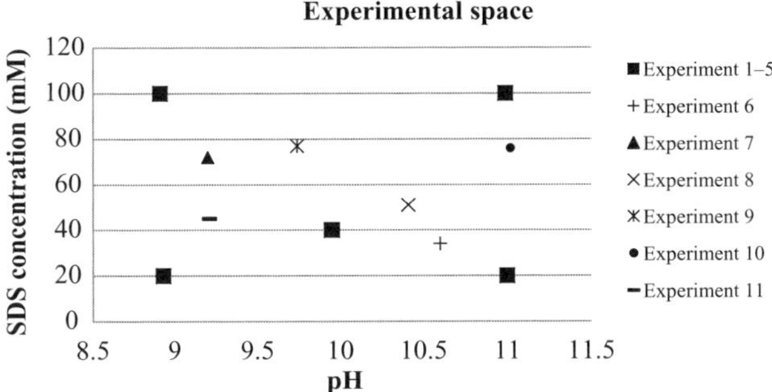

FIGURE 7.5. Experiments conducted for the optimization of the separation of the amino acids. Experiment 11 produced a global optimum.

magnitude to the training error, the model was considered to describe the experimental space adequately. This procedure was repeated until the predicted optimum and the experimentally determined optimum converged. It was necessary to conduct 11 experiments for our model to satisfy this criterion. These experiments are shown in Figure 7.5. The final network was an MLP with a training error of 0.1487 RMS and a verification error of 0.4165 RMS. These values indicate that the model had overlearned slightly, as the verification error was larger than the training error. However, the errors were of similar magnitude and the network was considered to have good predictive capabilities.

7.3.3. Construction of Resolution Response Surface

The response surface constructed from this model is shown as a surface plot in Figure 7.6. The resolution response surface was generated by multiplying the peak pair resolutions calculated from the amino acid mobilities predicted by the ANN. For this calculation, it was assumed that the peak widths were constant for each experiment. Therefore, the peak pair resolution calculation was reduced to the difference between the mobility of each of the adjacent peaks. The product resolution was chosen as it gives a simple measure of the overall resolution of the separation, with the largest value representing the greatest spread of the peaks.

7.3.4. Determination of Optimum

There were three optima evident: (1) 20 mM SDS and pH = 8.9; (2) 45 mM SDS and pH = 9.2; and (3) 55 mM SDS and pH = 9.9. Optima 1 indicated that the resolution was increasing below pH 8.9. This was outside the experimental space, demonstrating that extrapolation should not be relied upon. The scouting experiments showed that separations below 8.9 were poor (Fig. 7.3). The electropherogram obtained at optimum 2 is shown in Figure 7.7. Most of the

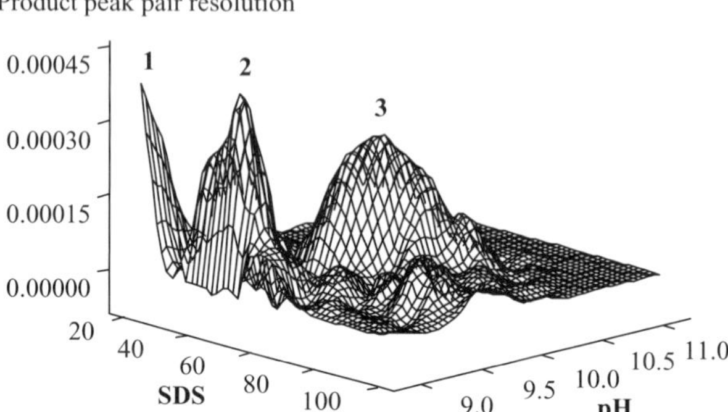

FIGURE 7.6. Response surface plot for the product peak pair resolution of amino acids. Three optimum are shown: (1) 20 mM SDS and pH = 8.9; (2) 45 mM SDS and pH = 9.2; and (3) 55 mM SDS and pH = 9.9.

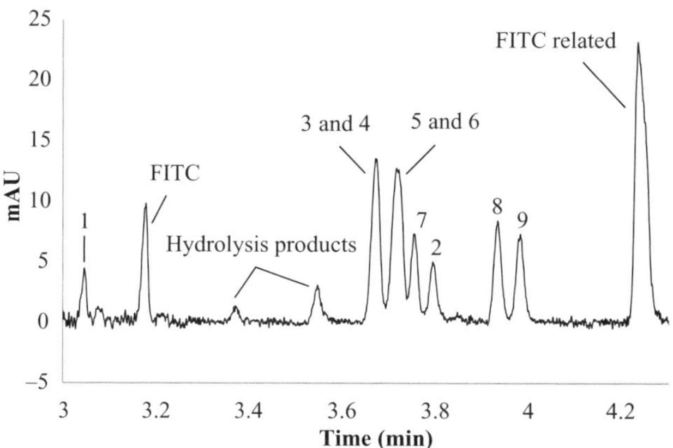

FIGURE 7.7. Separation of an amino acid mix containing 1 = ornithine-FITC; 2 = threonine-FITC; 3 = leucine-FITC; 4 = isoleucine-FITC; 5 = proline-FITC; 6 = histidine-FITC; 7 = phenylalanine-FITC; 8 = alanine-FITC; and 9 = serine-FITC. The background electrolyte consisted of 45 mM SDS, 10 mM borate, and a pH of 9.21.

amino acids were separated with the exceptions of leucine and isoleucine, as well as proline and histidine.

These data were reintroduced to the ANN as another verification point. The response surface was similar to that generated from experiment 11. This indicated that the network had sufficient data and had reached its operation limits. Therefore, optimum 2 provided the best separation that was possible using this separation electrolyte. It should be noted isoleucine and leucine, and histidine and proline comigrated under all experimental conditions that were investigated. Optimum 3 was also investigated and as the response surface indicated, the resolution was not as good as optimum 2 (Fig. 7.8).

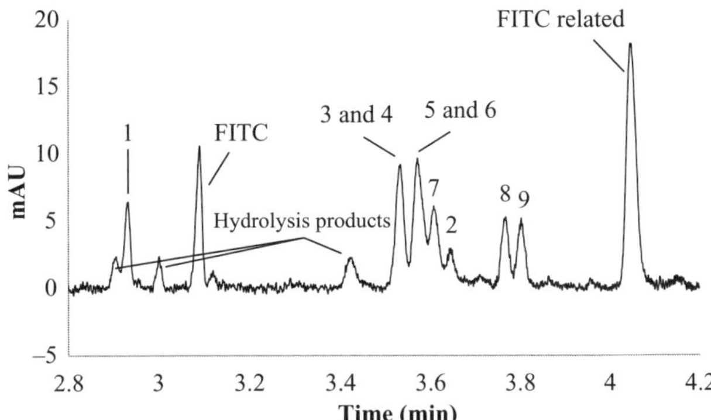

FIGURE 7.8. Separation of the amino acid mix containing 1 = ornithine-FITC; 2 = threonine-FITC; 3 = leucine-FITC; 4 = isoleucine-FITC; 5 = proline-FITC; 6 = histidine-FITC; 7 = phenylalanine-FITC; 8 = alanine-FITC; and 9 = serine-FITC. The background electrolyte consisted of 55 mM SDS, 10 mM borate, and a pH of 9.9.

The difficulties encountered in the separation of amino acids can be attributed to the altered structural profile of the amino acid once derivatized. Underivatized amino acids have been separated previously, but require contactless conductivity detection to identify all amino acids (11). The reaction is shown in Figure 7.1. The attachment of the fluorescent/UV label minimized the structural differences in side chains. Therefore, as the mass-to-charge ratios were similar, the addition of a surfactant such as SDS did not offer sufficient selectivity (13, 14). Nevertheless, the application of an ANN to the optimization of this separation rapidly arrived at the optimum conditions.

7.4. CONCLUSION

An ANN was successfully employed to optimize the separation of FITC-labeled amino acids employing a borate–SDS electrolyte. The optimization process required a total of 11 experiments. It was found that the electrolyte was not suitable for the complete separation of all of the derivatized amino acids under investigation. This conclusion was arrived at rapidly, and avoided further unnecessary experimentation.

ACKNOWLEDGMENT

The authors gratefully acknowledge the assistance of the staff at the University of Technology of Sydney who graciously offered their time and support toward this project.

REFERENCES

1. Harris, D. (2007) *Quantitative Chemical Analysis*, W.H. Freeman and Company, New York.
2. Bocaz-Beneventi, G., Latorre, R., Farková, M., and Havel, J. (2002) *Anal Chim Acta*, **452**, 47–63
3. Novotná, K., Havlis, J., and Havel, J. (2005) *J Chromatogr A*, **1096**, 50–57.
4. Madden, J.E., Avdalovic, N., Haddad, P.R., and Havel, J. (2001) *J Chromatogr A*, **910**, 173–179.
5. Casamento, S.G., Kwok, B.K., Roux, C.P., Dawson, M., and Doble, P.A. (2003) *J Forensic Sci*, **48**, 1075–1083.
6. Tran, A.T.K., Hyne, R.V., Pablo, F., Day, W.R., and Doble, P.A. (2007) *Talanta*, **71**, 1268–1275
7. Havel, J., Lubal, P., and Farková, M. (2002) *Polyhedron*, **21**, 1375–1384.
8. Andreas Ramseier, J.C.W.T. (1998) *Electrophoresis*, **19**, 2956–2966.
9. Havel, J., Breadmore, M., Macka, M., and Haddad, P.R. (1999) *J Chromatogr A*, **850**, 345–353.
10. Casamento, S., Kwok, B., Roux, C., Dawson, M. and Doble, P. (2003) *J Forensic Sci*, **48** 1075–1083.
11. Tuma, P., Samcová, E., and Andelová, K. (2006) *J Chromatogr B*, **839**, 12–18.
12. Doble, P., Sandercock, M., Du Pasquier, E., Petocz, P., Roux, C., and Dawson, M. (2003) *Forensic Sci Intern*, **132**, 26–39.
13. Lalljie, S.P.D. and Sandra, P. (1995) *Chromatographia*, **40**, 519–526.
14. Lalljie, S.P.D. and Sandra, P. (1995) *Chromatographia*, **40**, 513–518.

PART II

EXPLORATORY DATA ANALYSIS, PREDICTION, AND CLASSIFICATION

CHAPTER 8

DEVELOPMENT OF CAPILLARY ELECTROPHORESIS FINGERPRINTS AND MULTIVARIATE STATISTICS FOR THE DIFFERENTIATION OF OPIUM AND POPPY STRAW SAMPLES

RAYMOND G. REID, SUSANNE P. BOYLE, ANN S. LOW, and DAVID G. DURHAM
School of Pharmacy, The Robert Gordon University, Schoolhill, Aberdeen, UK

CONTENTS

8.1. INTRODUCTION

Capillary zone electrophoresis (CZE) is a technique that is being increasingly used in the separation of herbal medicinal products (1–23). It is based on the differential migration of ions in an electric field, either by attraction or repul-

Chemometric Methods in Capillary Electrophoresis. Edited by Grady Hanrahan and Frank A. Gomez

sion. A positive electrode (anode) and a negative electrode (cathode) are placed in separate solutions containing ions, connected by a fused silica capillary. The capillary is initially filled with buffer by applying a pressure to the inlet vial or a vacuum to the outlet vial. Samples are applied using either pressure or by applying a small voltage. When the running voltage is applied across the electrodes, solute ions of different charge move through the capillary. Figure 8.1 shows the main components of a standard capillary electrophoresis (CE) instrument.

The first objective was to develop a CZE separation (or fingerprint) using only migration times and peak areas for opium from different locations and poppy straw samples from different plants. Previously, a CZE method had shown poor separation for certain alkaloids in opium samples (24), and in other methods pH has had to be strictly controlled (25, 26). Some methods were unable to operate at the optimum detection wavelengths (25, 27, 28), while others have used various cyclodextrin additives to obtain separation (25, 29–31).

To improve the separation of the alkaloids extracted from opium and poppy straw samples, a combination of CZE and micellar electrokinetic chromatography (MEKC) was proposed. These methods allowed increased resolution and detection limits, using a modified stacking technique. This process has been named as sweeping (32–50). Sweeping is a concentration method that is based on interactions of a pseudostationary phase such as sodium dodecyl sulfate (SDS) with the compounds being separated. The capillary was normally filled with a low pH buffer at a specific concentration containing SDS and methanol. The sample was prepared in the same concentration of buffer without the SDS and methanol. The sample was injected for a specific time (300–500 s) hydrodynamically at a pressure of 50 mbar, and the sample was replaced by the buffer containing SDS and methanol. A negative voltage was applied and the large sample zone was swept to the injection side of the bound-

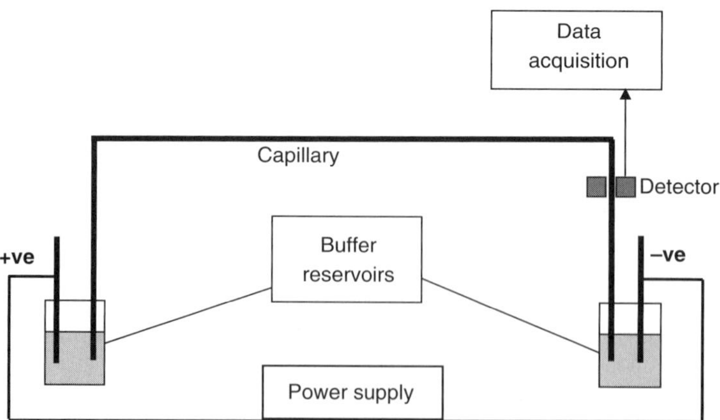

FIGURE 8.1. Representative diagram of the main components in a capillary electrophoresis system.

ary by the SDS micelles to form a very narrow concentrated zone. Samples then separated by the MEKC mechanism against the electroosmotic flow, which was very low because of the pH. The process was highly efficient and allowed increased detection limits for various compounds (24, 27, 51–54).

Multivariate data consist of many observations on variables for a large number of samples, such as the determination of metals in batches of honey (55) or wine samples (56) from different regions. It becomes difficult to visually see patterns within the samples so a statistical approach is used to analyze the data. For this type of pattern recognition, it is normally best to follow a decision tree (57). Figure 8.2 shows the decision tree that was followed for pattern recognition within the opium and poppy straw samples.

By following this decision tree, it became apparent that three different methods would have to be used, namely hierarchical cluster analysis (HCA), principal component analysis (PCA), and soft independent modeling of class analogy (SIMCA). Pattern recognition consists of two general areas, which are either supervised or unsupervised. HCA and PCA are two examples of unsupervised pattern recognition, where no prior knowledge of groupings is

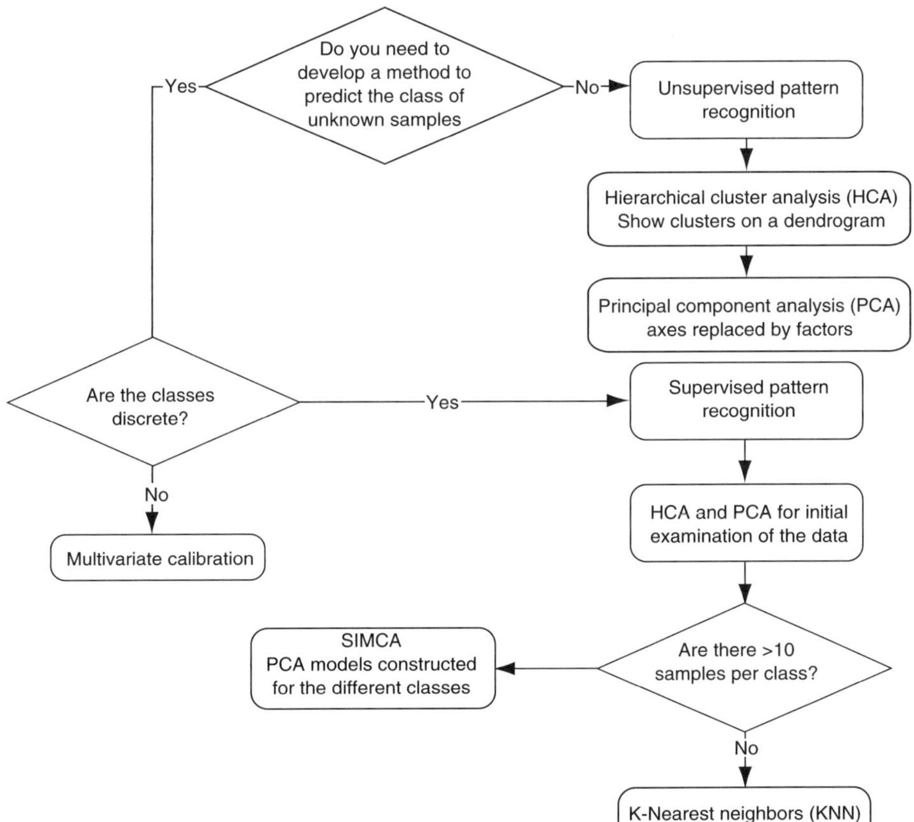

FIGURE 8.2. Illustration of the decision tree used for pattern recognition.

required. SIMCA is used for supervised pattern recognition where knowledge of the groupings is required.

All three statistical methods have been used previously in conjunction with chromatographic fingerprints to differentiate different types of samples. Most of these published methods use liquid chromatography (high performance liquid chromatography [HPLC]) for development of a fingerprint. HCA has been used in the analysis of *Pseudostellaria heterophylla* roots (58, 59), *Lidanpaishi* tablets (60), and in the study of medicinal *Taxus* species (61). PCA has been used in conjunction with HPLC fingerprints to determine the quality of various herbal products (62–67). SIMCA on the other hand has been used on a few occasions for differentiation of *Stephania lour* (68) and *Ganoderma lucidum* (69). CE in different formats has been used to develop fingerprints in various herbal products such as *Scutellaria* (70–71), *Hedera* (72), *Salvia miltiorrhiza Bunge* (73), *Flos carthami* (74), and *Echinacea purpurea* (75). These methods, however, have not incorporated multivariate statistical analysis.

In this chapter we have continued previous work (54) to investigate the use of the complete fingerprint developed using CZE. No quantitative data were used, and no attempt was made to identify any of the compounds present in the opium and poppy straw samples. All peaks found in the samples were measured and the resulting data were used in HCA, PCA, and SIMCA.

8.2. EXPERIMENTAL

8.2.1. Development of CE Fingerprint

The extraction method for the opium and poppy straw samples, and the separation method along with a suitable choice of internal standard (IS) have been described previously (54). Opium samples were available from four different locations, and poppy straw samples from five plants were also available for analysis. Samples were injected in triplicate for analysis.

8.2.2. Multivariate Statistical Analysis

The data produced from the fingerprint separations were used to create a spreadsheet containing relative migration times and relative peak areas for all peaks. The data were subjected to HCA, PCA, and SIMCA for evaluation.

8.3. RESULTS AND DISCUSSION

8.3.1. Development of CE Fingerprint

The number of peaks obtained using the CE separation of opium and poppy straw was quite large. Representative electropherograms for opium and poppy

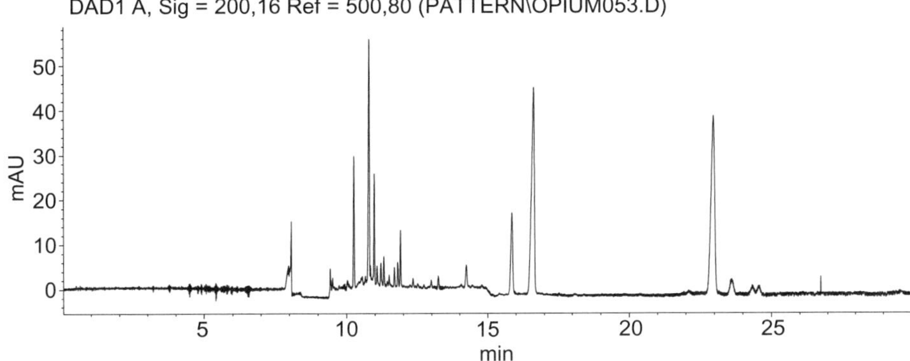

FIGURE 8.3. Representative electropherogram from a Yugoslavian opium sample. Capillary –60 cm × 0.5 μm fused silica, monitoring wavelength –200 nm, voltage applied –30 kV. Running buffer –50 mM disodium hydrogen orthophosphate, pH 2.5, 80 mM sodium dodecyl sulfate, 25% methanol.

straw are shown in Figures 8.3 and 8.4, respectively. The migration times and peak areas for all the peaks in the opium and poppy straw samples were measured. To evaluate the use of all the data, relative retention times and relative peak areas were calculated using the IS. Relative migration times were calculated by dividing a peak migration time by the migration time of the internal standard. Similarly, a peak area ratio of analyte peak area/IS peak area was determined. The relative migration times yielded 21 peaks as being representative of the opium samples and 15 for the poppy straw samples. In some samples, certain peaks were missing, and to allow all the options available in the software to work, they were allocated negligible values of 0.001 for peak area ratio.

8.3.2. HCA

HCA was used to check for clusters within data sets. It is normally visualized in a two-dimensional format, as a dendrogram, which was qualitative in nature (57). HCA initially measured the interpoint distance between all the samples, the Euclidean distance. This is the simplest and most frequently used method and can be represented as the square root of the sum of the squared differences between observations. An agglomerative algorithm was used to calculate the matrix of distances and begins by defining each point in the data as a separate cluster. Clusters are then merged at this point using a linkage method. There were three general linkage methods available, with some variations on each of them. To achieve the clustering the data were required to be preprocessed and various options were available. The Pirouette 4.0 Lite Classify software (Infometrix Inc., Bothell, WA) used for multivariate statistical analysis included seven linkage and five preprocessing methods. All of the preprocessing and linkage methods were investigated for the opium and poppy straw samples.

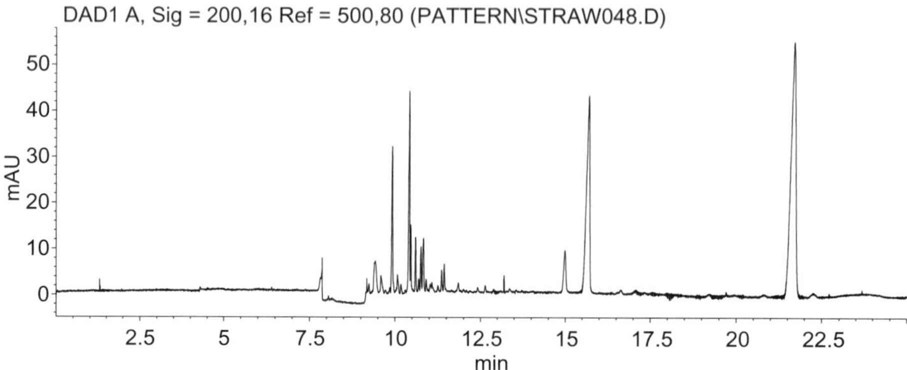

FIGURE 8.4. Representative electropherogram from straw sample S3. Capillary –60 cm × 0.5 μm fused silica, monitoring wavelength –200 nm, voltage applied –30 kV. Running buffer –50 mM disodium hydrogen orthophosphate, pH 2.5, 80 mM sodium dodecyl sulfate, 25% methanol.

The best results for the HCA analysis of the opium samples was obtained using auto-scale preprocessing with incremental linkage. The opium samples from different locations were clearly identified in separate clusters using a similarity value of 0.730 and can be seen in Figure 8.5(a). The optimum results for the poppy straw samples were obtained using range scale preprocessing again with incremental linkage. Figure 8.5(b) shows that the different types of plant species were clearly identified using a similarity of 0.733. This technique shows that it is possible to cluster opium samples from different locations and poppy straw samples from different plant species, but it was not possible to use this method to identify unknowns.

8.3.3. PCA

PCA is a tool that allows better visualisation of data in a three-dimensional (3D) environment for exploratory analysis. It is similar to HCA in that it graphically represents inter-sample and inter-variable relationships. PCA reduces the dimensionality by combining two variables into a single linear combination. These variables are called principal components or factors that are ordered so that the first few retain most of the variation present in all of the original variables. The Pirouette software uses five different types of preprocessing, along with two different types of validation and various rotation options. All the options were tested and the optimum results for the opium and poppy straw samples were obtained using auto-scale preprocessing with six factors, with no validation or rotation. Figure 8.6(a) and (b) show the 3D plots for the opium and poppy straw samples, respectively. While clear differentiation of both opium and poppy straw samples were evident, once again it was not possible to use PCA for predictions of unknown samples.

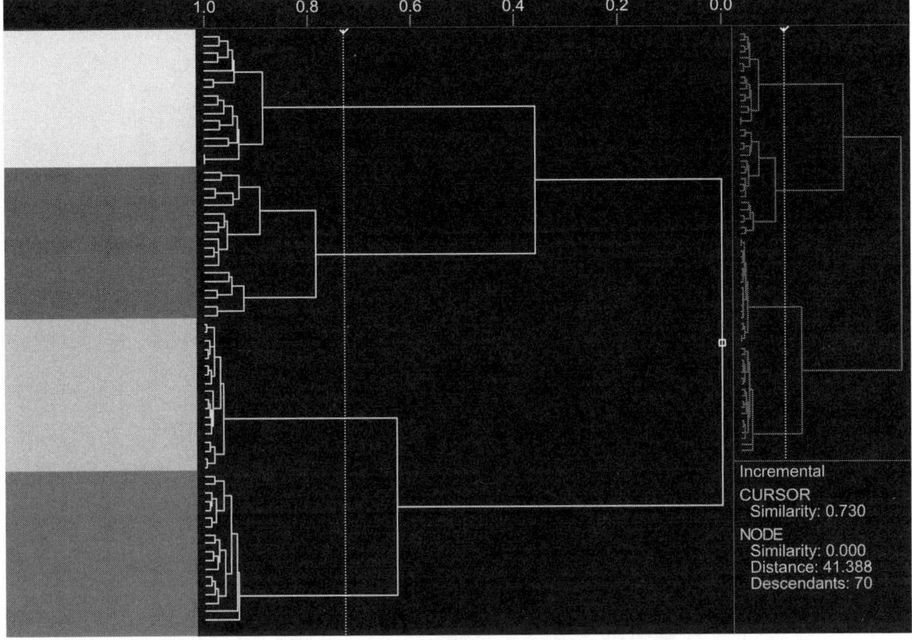

(a)

1.0
0.8
0.6
0.4
0.2
0.0
Incremental
CURSOR
Similarity: 0.733
NODE
Similarity: 0.000
Distance: 9.950
Descendants: 100

(b)

FIGURE 8.5. (a) Dendrogram obtained for HCA analysis of opium samples from different locations using auto-scale preprocessing and incremental linkage using a similarity value of 0.730. Cluster identification: yellow = Indian; red = Yugoslavian; green = Turkish; and purple = Persian. (b) Dendrogram obtained for HCA analysis of poppy straw samples using range scale preprocessing and incremental linkage using a similarity value of 0.733. Cluster identification: yellow = S4 and S5; red = S1; green = S2; and purple = S3. See color insert.

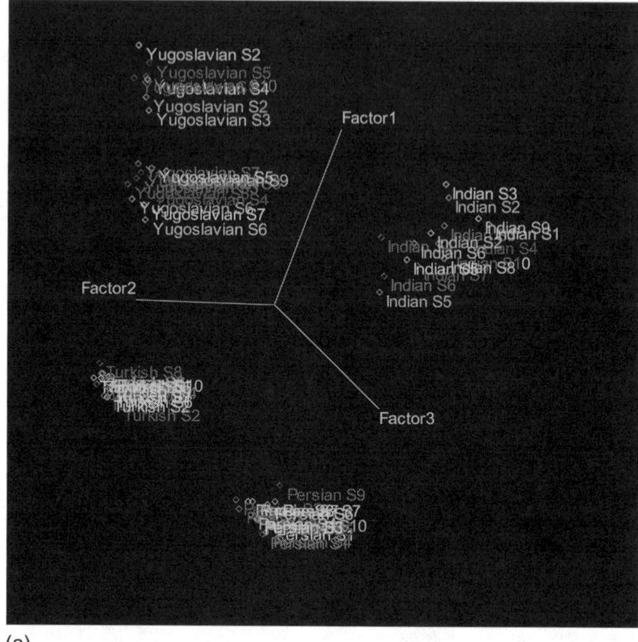

(a)

(b)

FIGURE 8.6. (a) 3D plot using the scores calculated from PCA analysis using auto-scale preprocessing with six factors for the opium samples from different locations. (b) 3D plot using the scores calculated from PCA analysis using auto-scale preprocessing with six factors for the poppy straw samples. See color insert.

8.3.4. SIMCA

SIMCA is a supervised pattern recognition technique, which needs to have the data classified manually or done using HCA. SIMCA then performs PCA on each class with a sufficient number of factors retained to account for most of the variation within classes. The number of factors retained is very important. If too few are selected, the information in the model set can become distorted. By using a procedure called cross validation, segments of the data are omitted during PCA, and the omitted data are predicted and compared to the actual value. This is repeated for every data element until each point has been excluded once from the determination. The PCA model that yields the minimum prediction error for the omitted data is retained.

After the SIMCA classification model has been created, there are three possible predictions for unknown samples: the sample fits only one predefined class, the sample does not fit any predefined class, or the sample fits into more than one predefined class. To check the model, the data set is divided into "training" and "unknown" sets to evaluate the SIMCA model.

The samples in the unknown set are then predicted using the SIMCA model, and the results are evaluated to determine if modification of the model is required. If the number of misclassifications (or wrongly identified) is unacceptable, then parameters can be adjusted and a new model is formed. The process is repeated to get acceptable classifications.

The opium training set was first subjected to HCA using auto-scale preprocessing with incremental linkage to define classes, as described previously. The data were then subjected to SIMCA with various decisions required on best method of preprocessing, the choice of scope (global or local), number of factors, and a suitable choice of a threshold value (0.99–0.01). The best results for the opium samples were obtained using global scope with a threshold of 0.95. Figure 8.7(a) shows the 3D plot obtained using the training set of data after SIMCA analysis using mean-centered preprocessing and five factors. Clear differentiation of the opium samples from different locations was obtained, and a model was built using these parameters. All the models were tested on the unknown samples, and the results of changing the preprocessing and the number of factors are shown in Table 8.1.

It was noted for the opium samples that increasing the number of factors decreased the percentage of correct predictions for four of the five preprocessing options. For mean-centered preprocessing, increasing the number of factors increased the percentage of correct predictions. Using mean-centered preprocessing with five factors matched 97.1% of the samples to the correct classification of the unknowns. One sample was erroneously classified.

The process was repeated for the straw samples. The straw training set was subjected to HCA using range scale preprocessing with incremental linkage to firstly classify samples. Global scope and a 0.95 threshold provided the best results, and the preprocessing and number of factors were tested. Once again,

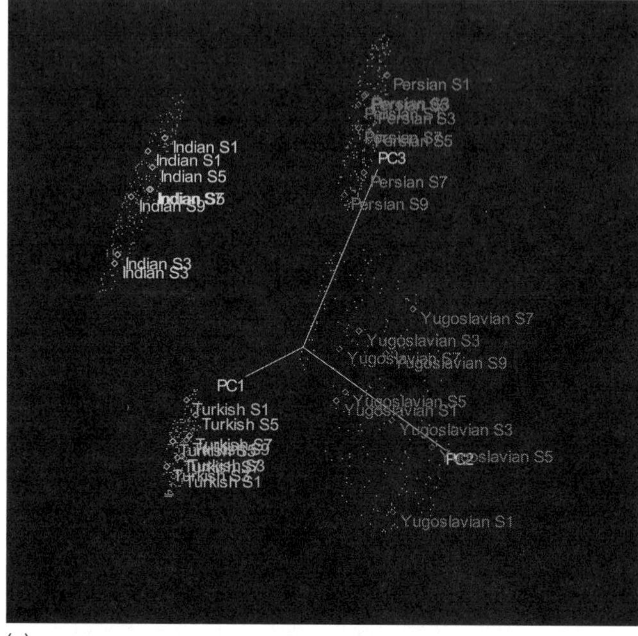

(a)

(b)

FIGURE 8.7. (a) 3D plot after SIMCA analysis using mean-centered preprocessing with five factors and a probability threshold of 0.95 for the opium samples from different locations. (b) 3D plot after SIMCA analysis using range scale preprocessing with three factors at a probability threshold of 0.95 for the poppy straw samples. See color insert.

TABLE 8.1. Summary of the effect of altering key preprocessing method parameters in SIMCA analysis on the level of % correct predictions achieved for the opium samples from different locations

Preprocessing	No. of Factors	% Correct
None	3	91.4
	5	88.6
	7	88.6
Auto-scale	3	94.3
	5	91.4
	7	91.4
Mean-centered	3	94.3
	5	97.1
	7	97.1
Range scale	3	94.3
	5	88.6
	7	88.6
Variance scale	3	94.3
	5	88.6
	7	88.6

the SIMCA variables would be used to create various models for use in determinations of the unknown set. Figure 8.7(b) shows the 3D plot obtained using the poppy straw training set after SIMCA analysis using range scale preprocessing and three factors. Once again, clear differentiation of plant types is observed. The unknown poppy straw samples were subjected to the SIMCA models. The models were tested on the unknown poppy straw, and the results are shown in Table 8.2.

For all the preprocessing methods, the number of correct predictions was reduced as the number of factors was increased. For range and variance scale preprocessing with three factors, 100% correct predictions were obtained. The results show that it was possible to predict the type of plant the poppy straw was originally from.

8.4. CONCLUSION

The results show that using a CZE fingerprint with multivariate statistical analysis, it was possible to differentiate opium samples from different locations and to be able to predict unknown samples with a high degree of reliability. Similar results were achieved for samples of poppy straw from different plants. This method needed no identification or quantitation of individual components, so it was less time-consuming. The results were similar to those reported previously using quantitative data for individual components (54).

TABLE 8.2. Summary of the effect of altering key preprocessing method parameters in SIMCA analysis on the level of % correct predictions achieved for the poppy straw samples from different plants

Preprocessing	No. of Factors	% Correct
None	3	92.0
	5	60.0
	7	40.0
Auto-scale	3	96.0
	5	66.0
	7	64.0
Mean-centered	3	92.0
	5	70.0
	7	38.0
Range scale	3	100.0
	5	86.0
	7	50.0
Variance scale	3	100.0
	5	90.0
	7	38.0

ACKNOWLEDGMENT

The authors acknowledge the donation of poppy straw samples from Dr. A.J. Fist, Tasmanian Alkaloids Pty Ltd. Westbury, Tasmania.

REFERENCES

1. Liu, Y.M. and Sheu, S.J. (1994) *J High Res Chromatogr*, **17**, 559–560.
2. Liu, Y.M. and Sheu, J.J. (1994) *Anal Chim Acta*, **288**, 221–226.
3. Li, K.W. and Sheu, S.J. (1995) *Anal Chim Acta*, **313**, 113–120.
4. Ganzera, M. (2008) *Electrophoresis*, **29**, 3489–3503.
5. Chen, J., Zhao, H., Wang, X., Lee, F.S., Yang, H., and Zheng, L. (2008) *Electrophoresis*, **29**, 2135–2147.
6. Xu, X., Ye, H., Wang, W., and Chen, G. (2005) *J Agric Food Chem*, **53**, 5853–5857.
7. Qi, S., Ding, L., Tian, K., Chen, X., and Hu, Z. (2006) *J Pharm Biomed Anal*, **40**, 35–41.
8. Li, Y., Qi, S., Chen, X., and Hu, Z. (2004) *Electrophoresis*, **25**, 3003–3009.
9. Wang, A., Zhou, Y., Wu, F., He, P., and Fang, Y. (2004) *J Pharm Biomed Anal*, **35**, 959–964.
10. Sun, Y., Guo, T., Sui, Y., and Li, F. (2003) *J Chromatogr B*, **792**, 147–152.
11. Okunji, C.O., Ware, T.A., Hicks, R.P., Iwu, M.M., and Skanchy, D.J. (2002) *Planta Med*, **68**, 440–444.

12. Cao, Y., Zhang, X., Fang, Y., and Ye, J. (2001) *Analyst*, **126**, 1524–1528.
13. Li, S.P., Li, P., Dong, T.T., and Tsim, K.W. (2001) *Electrophoresis*, **22**, 144–150.
14. Song, J.Z., Xu, H.X., Tian, S.J., and But, P.P. (1999) *J Chromatogr A*, **857**, 303–311.
15. Zhang, Y., Zhao, L., and Shi, Y.P. (2007) *J Chromatogr Sci*, **45**, 600–604.
16. Elosta, S., Gajdosova, D., and Havel, J. (2006) *J Sep Sci*, **27**, 1174–1179.
17. Che, A.J., Zhang, J.Y., Li, C.H., Chen, X.F., Hu, Z.D., and Chen, X.G. (2004) *J Sep Sci*, **27**, 569–575.
18. Sun, G., Wang, Y., Sun, Y., and Bi, K. (2003) *Anal Sci*, **19**, 1395–1399.
19. Ji, Y.B., Alaerts, G., Xu, C.J., Hu, Y.Z., and Vander Heyden, Y. (2006) *J Chromatogr A*, **1128**, 273–281.
20. Prinza, S., Singhubera, J., Zhub, M., and Koppa, B. (2006) *Planta Med*, **72**, 248–261.
21. Marchart, E., Krenn, L., and Kopp, B. (2003) *Planta Med*, **69**, 452–456.
22. Glockl, I., Veit, M., and Blaschke, G. (2002) *Planta Med*, **68**, 158–161.
23. Wu, M.H., Zhao, L.H., Song, Y., Zhang, W., Xiang, B.R., and Mei, L.H. (2005) *Planta Med*, **71**, 1152–1156.
24. Craige Trenerry, V., Wells, R.J., and Robertson, J. (1995) *J Chromatogr A*, **718**, 217–225.
25. Bjornsdottir, I. and Hansen, S.H. (1995) *J Pharm Biomed Anal*, **13**, 1473–1481.
26. Lurie, I.S. (1997) *J Chromatogr A*, **780**, 265–284.
27. Bjornsdottir, I. and Hansen, S.H. (1995) *J Pharm Biomed Anal*, **13**, 687–693.
28. Stockigt, J., Sheludk, Y., Unger, M., Gerasimenko, I., Warzecha, H., and Stockigt, D. (1997) *J Chromatogr A*, **767**, 263–276.
29. Lurie, I.S., Panicker, S., Hays, P.A., Garcia, A.D., and Geer, B.L. (2003) *J Chromatogr A*, **984**, 109–120.
30. Hindson, B.J., Francis, P.S., Purcell, S.D., and Barnett, N.W. (2007) *J Pharm Biomed Anal*, **43**, 1164–1168.
31. Taylor, R.B., Low, A.S., and Reid, R.G. (1996) *J Chromatogr B*, **675**, 213–223.
32. Zakaria, P., Macka, M., and Haddad, P.R. (2003) *J Chromatogr A*, **985**, 493–501.
33. Durham, D.G., Reid, R.G., Wangbooskul, J., and Daodee, S. (2002) *Phytochem Anal*, **13**, 358–362.
34. Kim, J.B., Quirino, J.P., Otsuka, K., and Terabe, S. (2001) *J Chromatogr A*, **916**, 123–130.
35. Monton, M.R., Quirino, J.P., Otsuka, K., and Terabe, S. (2001) *J Chromatogr A*, **939**, 99–108.
36. Quirino, J.P., Dulay, M.T., Bennett, B.D., and Zare, R.N. (2001) *Anal Chem*, **73**, 5539–5543.
37. Quirino, J.P., Kim, J.B., and Terabe, S. (2002) *J Chromatogr A*, **965**, 357–373.
38. Quirino, J.P., Otsuka, K., and Terabe, S. (1998) *J Chromatogr B*, **714**, 29–38.
39. Quirino, J.P. and Terabe, S. (1997) *J Capillary Electrop*, **4**, 233–245.
40. Quirino, J.P. and Terabe, S. (1999) *J Chromatogr A*, **856**, 465–482.
41. Quirino, J.P. and Terabe, S. (1999) *J Chromatogr A*, **850**, 339–344.

42. Quirino, J.P. and Terabe, S. (2000) *Anal Chem*, **72**, 1023–1030.
43. Quirino, J.P. and Terabe, S. (2000) *Electrophoresis*, **21**, 355–359.
44. Quirino, J.P. and Terabe, S. (2000) *J Chromatogr A*, **902**, 119–135.
45. Quirino, J.P., Terabe, S., and Bocek, P. (2000) *Anal Chem*, **72**, 1934–1943.
46. Quirino, J.P. and Terabe, S. (1997) *J Chromatogr A*, **781**, 119–128.
47. Taylor, R.B., Reid, R.G., and Low, A.S. (2001) *J Chromatogr A*, **916**, 201–206.
48. Wu, C.H., Chen, M.C., Su, A.K., Shu, P.Y., Chou, S.H., and Lin, C.H. (2003) *J Chromatogr B*, **785**, 317–325.
49. Quirino, J.P., Iwa, Y., Otsuka, K., and Terabe, S. (2000) *Electrophoresis*, **21**, 2899–2903.
50. Sun, S.W. and Tseng, H.M. (2005) *J Pharm Biomed Anal*, **37**, 39–45.
51. Olieman, C., Maat, L., Waliszewski, K., and Beyeerman, H.C. (1990) *J Chromatogr A*, **133**, 382–385.
52. Zhanpin, W. (1994) *Forensic Sci Int*, **64**, 103–106.
53. Li, S., He, C., Liu, H., Li, K., and Liu, F. (2005) *J Chromatogr B*, **826**, 58–62.
54. Reid, R.G., Durham, D.G., Boyle, S.P., Low, A.S., and Wangboonskul, J. (2007) *Anal Chim Acta*, **605**, 20–27.
55. Latorre, M.J., Pena, R., Pita, C., Botana, A., Garcia, S., and Herrero, C. (1999) *Food Chem*, **66**, 263–268.
56. Marengo, E. and Aceto, M. (2003) *Food Chem*, **81**, 621–630
57. Beebe, K.R., Pell, R.J., and Seasholtz, M.B. (1999) *Chemometrics: A Practical Guide*, John Wiley and Sons, New York.
58. Han, C., Shen, Y., Chen, J., Lee, F.S., and Wang, X. (2006) *J Sep Sci*, **29**, 2197–2202.
59. Han, C., Shen, Y., Chen, J., Lee, F.S., and Wang, X. (2008) *J Chromatogr B*, **862**, 125–131.
60. Xu, L., Han, X., Qi, Y., Xu, Y., Yin, L., Peng, J., Liu, K., and Sun, C. (2009) *Anal Chim Acta*, **633**, 136–148.
61. Ge, G.B., Zhang, Y.Y., Hao, D.C., Hu, Y., Luan, H.W., Liu, X.B., He, Y.Q., Wang, Z.T., and Yang, L. (2008) *Planta Med*, **74**, 773–779.
62. Xie, B., Gong, T., Tang, M., Mi, D., Zhang, X., Liu, J., and Zhang, Z. (2008) *J Pharm Biomed Anal*, **48**, 1261–1266.
63. Soares, P.K. and Scarminio, I.S. (2008) *Phytochem Anal*, **19**, 78–85.
64. Yang, J., Chen, L.H., Zhang, Q., Lai, M.X., and Wang, Q. (2007) *J Sep Sci*, **30**, 1276–1283.
65. Obradovic, M., Krajsek, S.S., Dermastia, M., and Kreft, S. (2007) *Phytochem Anal*, **18**, 123–132.
66. Xiaohui, F., Yi, W., and Yiyu, C. (2006) *J Pharm Biomed Anal*, **40**, 591–597.
67. Yan, S.K., Xin, W.F., Luo, G.A., Wang, Y.M., and Cheng, Y.Y. (2005) *J Chromatogr A*, **1090**, 90–97.
68. Huang, J.M., Guo, J.X., Qu, L.B., and Xiang, B.R. (1999) *J Asian Nat Prod Res*, **1**, 215–220.
69. Chen, Y., Zhu, S.B., Xie, M.Y., Nie, S.P., Liu, W., Li, C., Gong, X.F., and Wang, Y.X. (2008) *Anal Chim Acta*, **623**, 146–156.

70. Sun, G. and Shi, C. (2008) *J Chromatogr Sci*, **46**, 454–460.
71. Yu, K., Gong, Y., Lin, Z., and Cheng, Y. (2007) *J Pharm Biomed Anal*, **43**, 540–548.
72. Cianchino, V., Ortega, C., Acosta, G., Martinez, L.D., and Gomez, M.R. (2007) *Pharmazie*, **62**, 262–265.
73. Gu, M., Zhang, S., Su, Z., Chen, Y., and Ouyang, F. (2004) *J Chromatogr A*, **1057**, 133–140.
74. Sun, Y., Guo, T., Sui, Y., and Li, F. (2003) *J Chromatogr B*, **792**, 147–152.
75. Gotti, R., Fiori, J., Hudaib, M., and Cavrini, V. (2002) *Electrophoresis*, **23**, 3084–3092.

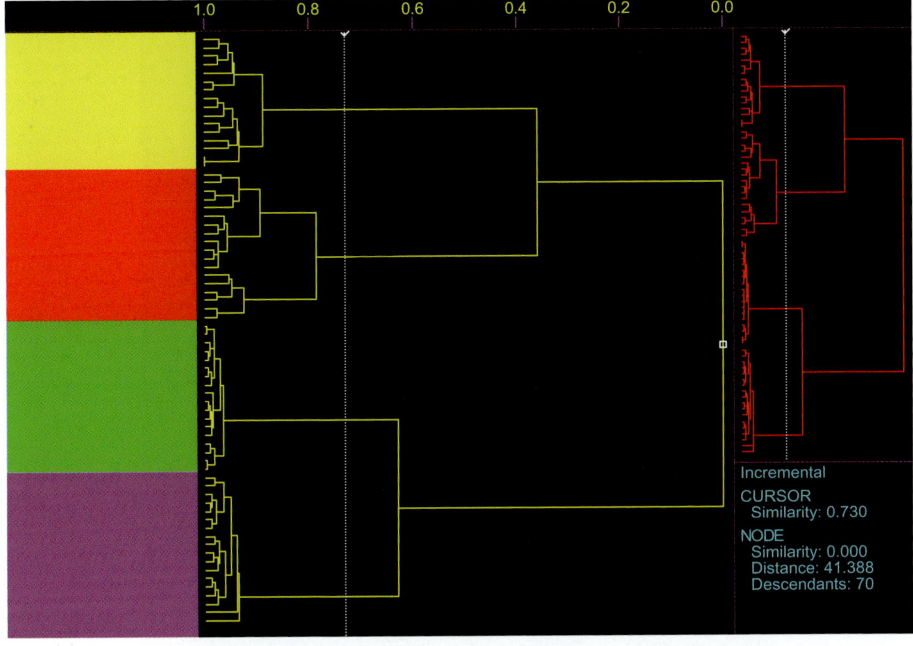

(a)

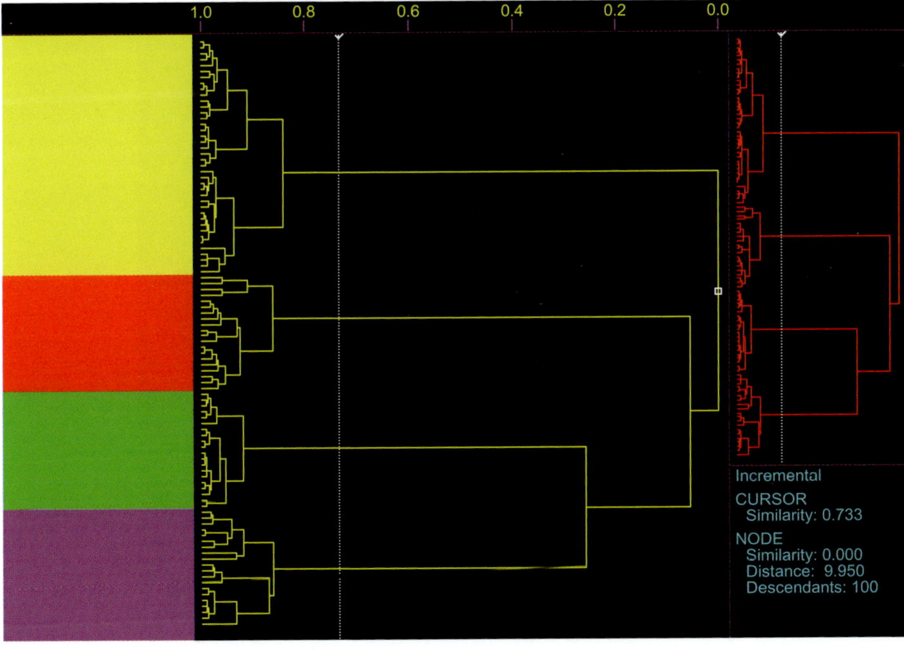

(b)

FIGURE 8.5. (a) Dendrogram obtained for HCA analysis of opium samples from different locations using auto-scale preprocessing and incremental linkage using a similarity value of 0.730. Cluster identification: yellow = Indian; red = Yugoslavian; green = Turkish; and purple = Persian. (b) Dendrogram obtained for HCA analysis of poppy straw samples using range scale preprocessing and incremental linkage using a similarity value of 0.733. Cluster identification: yellow = S4 and S5; red = S1; green = S2; and purple = S3.

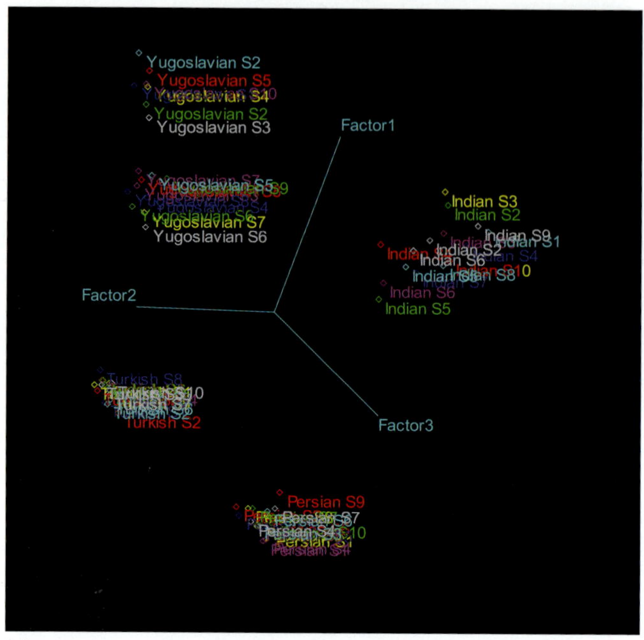

(a)

(b)

FIGURE 8.6. (a) 3D plot using the scores calculated from PCA analysis using auto-scale preprocessing with six factors for the opium samples from different locations. (b) 3D plot using the scores calculated from PCA analysis using auto-scale preprocessing with six factors for the poppy straw samples.

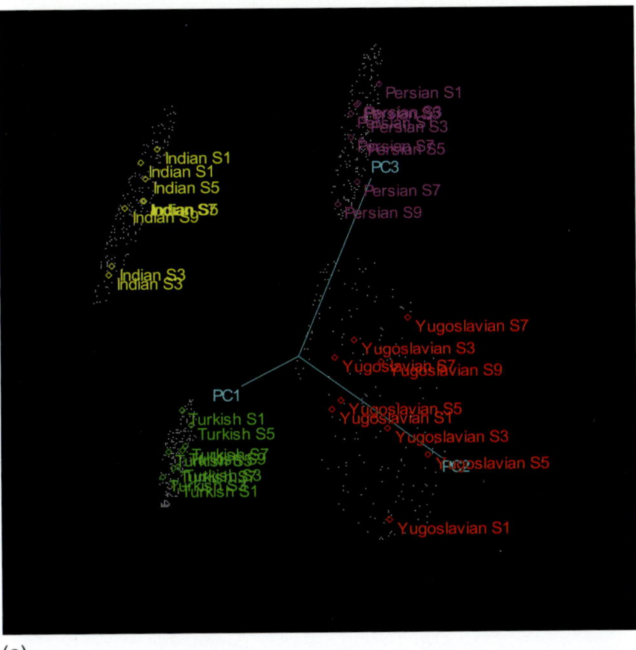

(a)

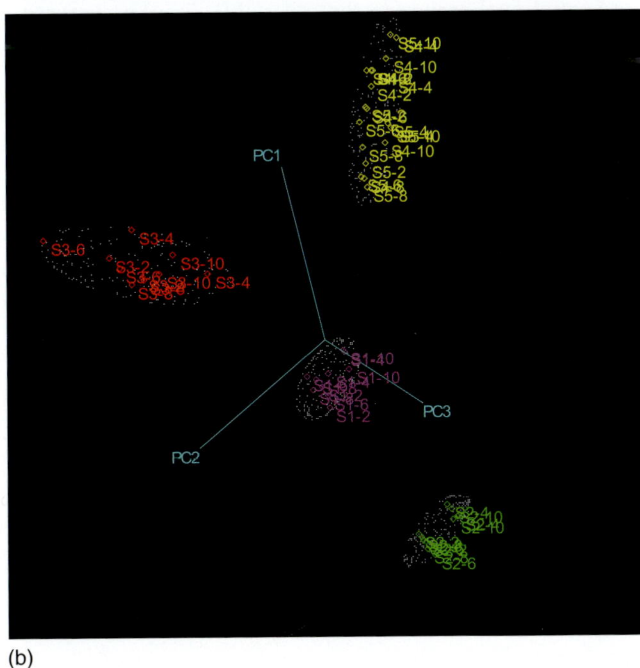

(b)

FIGURE 8.7. (a) 3D plot after SIMCA analysis using mean-centered preprocessing with five factors and a probability threshold of 0.95 for the opium samples from different locations. (b) 3D plot after SIMCA analysis using range scale preprocessing with three factors at a probability threshold of 0.95 for the poppy straw samples.

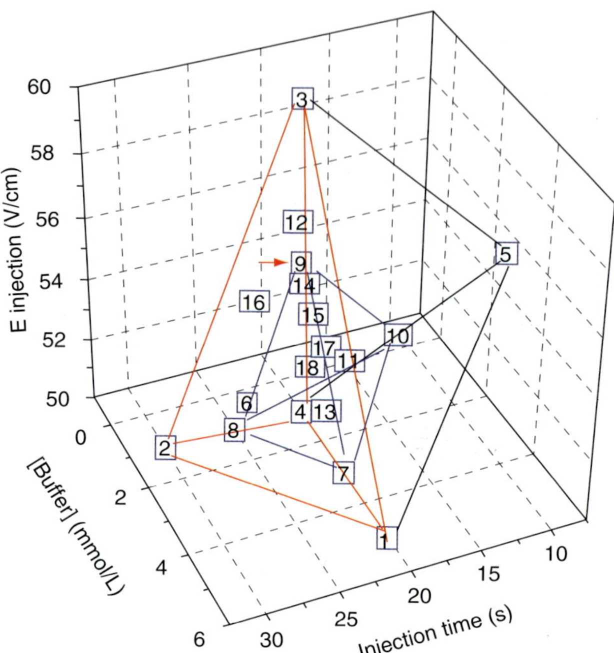

FIGURE 12.3. Spatial evolution of a three-variable simplex. The red lines link the initial conditions (vertices 1–4). The blue lines show the simplex figure after the radical contraction (vertices 4, 7–9) and the first reflection after contraction (vertex 10, blue lines). The arrow points to the best condition. Reprinted with permission from Reference 4.

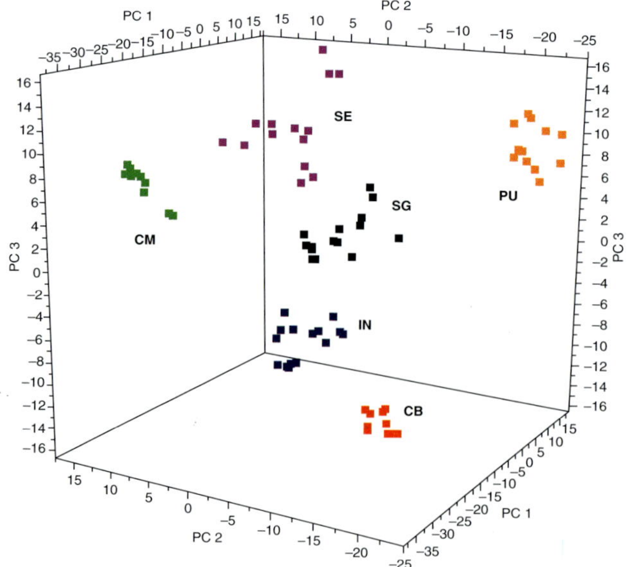

FIGURE 13.5. Three-dimensional score plot (PC1–PC2–PC3) of different *Corydalis* species electropherograms. Reproduced with permission from Sturm et al. (44).

CHAPTER 9

MULTIVARIATE CURVE RESOLUTION BASED ON ALTERNATING LEAST SQUARES IN CAPILLARY ELECTROPHORESIS

JAVIER SAURINA
Department of Analytical Chemistry, University of Barcelona, Barcelona, Spain

CONTENTS

9.1. INTRODUCTION

Capillary electrophoresis (CE) has proved to be a powerful separation technique increasingly utilized for the establishment of analytical methods in biochemical, clinical, pharmaceutical, and food fields (1–8). One of the most

Chemometric Methods in Capillary Electrophoresis. Edited by Grady Hanrahan and Frank A. Gomez

acclaimed features of CE is its great resolution capability leading to excellent separations of a wide variety of compounds. However, the performance of CE is obviously limited, and resolution problems may eventually arise. The similarities in the physicochemical characteristics of analytes, often belonging to the same family of compounds (i.e., sharing common structural features), hinder the separation. The sample matrix is an additional source of interferences and unexpected peaks, which may comigrate with the analytes.

Some preliminary aspects that cannot be underestimated to reach good separations involve, of course, the election of a suitable CE mode according to the physicochemical characteristics of analytes as well as the careful optimization of the experimental conditions. The introduction of micellar electrokinetic chromatography (MEKC) and electrochromatography has contributed to enlarge enormously the fields of application of CE, especially for dealing with neutral components (9–12). Regarding the optimization of the separation, it can be faced efficiently from a limited number of experiments with experimental design and multicriteria responses (13–15). However, after expending reasonable time and experimental efforts in optimization, in certain cases, the full electrophoretic resolution of all analytes might be not accomplished. Indeed, the occurrence of overlapping peaks is more common than we would desire, so approaches for solving this shortcoming are needed.

The study of comigrations is fundamental in order to be aware of the limitations of electrophoretic separation. Resolution deficiencies can be checked in a very simple way from the inspection of overlaid electropherograms of pure analytes and blanks. Besides, the appearance of partially resolved peaks, shoulders, tails, etc. is another sign of poor separation. In the case of minor components imbibed in a large peak as well as in the case of high overlapping, however, the detection of contamination may result in a more complex issue, and a mathematical evaluation of data may be required to ascertain the occurrence of comigration (16).

The problem of poor separation in CE has been addressed, mainly considering those strategies adopted in chromatography in analogous circumstances (17–20). One of the simplest approaches of mathematically increasing the resolution of peaks relies on working with derivative electropherograms (21). Derivative peaks display higher resolution than the original ones, and, thus, the chance of achieving a good separation is higher. However, the approach is hindered by a noticeable parallel loss of sensitivity and the inability to resolve strong overlapping or uncontrolled peaks. In conclusion, the significance of this treatment is actually limited to pseudo-academic examples while its application to "real-life" samples seems to be unreliable.

Other strategies for improving the resolution of comigrating components rely on the use of multiway detectors such as diode array detector (DAD) and charge-coupled devices (CCDs) as a way of getting spectral information over the entire electropherogram (16). If selective wavelengths are found for each overlapping species, they can be used to specifically monitor the corresponding components without interference. Unfortunately, in UV-visible spectroscopy, full spectral selectivity in multicomponent systems is hardly encountered since spectra of close species are likely similar.

Obtaining selectivity through the spectral domain is much more feasible in the case of mass spectrometry (MS). MS spectra currently contain mass-to-charge peaks characteristic of each component. Hence, the spectral selectivity can be exploited to resolve deficient separations by monitoring mass traces specific of each compound (22, 23). Additional advantages derived from the use of MS consist of the high sensitivity and the almost universal nature of the detection. The extensive analytical possibilities of MS detection have been confirmed previously in a multitude of high performance liquid chromatography–mass spectrometry (HPLC–MS) and gas chromatography–mass spectrometry (GC–MS) applications. However, in contrast to chromatography, CE–MS applications are still scarce due to the cost of the equipment and the difficulty of making compatible the current CE flow rates with MS requirements. Some pioneering attempts to couple CE and MS were from Sentellas et al. in the determination of drugs and metabolites in body fluids (24). In the referred work, authors proved the excellent performance of the technique even in the case of strong overlapping. The marked introduction of commercial CE–MS instruments is contributing to the rapid expansion of such techniques since the robustness of the hyphenation has been significantly improved.

And what happens if we are not able to get full selectivity through any measurement domain? Does it mean that we have to discard the method, then lose our efforts, time, and money? As we illustrate in this chapter, a solution to this problem can be obtained mathematically by using curve resolution methods. The combination of CE and curve resolution tools becomes, in general, greatly satisfactory as it takes advantage of synergisms between physicochemical and mathematical separations. Here, we should remark that the possibility of mathematically discriminating the components inevitably implies that such components should be slightly different either in the electrophoretic profiles or in the spectra. Conversely, if the electrophoretic and spectral behavior of such components is almost identical, the resolution will be impossible as they will be seen as an only component. In conclusion, even in the absence of full selective data, when the profiles of the components are sufficiently different, they can be treated mathematically using curve resolution methods to recover the underlying contributions of pure components (see section 9.2).

As shown in Figure 9.1, a scheme for dealing with electrophoretic data could be presented in a number of steps. First, one could inspect the electropherograms trying to find problematic peaks to be studied in more detail. Second, the purity of such suspicious peaks can be analyzed in order to confirm or discard the occurrence of imbibed contributions. In the case of comigrations, the next step aims at the resolution of underlying analyte profiles in the complex peak. Finally, analytes can be quantified by comparison of the recovered peak with those of the standards extracted under equivalent conditions.

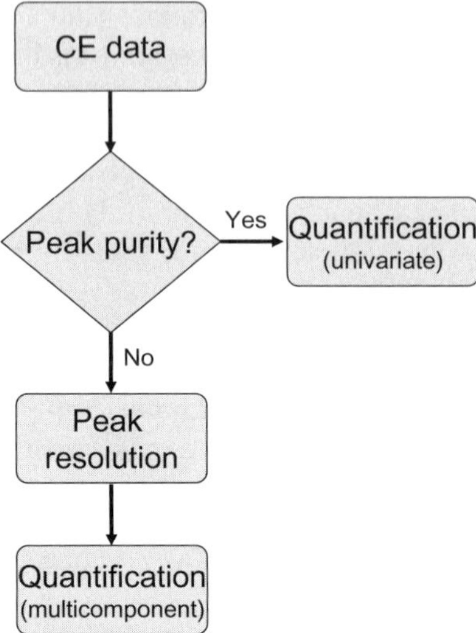

FIGURE 9.1. Scheme of the study of CE data.

9.2. MATHEMATICAL BACKGROUND

9.2.1. Preliminary Concepts

It is well known that the accuracy of CE determinations using univariate calibration models, such as linear regression, relies on the selectivity of the electrophoretic data. Peaks of analytes must be baseline resolved and the occurrence of comigrations and minor impurities should be avoided. Note that peak contaminations lead to wrong integrations, and, consequently, the concentrations estimated from these data may be unreliable.

The evaluation of the peak purity as a selectivity criterion is a fundamental issue deserving thorough attention. If peaks are found to be heterogeneous, chemometric methods based on curve resolution can be used to isolate the pure analyte contributions from a mixture system, thus making possible an accurate quantification of components (16).

Purity assays rely on the analysis of spectral information over the electropherogram so that the CE instrument must be compatible with this option. Fortunately, most of current commercial CE instruments are furnished with fast-scanning detectors, and they easily generate spectral data over the entire electropherogram. Note that if the shape of these spectra is constant from front to tail, reasonably, the peak should correspond to a single component. Conversely, a variation in the shapes may indicate a peak contamination (see scheme in Fig. 9.2). Although less common, a similar analysis can be carried out through the electrophoretic domain by comparing CE profiles recorded at different wavelengths.

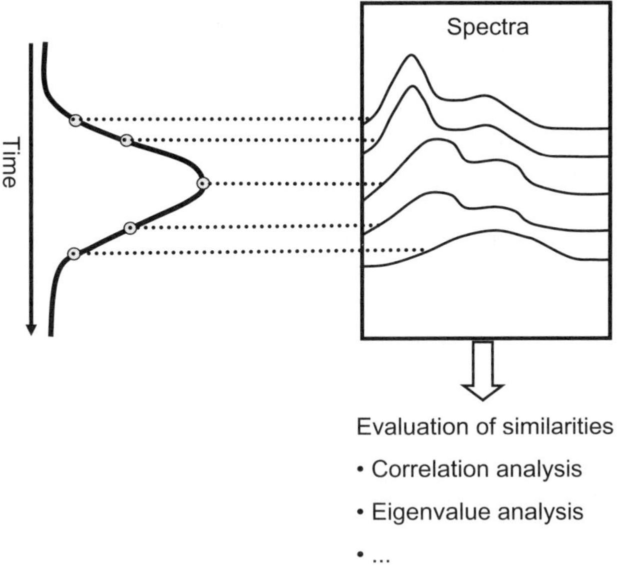

FIGURE 9.2. Evaluation of peak purity by spectral analysis.

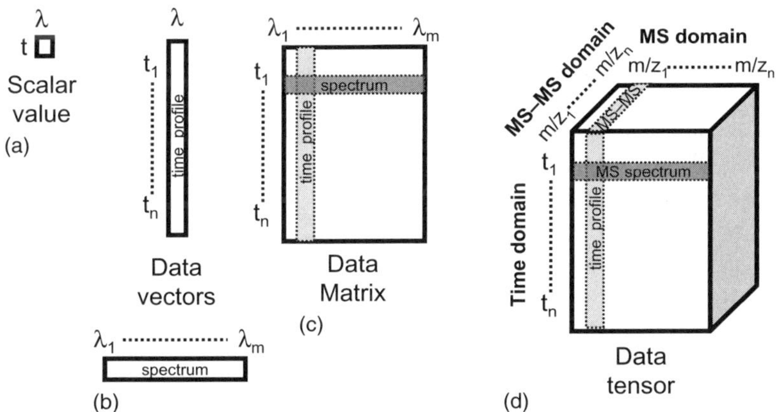

FIGURE 9.3. Types of data in CE.

9.2.2. Electrophoretic Data

As shown in Figure 9.3, CE provides data of different dimensionality that can be exploited for different qualitative and quantitative tasks (25, 26), namely:

Scalar data (zero-way data), such as peak areas, are used in quantitative determinations based on univariate calibration. As commented, the measured data have to be fully selective for the analyte of interest and interferences cannot be present.

Data arrays (one-way data) consist of spectral values taken at a given time point or electrophoretic responses at a given wavelength over time. One-way data can be used for sample characterization, classification, and quantification with multivariate calibration methods (e.g., principal component analysis and partial least square regression).

Data matrices (two-way data). Electrophoretic data resulting from multiway detectors, such as in CE–DAD and CE–MS techniques, can be arranged in a table of values or a data matrix. Data are structured over the two domains of measurement, in which each column corresponds to a wavelength (or m/q ratio) and each row corresponds to a time point. Two-way data can be exploited for studies of peak purity and mixture resolution using curve resolution and related factor analysis methods.

Data tensors (three-way data). Progressing on the complexity of the structure of data, three-way data sets involve three domains of measurement. As an example, CE with MS–MS detection could theoretically generate such type of data. In practice, however, the full spectral acquisition required for tensorial data is not technically available yet. Besides, mathematical tools dealing with data tensors are not fully established (27, 28).

9.2.2.1. Data Augmentation. The combination of data from different runs results in a valuable way to enrich the information content and expand the possibilities of CE methods (27–32). Matrices from various runs can be packed together in a tensor of superior dimensionality. Alternatively, as schematized in Figure 9.4, two-way data sets can be arranged in augmented data matrices in two ways:

Column-wise augmentation, in which matrices of various runs are joined one below the others in a structure that keeps common wavelengths (or m/q ratios) in the same column. According to MATLAB nomenclature, these matrices can be written as **[Run 1;Run 2;Run 3; … ;Run j]**.

Row-wise augmentation, in which matrices are joined one aside the others in a structure that keeps common times in the same row. Row-wise arrangements can be represented as **[Run 1,Run 2,Run 3, … , Run j]**.

Multivariate curve resolution can be used for the analysis of augmented sets as a way of reinforcing conclusions on peak purity, improving the resolution of overlapping compounds, and performing multicomponent determinations in the presence of interferences.

From the mathematical point of view, the construction of augmented arrangements assumes that a given species is characterized by the same profile in any run. Hence, in column-wise augmentation (i.e., wavelength-wise augmentation), each species is defined by a unique unit spectrum in any run. Analogously, in row-wise augmentation (time-wise augmentation), each species is characterized by a unique unit peak profile.

The simultaneous concurrence of equality in spectral and electrophoretic profiles, that is, each component is described by a unique dyad of vectors, leads to the so-called trilinearity. Trilinear data offer excellent possibilities for

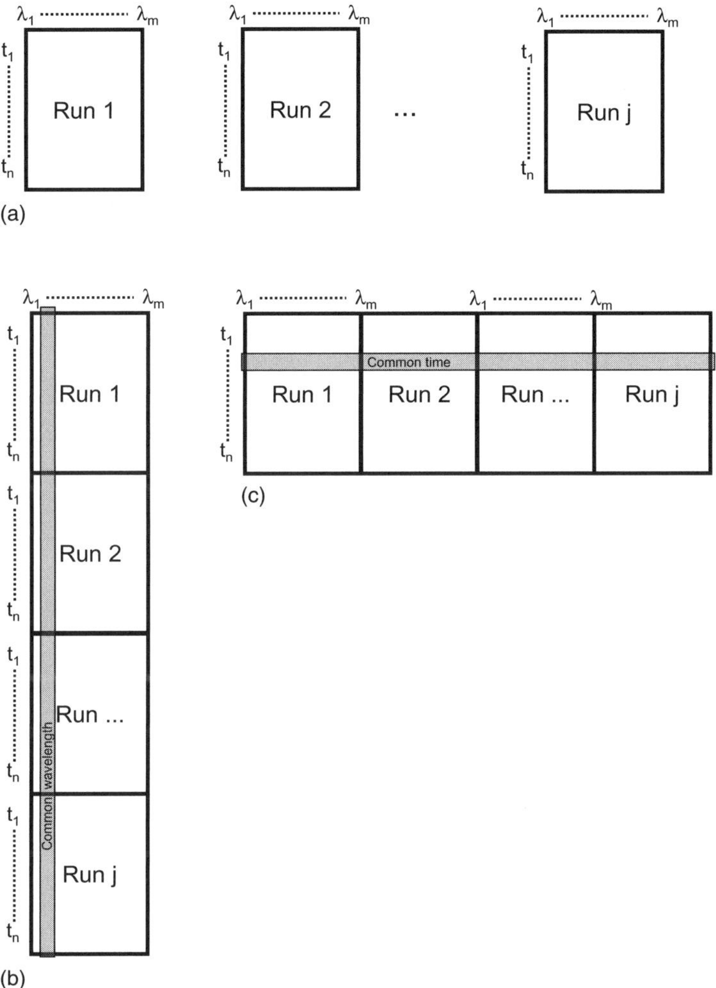

FIGURE 9.4. Matrix augmentation for the simultaneous analysis of CE runs. (a) Individual data sets; (b) Column-wise arrangement; and (c) Row-wise arrangement.

avoiding ambiguous resolutions, even in the presence of unknown interferences. This property, known as second-order advantage, opens up a wide variety of possibilities in resolution and quantification issues (25, 26).

9.2.3. Preprocessing CE Data

Preprocessing procedures are focused on improving the characteristics of CE data before proceeding with resolution and quantification tasks (16). Variations in the migration time of electrophoretic peak, often around 1%–2%, may be responsible for data desynchronization and lost of trilinearity. Peak shifting can be minimized with an alignment procedure based on the peak maximum position. Additional effects of peak broadening or sharpening may occur so that, if they are relevant, complementary peak width correction may be needed.

Another common treatment consists of background spectral correction by subtracting the baseline spectrum before the peak appearance. Drifts in the baseline of electropherograms can be circumvented by absorbance subtraction. Beyond these simple corrections, more sophisticated treatments for detrending and noise filtering using wavelets, artificial neural networks, and so on can be used.

9.2.4. Multivariate Curve Resolution

Curve resolution methods are focused on extracting information of the pure components in a mixture system through a suitable factorization of the experimental data matrix **D** into the product of two simpler matrices **C** and $\mathbf{S}^T$ that refer to pure peak profiles and pure spectra of components, respectively (33). Mathematically, the equation of the resolution process can be written as follows:

$$\mathbf{D} = \mathrm{C} \times \mathbf{S}^{\mathrm{T}} + \mathbf{E} \qquad \text{(Eq. 9.1)}$$

where **E** is the matrix of residuals not explained by the components recovered. Schematically, the resolution process is depicted in Figure 9.5.

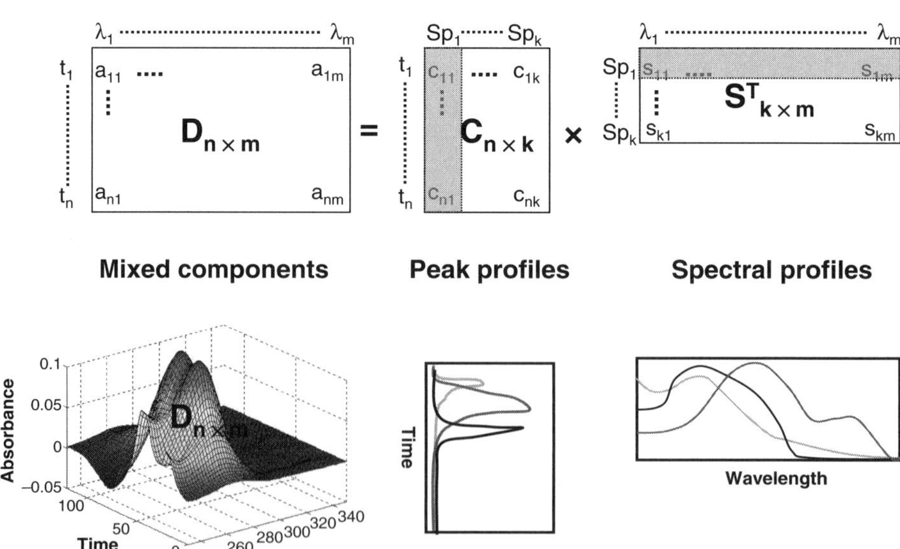

FIGURE 9.5. Scheme of the resolution of a mixture system into the spectral and peak profiles.

Various resolution methods have been proposed in the literature for dealing with the analysis of unresolved separation systems, including electrophoretic data (see section 9.3). Among them, the so-called multivariate curve resolution based on alternating least squares (MCR–ALS) method is used in this chapter (34, 35). MCR–ALS has proved to be highly efficient in a wide variety of chemical cases including kinetic processes (36–39), equilibrium modeling (40, 41), flow-injection analysis (42, 43), HPLC, and CE (20, 30–33, 44). A free version of MCR–ALS written in MATLAB environment can be downloaded from the web page of our working group at http://www.ub.edu/mcr/welcome.html. The principal steps of MCR–ALS (see scheme in Fig. 9.6) are described in the following sections.

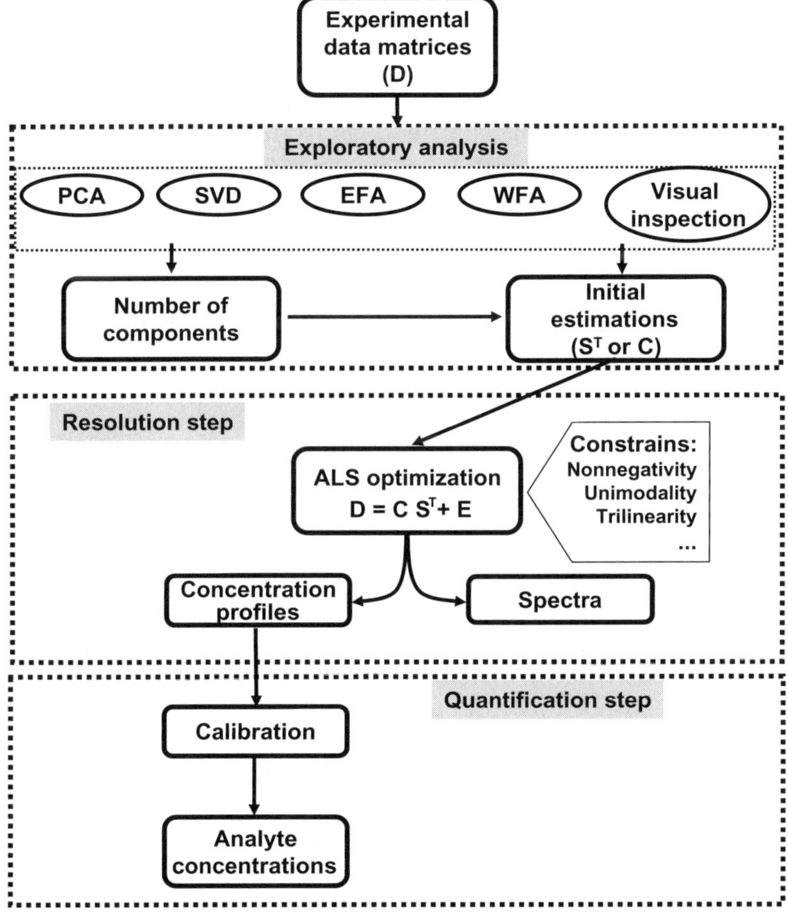

FIGURE 9.6. Scheme of steps of multivariate curve resolution based on alternating least squares (MCR–ALS procedure).
PCA = principal component analysis.

9.2.4.1. Exploratory Data Analysis. Before proceeding with the resolution process, exploratory studies are needed to determine the number of species of the system, to get a preliminary idea of the evolution of such components in the peak, and to obtain initial estimations of the species of interest.

9.2.4.1.1. Estimating the Number of Components in a Given Peak. The evaluation of the number of components in a given CE peak is synonymous with the diagnosis of its purity. Mathematical tools can be used to ascertain the number of relevant contributions or the rank of the experimental matrix **D**. The rank depends on the chemical species of the system as well as some physical factors that eventually may contribute to the response. For instance, baseline drifts, peak shifting, peak warping, changes in refractive index, and so on may be sometimes relevant components. Some of these factors can be removed totally or partially by means of appropriate preprocessing procedures (see section 9.2.3).

By far, singular value decomposition (SVD) is the most popular algorithm to estimate the rank of the data matrix **D**. As a drawback of SVD, the threshold that separates significant contributions from noise is difficult to settle. Other eigenvalue-based and error functions can be utilized in a similar way, but the arbitrariness in the selection of the significant factors still persists. For this reason, additional assays may be required, especially in the case of complex data sets.

We should note the importance of the correct selection of the number of components, as an erroneous number may lead to wrong qualitative and quantitative conclusions. Then, how do we proceed in the evaluation of the number of species? The number of species that we have deduced from exploratory tools and our chemical knowledge is often merely tentative. At this point, we should perform the resolution considering this number and evaluate the consistency of the recovered results. If the resolution is not satisfactory, additional models should be built with other numbers of species in order to achieve the best results.

9.2.4.1.2. Evaluating of the Distribution of Components in the Peak by Local Rank Analysis. Complementary information about the evolution of the components inside the CE peak system can be obtained from local rank analysis. In this case, instead of estimating the rank of the whole **D** matrix, a succession of smaller submatrices derived from **D** is analyzed to get the evolution of the mathematical factors throughout the system. The most widely used evolutionary methods are as follows:

Evolving factor analysis (EFA). This technique calculates the eigenvalues of submatrices gradually enlarged in the time direction (see scheme in Fig. 9.7) (45). Starting from the first spectrum of the system, that is, the first row of **D** matrix, the following spectrum in the forward direction is added and the eigenvalues of this submatrix are calculated. This process is repeated sequentially, adding each time the next spectrum up to the end of **D**. Subsequently,

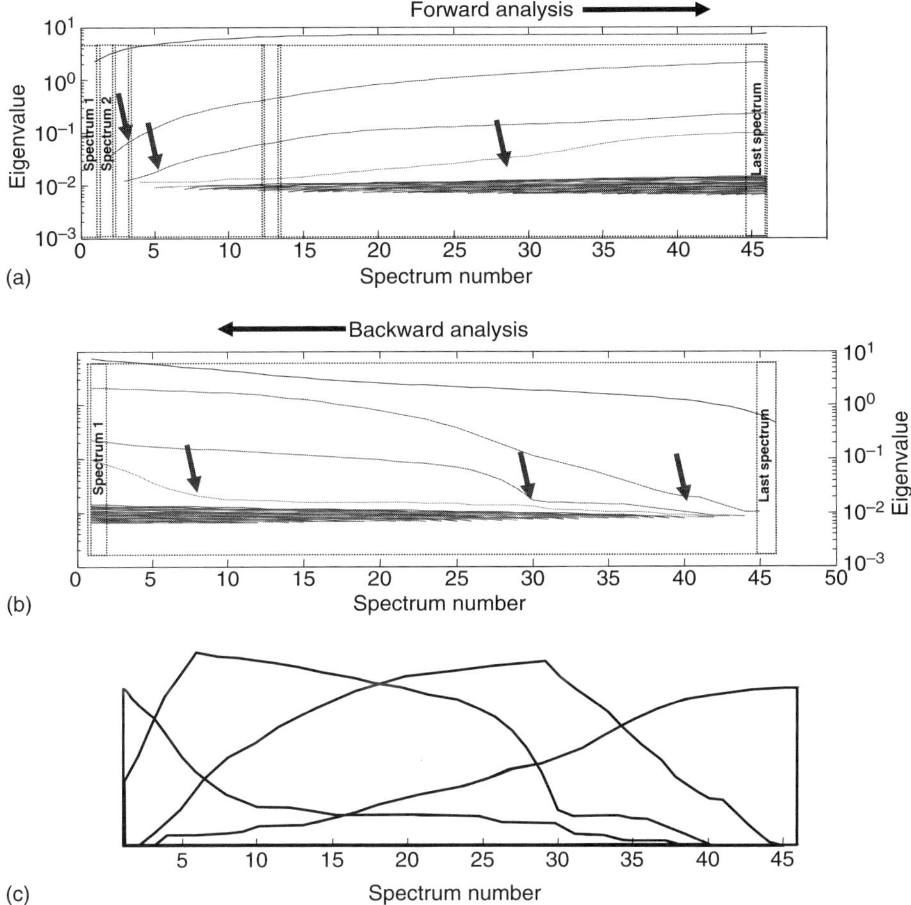

FIGURE 9.7. Scheme of the evolving factor analysis procedure. (a) Forward analysis; (b) Backward analysis; and (c) Reconstructed abstract profiles.

the evolution of the magnitude of eigenvalues is checked in the backward direction, starting from the last spectrum and going to the beginning of the system. From the evolution of eigenvalues, it is thus possible to detect the appearance of a new compound. Analogously, the disappearance of species can be followed in the backward direction. The reconstruction of the abstract profiles is based on two assumptions: the unimodal nature of peaks as only one maximum is expected (generally fulfilled in CE), and the fact that the first appearing factor is the first disappearing compound (not always true in CE).

Fixed-size moving-window–evolving factor analysis (FSMW–EFA). This technique, also called window factor analysis (WFA), is based on a window of a predefined number of rows or spectra, typically from three to five, which is moved over the time dimension of **D**, from the beginning to the end of the peak system (46). For each window submatrix, the eigenvalues are calculated and plotted as a function of time. The emergence and decay of factors is realized from the variation of eigenvalue lines.

Both EFA and WFA can be used to confirm conclusions on the rank of the system. Additionally, EFA and WFA provide relevant information on the occurrence of selective regions. The identification of such regions is extremely important in helping to solve resolution ambiguities via implementation of suitable constraints (see 9.2.4.1.4). Furthermore, these evolutionary methods may be useful for obtaining the initial estimations of components.

9.2.4.1.3. Obtaining Initial Estimations for Species. It is important to mention that the chance of reaching a good resolution increases when working with appropriate initial estimations since the convergence toward the underlying profiles may be more feasible. Both spectral and time domains are useful for obtaining the initial information in regard to the species to be modeled.

Several possibilities can be explored for obtaining the estimations of species, namely: visual inspection of the experimental data set, study of pure standards as a source of information, and, of course, application of mathematical tools. Among these possibilities, if pure standards are available, spectra (or peak profiles) of the desired analytes can be introduced directly in the optimization calculation.

Mathematically, EFA provides initial estimates of the time profiles, often quite rough and of poor quality. More frequently, algorithms searching for the purest variables of **D** (e.g., SIMPLISMA [SIMPLe-to-use Interactive Self-modeling Mixture Analysis]), applied either to the spectral or time domains, are used for finding the most characteristic profiles of the data set (47).

9.2.4.1.4. Alternating Least Square Optimization. The optimization process starts the iterative calculations from the initial estimates (spectral or electrophoretic profiles) of species to be modeled. If spectra are used as an input, the conjugated peak profile contributions **C** can be calculated as follows:

$$\mathbf{C} = \mathbf{D} \times (\mathbf{S}^{\mathbf{T}})^{+} \qquad \text{(Eq. 9.2)}$$

where the superindex + refers to the generalized inverse.

Subsequently, $\mathbf{S}^{\mathbf{T}}$ is updated using the expression:

$$\mathbf{S}^{\mathbf{T}} = (\mathbf{C})^{+} \times \mathbf{D} \qquad \text{(Eq. 9.3)}$$

where $(\mathbf{C})^{+}$ is the generalized inverse of **C**.

Analogously, if peak profiles **C** are used as an input, iterations start with the calculations of the conjugated spectra $\mathbf{S}^{\mathbf{T}}$. In any case, the iterative calculations of **C** and $\mathbf{S}^{\mathbf{T}}$ are repeated until reaching the optimum profiles. Three stopping criteria have been defined as follows: (i) reaching a convergence fitting error defined beforehand, (ii) exceeding a predefined number of iterations, and (iii) diverging in the fitting process 20 times consecutively.

A drawback inherent to all curve resolution methods is that optimized **C** and $\mathbf{S}^T$ profiles may present ambiguities in both intensity and shape (rotational ambiguity). The intensity ambiguity means that the recovered **C** and $\mathbf{S}^T$ can be multiplied (scaled) respectively by an unknown factor and its reciprocal without changing the result. The rotational ambiguity means that the recovered **C** and $\mathbf{S}^T$ profiles may be an unknown linear combination of the true profiles. These ambiguities may occur when the experimental data are not selective enough for some of the species present. Ambiguities can be solved, or at least minimized, when certain data features such as the occurrence of local selectivity and zero-concentration windows are met. However, as pointed out elsewhere (29), the most powerful way of reducing ambiguities relies on the simultaneous resolution of several related runs (see section 9.2.4.1.5).

Another weakness of resolution of CE overlapping data deals with the so-called rank deficiency caused by strong profile overlapping. In rank-deficient systems, the number of species detected mathematically is lower than the actual number of chemical components (48, 49). This may occur when two or more chemical species have equal or highly similar profiles in the two orders of measurement. When profiles are not exactly equal but slightly different, the addition of standard information from independent runs may contribute to facilitate the differentiation among species. Typically, the rank deficiency is solved by matrix augmentation as detailed below.

In order to get a better resolution of components of **D**, various natural constraints can be applied to restrict the mathematical solutions. Such constraints force a given spectral or time profile to fulfill a defined feature, thus reducing the ambiguity. In the case of CE data, the most relevant constraints are as follows:

Nonnegativity in the spectral and peak profiles. This constraint updates all negative values of peak and spectral profiles of species to zero. The restriction can be applied simultaneously to all species, or, alternatively, it can be implemented individually to selected compounds.

Unimodality. Such a concept relies on the fact that electrophoretic peaks have only one peak maximum. Hence, when a second peak is rising inside the profile of a given component, such a secondary peak is assumed to be due to a different species. The constraint cuts the secondary peak and sets the values in this range to zero. Note that this restriction is not applicable to spectra, as they may have several maxima.

Zero-concentration window. If a given species is absent in a given peak range, the corresponding values can be forced to be zero. The occurrence of zero concentration windows, often detected by EFA and WFA, is important since within this range the remaining species are expected to be better defined. For instance, in the case of two components, A and B, a zero-concentration window for A means that the region is selective to B. Hence, the information of B gained from this range should be free of rotational ambiguities.

The quality of the resolution results can be evaluated from a comparison between the actual spectral and time profiles of species with those recovered by MCR–ALS. Actual profiles can be found experimentally by recording CE runs of pure standards of components. The concordance between true and calculated profiles can be measured with correlation coefficients. Values close to 1 suggest that results are not affected by rotational ambiguities. Conversely, values significantly lower than 1 indicate that ambiguities still persist (43).

9.2.4.1.5. Simultaneous Analysis of Several Matrices: Resolution. As detailed in section 9.2.2.1, either column-wise or row-wise matrix augmentation can be considered to tackle the simultaneous analysis-related runs. The structure of the resolution process is shown in Figure 9.8 in which matrix **S** contains spectra of species and the augmented matrix **C** contains the concentration profiles in the different runs. Apart from those constraints implemented for the analysis of the individual data sets, additional restrictions can be used in this simultaneous analysis as follows:

Equal shape in the spectrum of each species. In general, each species is defined by a unique unit spectrum independently of the run. This constraint

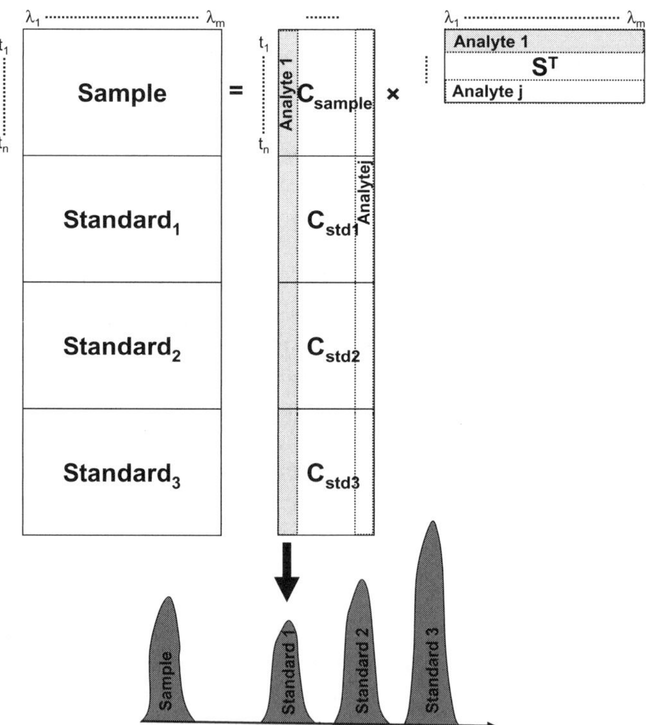

FIGURE 9.8. Scheme of the simultaneous resolution of several data sets and example of the quantification of analyte 1 from the extracted peak profiles.

is implicit to the construction of column-wise arrangements since the augmentation has no sense if unit spectra of species vary among runs.

Equal shape in the peak profile of each species. Similarly, when each species is defined by a unique peak profile shape in any run, this feature can be exploited to force the equality. The restriction is implemented as follows: For a given species, the peak profiles in all runs are analyzed together and the first principal component is taken as a representative shape in any run. The fulfillment of this constraint implies that peaks should be synchronized and interactions of comigrating species should be absent. These requirements are reasonably true if peak shifting in the time domain is minimized according to the data pretreatments described in section 9.2.3.

Trilinearity. The simultaneous achievement of equal shapes in the two domains of measurement leads to trilinear systems, and the so-called second-order advantage can be then exploited (25, 26). In these circumstances, in theory, the concentration of analyte(s) in unknown samples can be quantified using pure standards as a reference, even in the presence of unmodeled interferences. In practice, however, the occurrence of matrix effects altering the sensitivity may require the use of matrix-matched or standard additions (42, 50).

Partial trilinear systems involving equal shapes for certain component(s) have also been described. In the implementation of this constraint, the user can select the species to be restricted in this way, while the others can vary freely in the different processes (51).

9.2.4.1.6. Simultaneous Analysis of Several Matrices: Quantification. It is important to highlight that the quantitative information associated with the resolution is contained in **C**. As selectivity is, in theory, achieved mathematically after resolution of the augmented data set, the resulting peak profiles should be now free of interferences. Hence, analyte peak data such as areas or heights can be used for quantitative purposes in a very simple way (see Fig. 9.8).

In the simplest case, dealing with the simultaneous analysis of a sample mixture **M** with a standard **S** through the augmented system **[M;S]**, the quantification of the analyte in the unknown mixture is performed by comparison of peak areas as follows:

$$c_M = (a_M / a_S) c_S \qquad \text{(Eq. 9.4)}$$

where c_M and c_S are the concentrations of the analyte in the mixture and the standard, and a_M and a_S are the corresponding peak areas, respectively.

However, there is a wide variety of possibilities focused on the construction of augmented arrangements according to our needs. When standards of several compounds are added, the simultaneous determination of such analytes is then possible. For instance, in $[\mathbf{M};\mathbf{S}_A;\mathbf{S}_B;\mathbf{S}_C; \ldots]$, specific standard matrices of compounds A, B, and C are utilized for facilitating their resolution and making their quantifications in the sample M possible. In other cases, various standards of increasing concentration of a given analyte can be added to the system to get more robust modeling of profiles while improving the quantification possibilities. As an example, **[M;S_{A1};S_{A2};S_{A3}; …]** corresponds to a system focused on the quantification of A. Note that in this case, concentration calculations rely on linear regression as usual in univariate calibration. Other arrangements can be built considering several compounds with several standards simultaneously such as in **[M;S_{A1};S_{A2};S_{A3}; … ;S_{B1};S_{B2};S_{B3}; … ;S_{C1};S_{C2};S_{C3}; …]**.

Often, the incorporation of blank data from blank injections is a valuable way of enriching our knowledge about the background components of the system. As a result, factors such as electroosmotic flow (EOF) or micellar contributions can be more efficiently investigated. An example is represented in the analysis of **[M;B;S_A; …]** in which appropriate blank data **B** are added.

In complex samples containing multiple and unknown interferences, the study of blanks (if available) is extremely important to evaluate the rank and the distribution of such components over the peak system. In this case, moreover, the possible influence of the sample matrix on the sensitivity (i.e., the matrix effect) may result in an additional drawback to be taken into account. In CE, variations in the sensitivity between samples and standards may be due to multiple sources including differences in viscosity, differences in the intensity of stacking and sweeping phenomena, and other chemical factors. When dealing with matrix effects, the use of pure standards seems to be inappropriate and strategies based on matrix-matched standards and standard additions to the sample have to be followed (50). An example of matrix-matched systems is given in **[M;MS_{A1};MS_{A2};MS_{A3}; …]**, which represents a case including various standards of analyte A, namely MS_{A1}, MS_{A2}, MS_{A3}, …, prepared in a matrix of characteristics similar to that of the sample. Analogous arrangements can be constructed relying on the standard addition method, such as **[M;M_{A1};M_{A2};M_{A3}; …]**, where appropriate amounts of A are added to the sample M, thus resulting in successive additions M_{A1}, M_{A2}, M_{A3}, etc. Apart from the analysis of raw matrices, mathematical transformations concerning blank, analyte(s), or sample subtraction could be used.

9.3. APPLICATION OF CURVE RESOLUTION TO CE DATA

Recently, various papers have been published in the scientific literature dealing with the application of curve resolution and other factor analysis techniques to CE data. Lilley et al. have analyzed the peak purity of drugs and their metabolites in urine and pharmaceutical preparations using iterative target transformation factor analysis (ITTFA) (52, 53). Complementarily, ITTFA has been used for deconvoluting comigrations and tracking the individual sample components across the electropherogram. ITTFA and other factor analysis assays have also been applied to resolve benzodiazepines in a complex peak system from the simultaneous analysis of HPLC–DAD and

MEKC–DAD data (54). Studies from Kaniansky and coworkers have focused on using factor analysis, including ITTFA, WFA, and orthogonal projection approach (OPA), for the feasible identification of orotic acid at low concentration level in urine matrices (55, 56). The mathematical resolution of anionic surfactants that cannot be separated electrophoretically has been accomplished by OPA–ALS (57).

In a related study, Latorre et al. applied exploratory rank analysis to ascertain the number of components of complex nonresolved electrophoretic peaks of some amino acid derivatives (32). The performance of EFA, WFA, and MCR–ALS for following the evolution of overlapping species in the system was compared. It was found that MCR–ALS provided the best results in the case of strongly overlapping contributions. The simultaneous treatment of the sample mixture with data from standards of interest permitted the analytes to be successfully quantified.

Sentellas and coworkers described the resolution of species comigrating with the EOF (30, 31). The principal difficulty of these systems arises from the high similarity of the electrophoretic behaviors of analytes, thus hindering the resolution. Part of the material presented in Example 9.3.3 (below) has been adapted from these examples.

Hua Li and coworkers have presented numerous studies on the application of curve resolution to recover the underlying contributions components in overlapping peaks. In one such case, the qualitative performance of various curve resolution methods, including heuristic evolving latent projections (HELP), EFA, WFA, and MCR–ALS was compared (58). Authors have also evaluated strategies for constructing the augmented arrangements and their implications in the quantitative predictions (50–61). The improvement of the determination by using internal standards for the standardization of multivariate data has also been assayed (62). Apart from these brief bibliographic references on the application of curve resolution to CE, in the following section, various examples of different complexity are resolved and discussed in detail.

9.3.1. Example 1: Evaluation of Peak Purity: Study of the Tryptamine Peak

This example illustrates the application of exploratory methods to evaluate the homogeneity of CE peaks. Data chosen correspond to a method for the determination of biogenic amines in wines by field-amplified sample stacking and in-capillary derivatization (63). 1,2-naphthoquinone-4-sulfonate (NQS) has been used as a labeling agent. Reagent and buffer solutions are introduced hydrodynamically into the capillary, whereas the sample is injected electrokinetically, thus allowing an effective preconcentration of positively charged analytes. After injection, both separation and reaction processes occur simultaneously inside the capillary using a zone-passing derivatization approach in mixed tandem mode.

Although separation and derivatization conditions have been optimized thoroughly using experimental design and multicriteria functions, certain peaks are suspected to contain impurities from side products. In particular, the peak of tryptamine derivative shows a shoulder that might be due to the presence of one or various comigrating impurities (see Fig. 9.9a). A time window of 120 s centered in the peak maximum has been taken for a deeper study of peak purity.

The visual inspection of spectra in the front, center, and tail of the peak shows differences in shapes that could be attributed to contamination. SVD has been applied to the study of the number of significant components of the data set. Although the interpretation of SVD graphs and the extraction of conclusions require caution, Figure 9.9b suggests that probably three species might be relevant. Therefore, apart from the main analyte peak, two additional contributions seem to be imbibed in the gross signal. Complementary analyses relying on EFA and WFA have detected the emergence of two residual peaks adjacent to the principal tryptamine peak (see Fig. 9.9c). The remaining factors are clearly irrelevant. These preliminary studies are consistent with the presence of two impurities. It is important to remark that such interferences correspond to degradation products of derivatives or side reactions. These products do not appear in blank electropherograms, so the only way to detect their presence is by studying the sample electropherograms. As a final comment, note that the determination of tryptamine without removing interferences may be inaccurate so that the pure analyte contribution should be first isolated from the side products before proceeding with the quantification.

9.3.2. Example 2: Resolution of Poorly Separated Peaks: Putrescine + Tryptamine System

Another example bringing an additional degree of complexity is presented in the study of a partially resolved system involving putrescine and tryptamine derivatives (63). The method utilized is the same as in Example 9.1 above. CE data corresponding to a working time window of ±100 s centered on the peak maximum are shown in Figure 9.10a. It can be seen that the principal peaks of putrescine and tryptamine are not baseline resolved. The strategy for analyzing this system is analogous to that described above for checking the homogeneity of tryptamine peak. SVD results suggest that four relevant factors are present in this data set (Fig. 9.10b). According to the information recovered in Example X.1, apart from the two components of each amine derivative, the two additional contributions due to peak contaminations are also observed.

The following steps should be addressed for the resolution of species with MCR–ALS. Initial estimates to be used as an input for the optimization process have been extracted from the experimental data sets as follows: the spectra taken at the two peak maxima and two more spectra at the beginning and tail of tryptamine peak (approximately at those times corresponding to

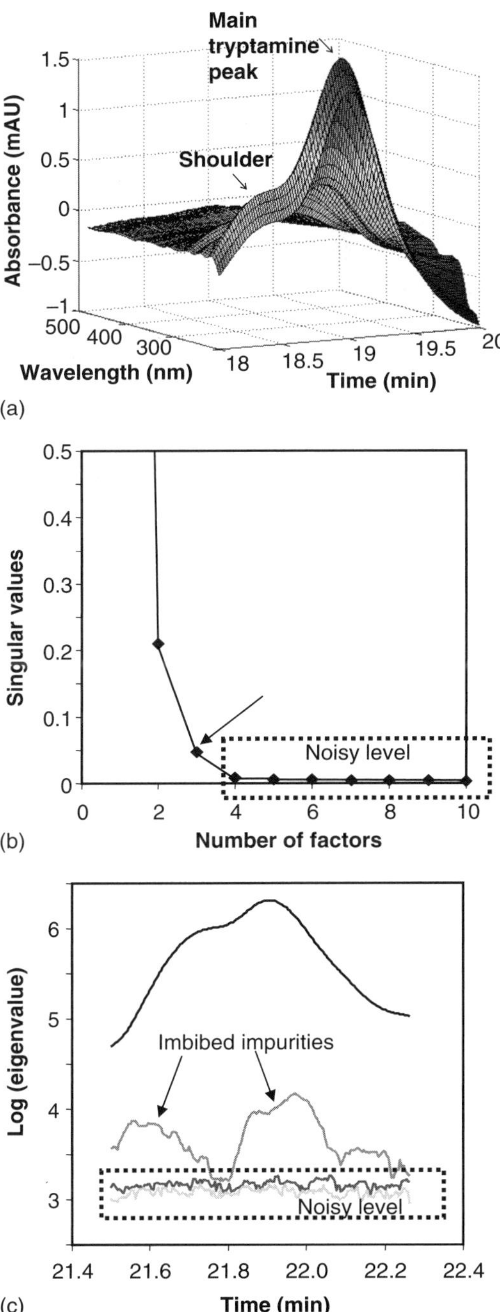

FIGURE 9.9. Evaluation of the peak purity of the tryptamine system. (a) Experimental data set; (b) determination of the number of components by SVD; and (c) study of impurities by window factor analysis.

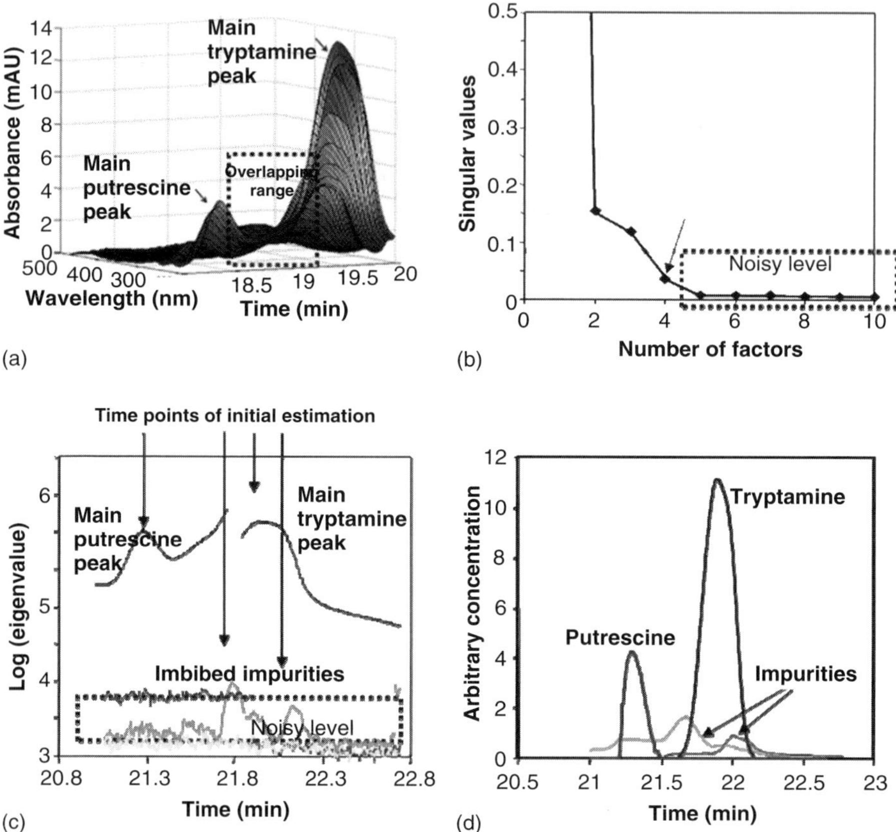

FIGURE 9.10. Evaluation of the peak purity of the putrescine-tryptamine system. (a) Experimental data set; (b) Determination of the number of components by SVD; (c) Study of impurities by window factor analysis (arrows indicate the time points at which spectra have been taken to be used as initial estimations; and (d) Results of the resolution of the data set by MCR–ALS.

the maximum of each emerging species, Fig. 9.10c). Concentration profiles resolved by MCR–ALS are shown in Figure 9.10d. Note that profiles of amine derivatives are apparently fully separated while imbibed interferences are responsible for the lack of baseline resolution between putrescine and tryptamine peaks.

9.3.3. Example 3: Simultaneous Resolution of Several Electrophoretic Runs

This section illustrates the resolution of components in strongly overlapping CE peak is described. Data correspond to a capillary zone electrophoresis (CZE) method for the determination of an antihistaminic drug and its metabolites (24). The main problem of this method arises in the occurrence of various poorly ionizable compounds, namely, 4-bromobenzensulfonamide, N-(2-methylsulfonyl-ethylamin-methylen)-4-bromobenzensulfonamide, and N-(2-methylsulfinyl-ethylamin-methylen)-4-bromobenzensulfonamide, here referred to as compounds A, B, and C, respectively. These species cannot be separated sufficiently by CZE and thus comigrate with the EOF. The use of micellar buffers could improve the resolution slightly, but, even in this case, the high similarity of the physicochemical characteristics of some of these compounds hinders the full separation.

Preliminary information gained from independent injections of blanks and pure standards indicates that spectral and peak profiles of metabolites are rather similar, with correlation coefficients between some species higher than 0.95. In these circumstances, the resolution of underlying contributions of components is expected to be difficult. As commented in the theory section, a powerful way of improving the resolution relies on the analysis of augmented arrangements, including standard(s) of the component(s) of interest. In this example, the unknown sample matrix **M** will be treated simultaneously with a blank (giving the EOF behavior) and a standard of metabolite A, referred to as matrices **B** and **SA**, respectively. Due to the higher stability of spectral data, the column-wise (wavelength-wise) matrix augmentation seems to be more convenient. Hence, the system to be analyzed could be written as **[M;B;SA]**.

Estimating the Number of Components A picture of the experimental data matrix **M** obtained from the injection of a mixture of metabolites A and B is shown in Figure 9.11a. The time window chosen in the study corresponds to the migration range of the neutral components. The exploratory analysis starts with the visual inspection of the original data. The comparison of spectra at different time points suggests the presence of various contributions. Mathematically, the number of components deduced from the SVD (Fig. 9.11b) of **M** seems to be two. This number does not agree with the presence of three chemical components corresponding to two metabolites plus an electroosmotic marker. The high similarity in the profiles of compounds A and B indicates that these two substances are hardly distinguishable.

As the addition of standard information of one or several components may facilitate the discrimination among species, the augmented arrangement consisting of **[M;B;SA]** has been analyzed. In this case, the number of components detected from **[M;B;SA]** is three, indicating that all chemical species can be seen, and thus, the rank deficiency due to profile overlapping has been solved.

Comparison of Strategies for Obtaining Initial Estimations The performance of the visual inspection of the experimental data set, SIMPLISMA and EFA, for obtaining initial estimations of peak profiles of components of **M** is compared here (Fig. 9.12). In general, EFA efficiently finds the time points of emergence and disappearance of factors, but the resulting profiles are just a

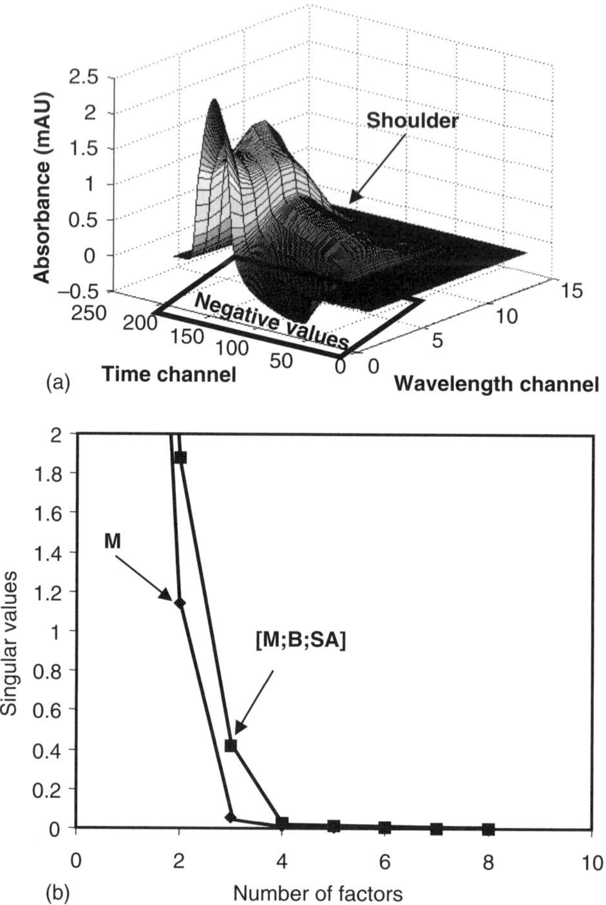

FIGURE 9.11. Study of electrophoretic data corresponding to a mixture of compounds A and B (see text for details). (a) Experimental data set; (b) SVD analysis of the individual data matrix **M** and the augmented system **[M;B;SA]**, being **B** and **SA** the matrices of blank and standard A.

poor approximation to electrophoretic peaks. The SIMPLISMA method looks into the data set to identify the purest variables of the system, for example, the less correlated variables. In this case, one of the estimations is clearly attributable to the EOF profile while the others may correspond to metabolites A and B, which mutually interfere. Finally, time profiles selected from the inspection of CE data seem to provide a more realistic approximation to the actual components. This option, shown in Figure 9.12c, is finally chosen to be used in the resolution process.

Resolution The simultaneous resolution of **[M;B;SA]** has been tackled, taking into account the specific features of this data system for the selection of constraints to be applied. Due to the particular shape of EOF contribution,

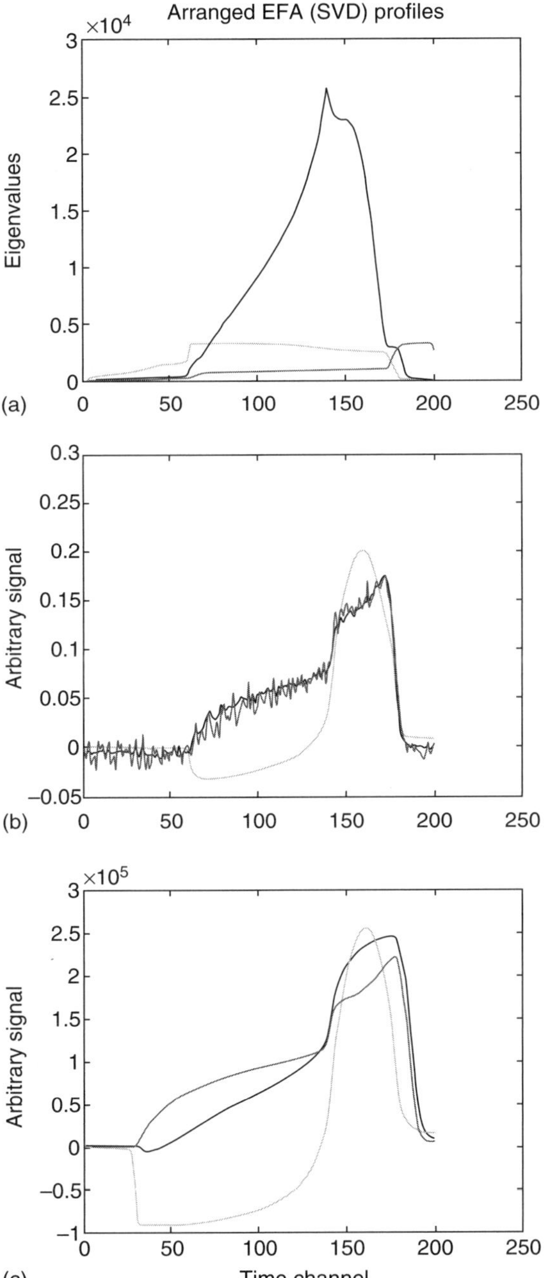

FIGURE 9.12. Study of approaches for obtaining initial estimations. (a) EFA; (b) SIMPLISMA; and (c) Estimations from the visual inspection of the data set.

the nonnegativity constraint cannot be utilized for this species. Conversely, the restriction is applicable to the rest of spectra and peak profiles. In the case of unimodality, this constraint can only be used in the peak profiles of A and B. The condition of equal shapes in both spectral and peak profiles has been considered for analyte A. EOF profiles cannot be constrained in this way since their shapes may vary in the runs. Optimized resolution results are summarized in Figure 9.13. It can be seen that the spectrum of EOF, associated with acetonitrile solvent, is clearly different from those of the metabolites. The shapes of the peak profiles are certainly peculiar, with a big shoulder next to an asymmetric main peak. This unusual profile has been attributed to the interaction between acetonitrile and analytes.

Some authors often consider the lack of fit as a good criterion to evaluate the goodness of the resolution. The lack of fit gives the error in the reproduction of the experimental data with the recovered components, but, in our opinion, this parameter says nothing about the reliability of profiles of analytes. Instead, we propose the comparison of actual and recovered spectra of analytes through the calculation of the correlation values as a more realistic way of proving the reliability of the resolution. It is thought that, in general, a good recovery of profiles is a reasonable guarantee of the quality of results. In the example of Figure 9.13, the similarities between actual and calculated profiles, in terms of correlation, are better than 0.98, thus demonstrating the success of the resolution (data not shown here).

Quantification From a quantitative point of view, the study of system **[M;B;SA]** corresponds to the determination of compound A in an unknown mixture M. Information regarding A has been included in the arrangement from **SA**, while no standard of compound B has been used; thus, B is acting as an unknown interference. The comparison of peak areas of compound A in the mixture and in the standard has been exploited to its quantification. The concentration predicted in this way is sufficiently accurate with a determination error below 5%.

This example tries to illustrate a representative case of simultaneous resolution and quantification. Obviously, other cases are also analytically relevant. For instance, if our interest is focused on the quantification of B, the system to be resolved should be **[M;B;SA]**. Additional arrangements can be built including A and B standards for the simultaneous determination of the two metabolites. All these possibilities cannot be treated here to avoid unnecessarily enlarging the chapter. Detailed information about other cases can be found in the literature (30, 31).

9.4. CONCLUSIONS

In conclusion, we should remark that the application of chemometrics to CE cannot be indiscriminate, and the most elemental CE fundamentals have to

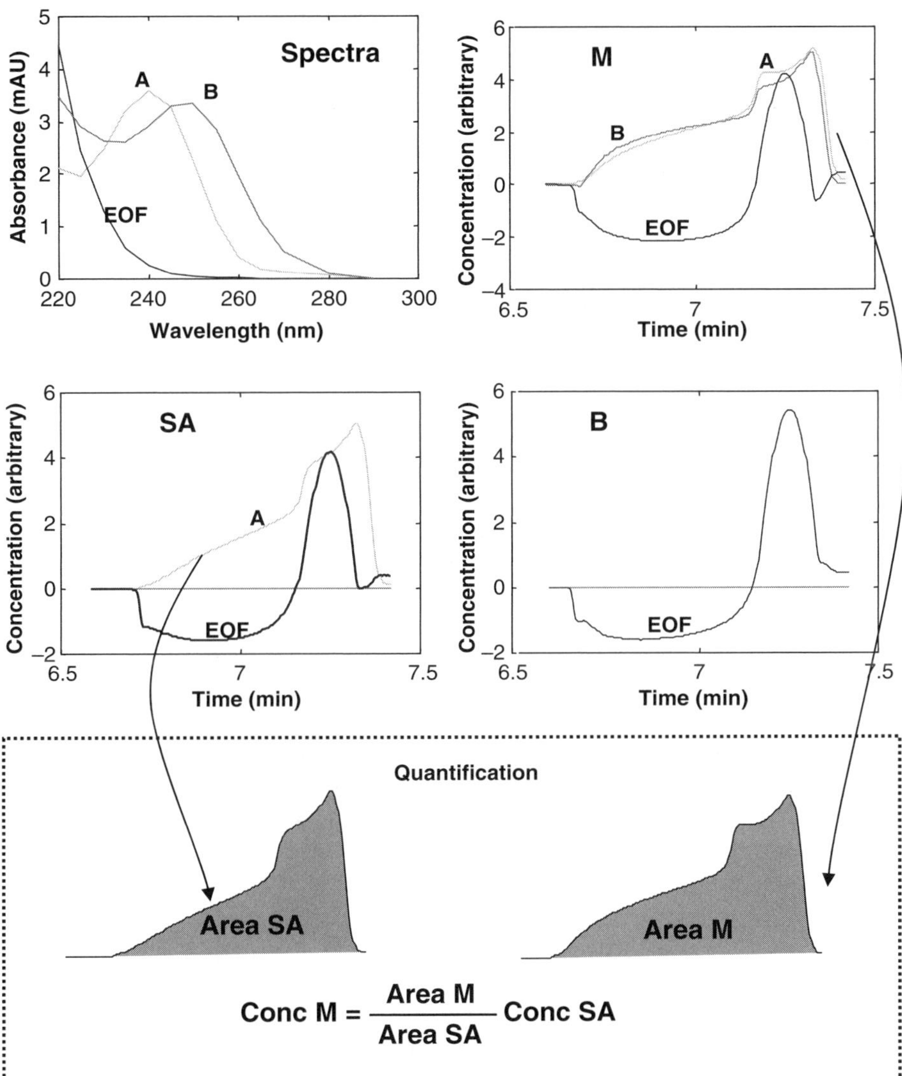

FIGURE 9.13. Results of the resolution of system **[M;B;SA]** by MCR–ALS and quantification of compound A from the comparison of peak areas.

be kept in mind. Hence, the choice of the most convenient CE mode, according to the characteristics of sample and analytes, and the careful optimization of the experimental conditions cannot be forgotten at the expense of further mathematical treatment of data. Sometimes we realize that certain separation methods have not been optimized correctly to generate, on purpose, overlapping systems that are resolved chemometrically. In our opinion, these practices may dissuade the potential users against the application of chemometrics.

As shown in this chapter, once the method has been optimized, we should check the electrophoretic separation of the sample in order to ascertain the presence of some problematic peaks and poor separations. The analysis of spectroelectrophoretic data obtained by CE–DAD or CE–MS may provide significant information about peak purity, analyte overlapping, and occurrence of imbibed peaks. As shown in Figure 9.1, curve resolution and related factor analysis methods can be applied to isolate the underlying analyte profiles from mixture systems. In many cases, extracting feasible conclusions from the exploratory analysis of CE data may be difficult, especially when dealing with biological, clinical, and food samples, due to the presence of multiple interfering components from the sample matrix. Besides, in peak purity assays and determination of the number of components, a certain degree of arbitrariness may occur. Resolution results may also be affected by ambiguities that can be solved or reduced under the application of suitable constraints. It has been proven that the most successful strategy for improving the resolution and minimizing ambiguities is based on the simultaneous analysis of various correlated runs sharing spectral or time information. In addition, the simultaneous analysis of samples and standards is the basis of the quantification by comparison of resolved peak profiles of analytes. There is a wide variety of quantification possibilities that can be treated depending on the number of analytes to be considered, number of standards of each analyte to be included in the arrangement, addition of blanks, etc. Furthermore, in the case of matrix effects on the sensitivity, strategies based on a generalization of the standard addition method or the use of matrix-matched standards could be followed.

Finally, commercial instruments progressively incorporate software for checking, for instance, the peak purity in a simple way and without needing solid chemometric skills. This may be a first step toward a progressive implementation of other algorithms to be used as standard processing tools.

REFERENCES

1. Frazier, R.A. (2001) *Electrophoresis*, **22**, 4197–4206.
2. Frazier, R.A., Ames, J.M., and Nursten, H.E. (1999) *Electrophoresis*, **20**, 3156–3180.
3. Issaq, H.J. (2000) *Electrophoresis*, **21**, 1921–1939.
4. Petersen, J.R., Okorodudu, A.O., Mohammad, A., and Payne, D.A. (2003) *Clin Chim Acta*, **330**, 1–30.
5. Jin, L.L., Ferrance, J., and Landers, J.P. (2001) *BioTech*, **31**, 1332–1353.
6. Dolnik, V. and Hutterer, K.M. (2001) *Electrophoresis*, **22**, 4163–4178.
7. Rochu, D. and Masson, P. (2002) *Electrophoresis*, **23**, 189–202.
8. Kasicka, V. (2001) *Electrophoresis*, **22**, 4139–4162.
9. Pyell, U. (2001) *Fresenius J Anal Chem*, **371**, 691–703.
10. Molina, M. and Silva, M. (2002) *Electrophoresis*, **23**, 3907–3921.

11. Rathore, A.S. (2002) *Electrophoresis*, **23**, 3827–3846.
12. Mistry, K., Krull, I., and Grinberg, N. (2002) *J Sep Sci*, **25**, 935–958.
13. Hanrahan, G., Montes, R., and Gomez, F.A. (2008) *Anal Bioanal Chem*, **390**, 169–179.
14. Sentellas, S. and Saurina, J. (2003) *J Sep Sci*, **26**, 875–885.
15. Siouffi, A.M. and Phan-Tan-Luu, R. (2000) *J Chromatogr A*, **892**, 75–106.
16. Sentellas, S. and Saurina, J. (2003) *J Sep Sci*, **26**, 1395–1402.
17. Duarte, A.C. and Capelo, S. (2008) *J Liq Chromatogr Rel Technol*, **29**, 1143–1176.
18. Daszytowski, M. and Walczak, B. (2006) *Trends Anal Chem*, **25**, 1081–1096.
19. Pierce, K.M., Hoggard, J.C., Mohler, R.E., and Synovec, R.E. (2008) *J Chromatogr A*, **1184**, 341–352.
20. de Juan, A. and Tauler, R. (2007) *J Chromatogr A*, **1158**, 184–195.
21. Dohnal, V., Zhang, F., Li, H., and Havel, J. (2002) *Electrophoresis*, **24**, 2462–2468.
22. Finehout, E.J. and Lee, K.H. (2004) *Biochem Mol Biol Educ*, **32**, 93–100.
23. Hoffmann, E. and Stroobant, V. (2003) *Mass Spectrometry: Principles and Applications*, John Wiley & Sons, West Essex.
24. Sentellas, S., Puignou, L., Moyano, E., and Galceran, M.T. (2000) *J Chromatogr A*, **888**, 281–292.
25. Booksh, K.S. and Kowalski, B.R. (1994) *Anal Chem*, **66**, 782A–791A.
26. Faber, K., Lorber, A., and Kowalski, B.R. (1997) *J Chemom*, **11**, 419–461.
27. Escandar, G.M., Olivieri, A.C., Faber, N.M., Goicoechea, H.C., Muñoz de la Peña, A., and Poppi, R.J. (2007) *Trends Anal Chem*, **26**, 752–765.
28. Gomez, V. and Callao, M.P. (2008) *Anal Chim Acta*, **627**, 169–183.
29. Tauler, R., Smilde, A.K., and Kowalski, B.R. (1995) *J Chemom*, **9**, 31–58.
30. Sentellas, S., Saurina, J., Hernández-Cassou, S., Galceran, M.T., and Puignou, L. (2001) *Electrophoresis*, **22**, 71–76.
31. Sentellas, S., Saurina, J., Hernández-Cassou, S., Galceran, M.T., and Puignou, L. (2001) *Anal Chim Acta*, **431**, 49–58.
32. Latorre, R.M., Saurina, J., and Hernández-Cassou, S. (2000) *Electrophoresis*, **21**, 563–572.
33. Lawton, W.H. and Sylvestre, E.A. (1971) *Technometrics*, **13**, 617–633.
34. de Juan, A. and Tauler, R. (2006) *Crit Rev Anal Chem*, **36**, 163–176.
35. de Juan, A., Casassas, E., and Tauler, R. (2000) *Encyclopedia of Analytical Chemistry: Instrumentation and Applications* (ed. R.A. Meyers), John Wiley & Sons, Chichester, pp. 9800–9837.
36. Argemí, A. and Saurina, J. (2007) *Talanta*, **74**, 176–182.
37. Ruckebusch, C., Duponchel, L., Huvenne, J.P., and Saurina, J. (2004) *Anal Chim Acta*, **515**, 183–190.
38. Mas, S., de Juan, A., Lacorte, S., and Tauler, R. (2008) *Anal Chim Acta*, **618**, 18–28.
39. Culzoni, M.J., Goicoechea, H.C., Ibáñez, G.A., Lozano, V., Marsili, N.R., Olivieri, A.C., and Pagani, A.P. (2008) *Anal Chim Acta*, **614**, 46–57.

40. del Toro, M., Gargallo, R., Eritja, R., and Jaumot, J. (2008) *Anal Biochem*, **379**, 8–15.
41. Argemí, A. and Saurina, J. (2007) *J Pharm Biomed Anal*, **44**, 859–866.
42. Checa, A., Oliver, R., Saurina, J., and Hernández-Cassou, S. (2007) *Anal Chim Acta*, **592**, 173–180.
43. Checa, A., Oliver, R., Saurina, J., and Hernández-Cassou, S. (2006) *Anal Chim Acta*, **572**, 155–161.
44. Peré-Trepat, E., Lacorte, S., and Tauler, R. (2007) *Anal Chim Acta*, **595**, 228–237.
45. Maeder, M. and Zuberbühler, A.D. (1986) *Anal Chim Acta*, **181**, 287–291.
46. Keller, H.R. and Massart, L.D. (1991) *Anal Chim Acta*, **246**, 379–390.
47. Windig, W. and Stephenson, D.A. (1992) *Anal Chem*, **64**, 2735–2742.
48. Amrhein, M., Srinivasan, B., Bonvin, D., and Schumacher, M.M. (1996) *Chemom Intell Lab Syst*, **33**, 17–33.
49. Saurina, J., Hernández-Cassou, S., Tauler, R., and Izquierdo-Ridorsa, A. (1998) *J Chemom*, **12**, 183–203.
50. Saurina, J. and Tauler, R. (2000) *Analyst*, **125**, 2038–2043.
51. Saurina, J., Hernández-Cassou, S., and Tauler, R. (1995) *Anal Chem*, **67**, 3722–3727.
52. Lilley, K.A. and Wheat, T.E. (1996) *J Chromatogr B*, **683**, 67–76.
53. Wheat, T.E., Chiklis, F.M., and Lilley, K.A. (1995) *J Liq Chromatogr*, **18**, 3643–3657.
54. van Zomeren, P.V., Metting, H.J., Coenegracht, P.M.J., and de Jong, G.J. (2005) *J Chromatogr A*, **1096**, 165–176.
55. Danlová, M., Strasik, S., and Kaniansky, D. (2003) *J Chromatogr A*, **990**, 121–132.
56. Strasik, S., Danlová, M., Molnárová, M., Ölvecká, E., and Kaniansky, D. (2003) *J Chromatogr A*, **990**, 23–33.
57. Bernabé Zafón, V., Torres Lapasió, J.R., Ortega Gadea, S., Simó Alfonso, E.F., and Ramos, G. (2004) *J Chromatrogr A*, **1036**, 205–216.
58. Li, H., Hou, J., Wang, K., and Zhang, F. (2006) *Talanta*, **70**, 336–343.
59. Li, H., Zhang, J.F., and Havel, J. (2003) *Electrophoresis*, **24**, 3107–3115.
60. Zhang, F. and Li, H. (2005) *Electrophoresis*, **26**, 1692–1702.
61. Zhang, F., Chen, Y., and Li, H. (2007) *Electrophoresis*, **28**, 3674–3683.
62. Zhang, F. and Li, H. (2006) *Chemom Intell Lab Syst*, **82**, 184–192.
63. García Villar, N., Saurina, J., and Hernández Cassou, S. (2006) *Electrophoresis*, **27**, 474–483.

CHAPTER 10

APPLICATION OF CHEMOMETRICS IN CAPILLARY ELECTROPHORESIS ANALYSIS OF HERBAL MEDICINES

SHAO-PING LI, XIAO-JIA CHEN, and FENG-QING YANG
Institute of Chinese Medical Sciences, University of Macau, Macao SAR, China

CONTENTS

10.1. INTRODUCTION

Herbal medicines, plant-derived materials, or products with therapeutic or other human health benefits that contain either raw or processed ingredients from one or more plants (1) have been utilized to treat various diseases for thousands of years, especially in Far Eastern countries. It is estimated that traditional herbal preparations account for 30%–50% of the total medicinal consumption in China (2). However, "The quantity and quality of the safety and efficacy data on traditional medicine are far from sufficient to meet the criteria needed to support its use worldwide. The reasons for the lack of research data are due not only to health-care policies, but also to a lack of adequate or

Chemometric Methods in Capillary Electrophoresis. Edited by Grady Hanrahan and Frank A. Gomez

accepted research methodology for evaluating traditional medicine" (3). According to the Chinese *Pharmacopoeia* (4), there are more than 400 crude drugs used widely. Each of these herbs usually contains hundreds of chemical constituents, but only a few compounds are responsible for the beneficial and/or hazardous effects. Therefore, efficient and selective methods are required for qualitative and quantitative analysis of their bioactive compounds.

The popularity of capillary electrophoresis (CE) continuously increased so that high performance CE instruments are now rapidly available, since the publication of Professor James W. Jorgenson's groundbreaking paper, "Free Zone Electrophoresis in Glass Capillaries" (5). At present, CE represents one of the most attractive analytical techniques for the rapid qualitative and quantitative analysis of molecules with a wide range of polarity and molecular weight, including not only small molecules such as drugs, but also macromolecules such as proteins or nucleic acids. Because of its versatility and high separation efficiency, CE is an interesting alternative to the widely used reverse-phase high performance liquid chromatography (RP-HPLC) (6) and gained much interest for the analysis of herbal extracts, pharmaceutical formulations, or food supplements (7–9). Generally, several chemical (buffer ionic strength or concentration and pH, organic solvents, and additives) and instrumental parameters (separating voltage and temperature) can be manipulated to obtain the optimum CE separation. Traditionally, the optimization is performed by varying one factor at a time, while other parameters are kept unchanged (univariate approach). This approach is the simplest and most commonly used, but it is time-consuming, and importantly, it does not reveal the interactions of all investigated factors (10). The larger the interaction effects, the greater the error will be found (Fig. 10.1). In addition, modern automatic analysis methods provide opportunities to collect large amounts of data very easily. To find the patterns and relationships of these data, multivariate analysis is necessary.

Chemometrics, first coined in 1971, is an interdisciplinary field that involves multivariate statistics, mathematical modeling, computer science, and analyti-

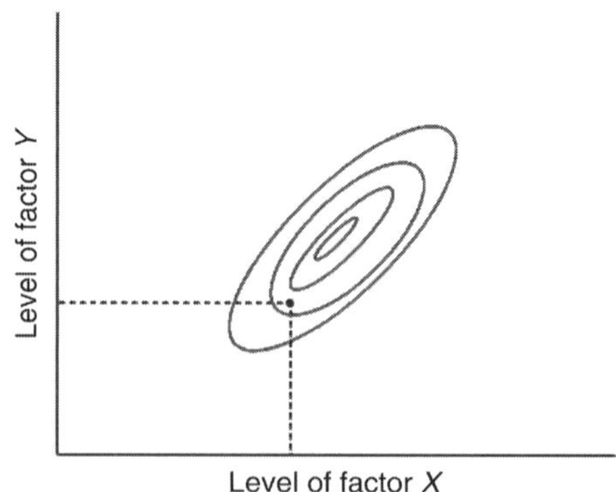

FIGURE 10.1. Simplified contour diagram shows significant X–Y interaction in which the univariate approach fails to locate the maximum.

cal chemistry. Some major application areas of chemometrics include (i) calibration, validation, and significance testing; (ii) optimization of chemical measurements and experimental procedures; and (iii) the extraction of maximum chemical information from analytical data (11). Recently, chemometrics has attracted the interest of analysts, and the application in CE method development and data processing has been reviewed (12–15) and/or reported (16–18). However, there has been no review on the application of chemometrics during CE analysis of herbal medicines. In this chapter, the application of chemometrics in optimization of sample preparation, separation condition, and data handling in CE analysis of herbal medicines will be reviewed and discussed.

10.2. TERMS AND PROCEDURES FOR MULTIVARIATE DESIGN

Selected terms involved in multivariate design are as follows (15):

Experimental domain is the level range of investigated variable, which is defined by the minimum and maximum limits of the experimental variables studied.

Experimental design is a specific set of experiments defined by a matrix composed of the different level combinations of the variables studied.

Factors or independent variables are experimental variables that can be changed independently of each other. Typical independent variables comprise the pH, temperature, reagent concentration, and voltage for CE analysis.

Levels of a variable are different values of a variable at which the experiments must be carried out.

Responses or dependent variables are the measured values of the results from experiments. Typical responses are the analytical signal (absorbance, abundance and potential, etc.), recovery of an analyte, and resolution among peaks for CE analysis.

Response surface methodology (RSM) is considered among the most relevant multivariate techniques used in analytical optimization. RSM consists of a group of mathematical and statistical techniques that are based on the fit of empirical models to the experimental data obtained in relation to experimental design. The procedures in the application of RSM as an optimization technique are as follows (14, 15): (i) determine the overall goals and objectives of the experiment; (ii) define the overall outcome (response) of the experiment; (iii) select independent variables of major effects on the system through screening studies and the delimitation of the experimental region; (iv) choose the experimental design and carry out the experiments according to the selected experimental matrix; (v) treat the obtained experimental data through the fit of a polynomial function using mathematic-statistical approaches; (vi) evaluate the model's fitness; (vii) verify the necessity and possibility of performing a displacement in direction to the optimal region; and (viii) obtain the optimum values for each studied variable.

10.3. OPTIMIZATION OF SAMPLE PREPARATION USING CHEMOMETRIC APPROACHES

CE analysis plays an important role in quality control of herbal medicines, which can be designed to provide qualitative data and quantitative measurement. During the process, sample preparation is one of the key steps that greatly influences the repeatability and accuracy of the analysis. It is reported that 70%–80% of analysis time is spent on sample preparation and more than 60% of analysis error is derived from nonstandard sample pretreatment. Therefore, a proper sample preparation approach is very important for analysis. Generally, extraction of active ingredients from herbal medicines is usually approached by systematic alteration of one variable affecting the recovery while keeping the other variables constant (19–22). However, this method may miss the solution even if the problem is apparently simple because it does not thoroughly explore the space of possible solutions. Thus, experimental design is a strategy that ensures efficient progress toward a solution using a series of small, carefully designed experiments. Actually, experimental design in sample preparation has been already used for optimizing liquid extraction (23), microwave extraction (24), pressurized liquid extraction (25), solid phase extraction (26), and solid phase microextraction (27).

Gotti et al. optimized two major factors, temperature and percentage of ethanol, for ultrasonic extraction of catechins from *Theobroma cacao* beans using a central composite design (CCD) (28). The investigated experimental domain was defined by the temperature of extraction ranging from 49 to 76 °C and by the percentage of ethanol ranging from 29% to 61% according to preliminary experiments. The results showed that a minimization of the response was obtained at the center of the experimental domain; in addition, a low level of percentage of ethanol seemed more suitable for an increase of the response. Finally, among the several possibilities, the optimized conditions were chosen.

Optimization of solid-phase extraction for determination of resveratrol in wines was also performed using artificial neural networks (ANN) in combination with CCD (29). Three factors (volume of sample, flow rate, and volume of methanol) and five levels (0.35–2.45 mL, 0.3–1.2 mL/min, and 0.68–4.92 mL, respectively) each were tested according to a CCD. The data obtained from experimental measurements were used for modeling using ANN. The variables were used as inputs for ANN. As output, the value of efficiency of extraction was used. Back propagation in combination with quick propagation as a training algorithm for multilayer perceptrons was applied for suitable network searching. The optimal structure of the network with three neurons in the hidden layer (3:3:1) was applied for prediction of efficiency with error up to 5%.

In addition, microwave power and radiation time of focused microwave-assisted extraction for the quantitative extraction of cocaine and benzoylecgonine from coca leaves were also optimized using CCD (24).

10.4. OPTIMIZATION OF SEPARATION CONDITIONS USING CHEMOMETRIC APPROACHES

Various chemometrics-based techniques including factorial designs, multivariate experimental design (e.g., RSM), and multivariate sequential optimization methods (e.g., simplex) have been devised to aid in the optimization of CE methods (13, 14, 30). Generally, the main effects and interactions can be statistically evaluated by factorial designs first. The variables that are significant for the separation can be selected and further optimized. Second, when factor interactions are found to be relevant, multivariate experimental designs or multivariate sequential optimization methods should be used for further optimization. To date, few analyses of herbal medicines using chemometrics-aided experimental designs were reported (14, 30), although the approaches have been intensively used for optimization of CE methods.

CCD is one of the most common designs generally used in response surface modeling, which allows for the determination of both linear and quadratic models. Full uniformly routable CCDs present the following characteristics: (i) they require an experiment number according to $N = k^2 + 2k + c_p$, where k is the factor number and (c_p) is the replicate number of the central point; (ii) all factors are studied in five levels ($-\alpha$, -1, 0, $+1$, $+\alpha$); (iii) the α-values depend on the number of variables and can be calculated by $\alpha = 2^{k/4}$. For two, three, and four variables, they are, respectively, 1.41, 1.68, and 2.00. Table 10.1 presents a comparison among the efficiencies of the CCD and other response surface designs for the quadratic model (31).

In order to find the optimum resolution for determination of six main nucleosides (adenine, uracil, adenosine, guanosine, uridine, and inosine) in *Cordyceps* by CE, Gong et al. (32) employed chemometric optimization based on CCD. Initial experiments were run in which the effects of five factors were examined. Three factors (buffer concentration, pH, and proportion of acetonitrile [ACN]) were chosen that displayed the most pronounced effect on the

TABLE 10.1. Comparison of efficiency of central composite design (CCD), Doehlert design (DM), and Box–Behnken design (BBD) (cited from Reference 31 with permission from Elsevier)

Factors (k)	Number of Coefficients (p)	Number of Experiments (f)			Efficiency (p/f)		
		CCD	DM	BBD	CCD	DM	BBD
2	6	9	7	—	0.67	0.86	—
3	10	15	13	13	0.67	0.77	0.77
4	15	25	21	25	0.60	0.71	0.60
5	21	43	31	41	0.49	0.68	0.61
6	28	77	43	61	0.36	0.65	0.46
7	36	143	57	85	0.25	0.63	0.42
8	45	273	73	113	0.16	0.62	0.40

separation expressed as resolution. Finally, a good separation was achieved based on CCD-aided optimization (Fig. 10.2). In addition, CCD was also used for optimization of buffer pH, percentage of ACN, and separation voltage during the determination of 11 nucleosides and nucleobases in *Cordyceps* by capillary electrochromatography (CEC), and resolution (*Rs*) of inosine with guanosine and analytical time (T_R) were considered as responses (33). The results showed that both *Rs* and T_R increased with reduction of the proportion of ACN and voltage. Therefore, the optimum conditions should be chosen

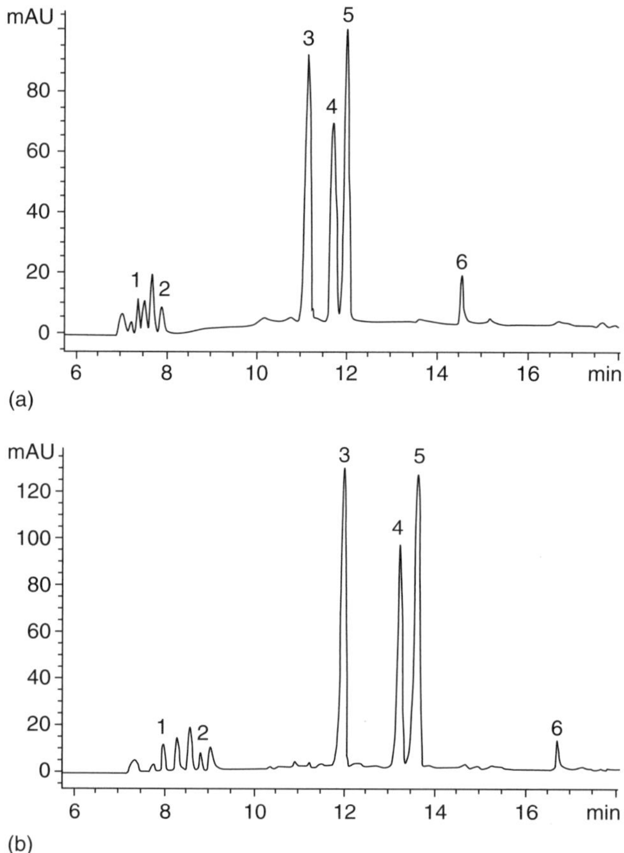

FIGURE 10.2. Electrophoretic profiles of *Cordyceps* before and after optimization. Conditions: pressure injection 50 mbar for 10 s, 56 cm × 75 µm i.d. capillary (48 cm effective length, Agilent fused-silica [Agilent Technologies, Waldbronn, Germany]), voltage 20 kV at temperature 20 °C, detected at 254 nm. (a) Running buffer 0.5 M boric acid-sodium hydroxide (pH 8.6) without acetonitrile as organic modifier. (b) Running buffer 0.5 M boric acid-sodium hydroxide (pH 8.6) with 12.2% acetonitrile as organic modifier.
1 = adenine; 2 = uracil; 3 = adenosine; 4 = guanosine; 5 = uridine; 6 = inosine.
Cited from Reference 32 with permission of Elsevier.

carefully in order to get higher *Rs* (≥1.5) and shorter T_R (≤20 min) because the two responses are incompatible. According to the response model, to obtain *Rs* ≥ 1.5, the conditions should be pH = 5.3, ACN% ≤ 5%, 10 kV ≤ voltage ≤ 22 kV. Similarly, to get $T_R \leq 20$ min, the conditions should be pH = 5.3, ACN% ≥ 3%, voltage ≥ 17 kV. Considering both conditions, in order to get better resolution in shorter analytical time, appropriate conditions were: pH = 5.3, 3% ≤ ACN% ≤ 5%, 17 kV ≤ voltage ≤ 22 kV. Herein, the proportion of ACN and voltage were optimum at 3% and 22 kV, respectively. Under the optimized conditions, baseline separation of 12 analytes (including internal standard [IS]) by CEC could be achieved in 20 min (Fig. 10.3). Indeed, CCD is a powerful tool for optimization of CE conditions. CE analysis of four flavonoids in *Epimedium* was also optimized by CCD (34), which was further confirmed as the optimum CE conditions for analysis of multiple flavonoids in *Epimedium* (35).

The Box–Behnken design (BBD) is a rotatable or nearly rotatable second-order design based on three-level incomplete factorial designs, while Doehlert matrices (DM) or Doehlert design describes a spherical experimental domain that stresses uniformity in space filling. Both have their specific characteristics. DM is considered the most efficient of the three commonly used designs: CCD, BBD, and DM (see Table 10.1). DM is also more efficient in mapping space and has potential for sequential design (36), where experiments can be reused when the boundaries have not been well chosen at first. The applications of BBD and DM in analytical chemistry have been well reviewed, although there are only a few cases for CE analysis of herbal medicines (31, 36).

Recently, ANN have been incorporated, either separately or in combination with the experimental design techniques discussed above, into CE optimization methods (29, 37, 38). ANN, which are computational models based

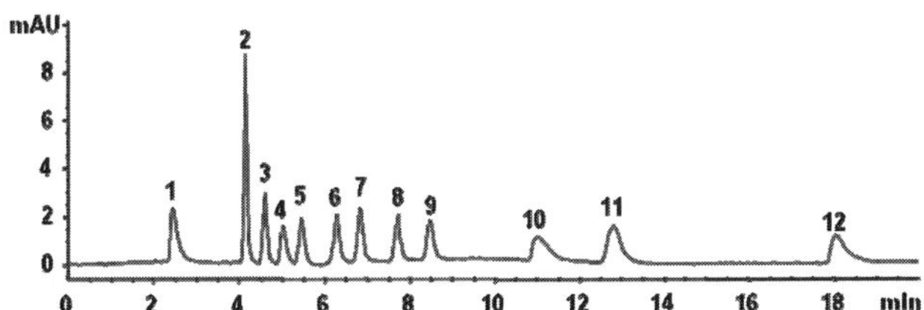

FIGURE 10.3. CEC profiles of 11 investigated compounds and internal standard (IS) after optimization. Conditions: CEC capillary Hypersil C18, 3 μm, 100 μm/25 cm column (Agilent Technologies, Waldbronn, Germany); electrokinetic injection (10 kV for 5 s); running buffer: 4 mM ammonium acetate-acetic acid contained 2 mM triethylamine (TEA) (pH 5.3) with 3% ACN as organic modifier; temperature, 20 °C; voltage, 22 kV. 1 = cytosine; 2 = uracil; 3 = uridine; 4 = hypoxanthine; 5 = 2′-deoxyuridine; 6 = inosine; 7 = guanosine; 8 = IS (5-chlorocytosine arabinoside); 9 = thymidine; 10 = adenine; 11 = adenosine; 12 = cordycepin. Adapted from Reference 33 with permission of Wiley-VCH.

on biological neural networks consisting of an interconnected group of artificial neurons and process information using a connectionist approach to represent the strengths (weights) of the connections (39), offer attractive possibilities for providing nonlinear modeling of response surfaces and optimization in CE analyses.

A combination of experimental design and ANN has been used for the optimization of capillary zone electrophoresis (CZE) separation of *Ginkgo biloba* leaf extract components (40). Generally, the approach has three stages: (i) performing a selected number of experiments using suitable experiments, (ii) a learning stage, where selection of ANN architecture can approximate the data, and (iii) prediction of the optimal experimental conditions under which the highest sensitivity of the determination can be reached (40). For ANN, the background electrolyte (BGE) concentration and the separation voltage were used as input parameters. The number of peaks and the differences in migration times (Δt) for neighboring peaks were the output parameters in the architecture of the neural network. Finally, a 12 kV separation voltage, 36 mM borate buffer as BGE, pH 9.2, at 35 °C, and an injection time of 2 s were chosen as the ANN optimal conditions. Experimental design (two-factor CCD) combined with ANN has also been applied to search for the optimal CE parameters for analysis of polyphenols in *Salvia officinalis* plant extracts (41). Table 10.2 showed some applications of chemometric approaches in CE analysis of herbal medicines.

10.5. DATA PROCESSING USING CHEMOMETRIC APPROACHES

A variety of methods for learning from data by inductive learning methods are being used in chemistry, for example, statistics, pattern recognition methods, ANNs, and genetic algorithms.

10.5.1. Principal Component Analysis (PCA)

The aim of PCA is to reduce the dimensionality of a data set that consists of a large number of interrelated variables, and replace them by new sets called principal components, while retaining as much as possible of the variation present in the original data set. PCA is easily performed using statistical software, such as SPSS, MATLAB, and STATISTICA, but the algorithms for PCA calculations can be found in chemometric-based books (48, 49). Because of its simplicity and versatility, PCA has been widely employed for evaluation of molecular physicochemical parameters (50, 51), quantitative structure–retention relationships (52–54), diagnosis of diseases (55, 56), and identification of food or medicines of different biological and geographical origins (57–62). During CE analysis, PCA is also used for origin authentication (18, 63–65), metabolic fingerprinting (66, 67), and selection of pseudostationary

TABLE 10.2. Selected applications of chemometric experimental design for CE analysis in herbal medicines (2004–2008)

Samples	Analytes	Mode	Optimization Method	Variables	Response	References
Soy capsules	Six isoflavones (glycitein, daidzein, genistein, daidzin, glycitin, genistin)	MEKC	3^2 factorial design	(SDS), MeOH %	A defined response function	(42)
Tobacco	Three acidic herbicides (2,4-D, dicamba, 2,4,5-T)	NACE	3^4 orthogonal design	(NH_4Ac), ACN %, apparent pH, voltage	Resolution, analysis time	(43)
Ginkgo biloba extracts	Fingerprint	CZE	$8^3 \times 4^2$ three-step sequential pseudo-level uniform design	(Borate), pH, MeOH %, temperature, voltage	Local overlap index, number of peaks	(44)
Salvia officinalis	Six polyphenols (epicatechin, catechin, vanillic acid, rosmarinic acid, caffeic acid, galllic acid)	CZE	Two-factor CCD and ANN	(Borate), voltage	Resolution, migration time	(41)
Herba Epimedii	Four flavonoids (icariin, epimedin A, epimedin B, epimedin C)	CZE	Three-factor CCD	(Borate), ACN %, pH	Resolution	(34)
Cordyceps	11 nucleosides and nucleobases (cytosine, uracil, uridine, hypoxanthine, 2′-deoxyuridine, inosine, guanosine, thymidine, adenine, adenosine, cordycepin)	CEC	Three-factor CCD	Voltage, pH, ACN %	Resolution between inosine and guanosine, entire run time	(33)

TABLE 10.2. *Continued*

Samples	Analytes	Mode	Optimization Method	Variables	Response	References
"SHUANGDAN" granule	Protocatechuic aldehyde, paeonol, danshensu, salvianolic acid B	MEKC	Four-factor CCD	(Borate), (phosphate), (SDS), ACN %	Modified chromatographic exponential function, resolution of three mark peaks, mobility time of final peak	(45)
Scutellaria baicalensis	Three flavonoids (baicalin, baicalein, wogonin)	MEKC	Five-factor CCD	(Borate), (phosphate), (SDS), ACN %, 2-propanol %	Resolution of six critical peak pairs	(46)
Cordyceps	Six nucleosides and bases (adenine, uracil, adenosine, guanosine, uridine, inosine)	CZE	Three-factor CCD	(Boric acid), pH, ACN %	Resolution	(32)
Ginkgo biloba extracts	Rutin, catechin, quercetin, epicatechin	CZE	Experimental design and ANN	(Borate), voltage	Number of peaks, difference in migration time for neighboring peaks	(40)
Resvis XR effervescent tablets (Biofutura Pharma, Milan, Italy)	Resveratrol, hesperidin, narirutin, L-ascorbic acid, vitamin B_2, *p*-coumaric acid, caffeic acid, ferulic acid, sinapic acid, flufenamic acid (IS)	CZE	Three-factor CCD	(Borate), ACN %, voltage	Resolution, analysis time	(47)
Commercial tablets of huperzine	(−)-Huperzine A	CZE	Two-factor CCD	(Acetate), voltage	Peak area, peak height, analysis time	(38)

MEKC = micellar electrokinetic chromatography; SDS = sodium dodecyl sulfate; NACE = nonaqueous capillary electrophoresis.

phases (68). The main application of PCA is differentiation of different species and locations for CE analysis of herbal medicines.

Central European *Corydalis* species, namely *Corydalis cava*, *Corydalis intermedia*, *Corydalis pumila*, and *Corydalis solida*, were investigated by nonaqueous CE–electrospray ion trap mass spectrometry. Application of PCA to the complete data set of 39 analytes and 79 samples allowed the identification of eight analytes responsible for lot discrimination. Hierarchical cluster analysis (HCA) also confirmed the findings of the explorative PCA (69). PCA was also applied to 65 *Glycyrrhiza* samples from different species and locations in order to investigate and visualize the chemical relationships to each other, which was performed using STATISTICA software on the basis of CZE peak area data of glycyrrhizin, glabridin, glycyrrhetic acid, liquiritin, and licochalcone A, and *Glycyrrhiza* samples from Europe and China were distinguished. Contribution of first and second principal components obtained, using the peak area data of the five compounds mentioned above, contributed 60.3%, representing variation within the data (70). Reid et al. (71) employed PCA to differentiate opium and poppy straw based on the contents of eight alkaloids. The first three principal components described 93.9% and 97.2% of the total variation for the opium and poppy straw samples, respectively. By applying PCA, opium samples from different locations and poppy straw samples from four plants of the same general genus were clearly differentiated, which were consistent with those established using HCA.

PCA is an efficient chemometric method, but it also has limitations. Above all, only the first few principal components are used, which may not reflect the whole information of the original data set. In addition, PCA does not define the principal components as concrete physical or physicochemical entities. Thus, rational explanation of principal components may be difficult on occasion.

10.5.2. HCA

HCA is one of the most commonly employed pattern recognition methods. The main objective of HCA is to find clusters of observations within a data set. The results are presented in a dendrogram, where the Euclidean distances among samples or variables are transformed into similarity indices. A small distance corresponds to a large index and means a large similarity. HCA has been applied for data analysis in several fields such as medical research (72), family psychology research (73), biogeographical classifications (74), and microarray data clustering (75). HCA has also been used for the discrimination of herbal medicines from different species or the same species of samples from different locations (76–78).

Cordyceps is an expensive traditional Chinese medicine, which is commonly sold in capsule form as a health food product. Because of the price difference, some manufacturers claim their products are derived from cultured *Cordyceps* mycelia from a natural source. In order to distinguish among various types of

TABLE 10.3. Selected applications of chemometrics in data processing for CE analysis of herbal medicines (2004–2008)

Samples	Approaches	Differentiation	References
Corydalis species	PCA and HCA	Species, location	(69)
Radix Glycyrrhizae (licorice)	PCA	Species, location	(70)
Opium and poppy straw	PCA and HCA	Location, species	(71)
Codyceps	HCA and PCA	Species	(32)
	HCA	Species	(79)
Sophora tonkinensis	HCA	Chemical characteristics	(80)

Cordyceps in the market, the profiles of water-soluble constituents derived from different sources of *Cordyceps* were determined by CE. By using the peak characteristics of CE profiles of different *Cordyceps* samples, HCA was performed. The result showed that those samples of natural *Cordyceps* grouped together were distinct from the cultured one (79). Therefore, natural and cultured *Cordyceps* could be distinguished based on their CE profiles of water-soluble constituents. Furthermore, HCA was performed based on 32 peak characteristics from electrophoretic profiles of 12 tested natural and cultured *Cordyceps* samples using a method named as average linkage between groups, and squared Euclidean distance as measurement. The natural and cultured *Cordyceps* were also grouped into two main clusters, cultured *Cordyceps* and natural *Cordyceps*. Among the peaks of electropherograms for cluster analysis of samples, two typical peaks of adenosine and inosine were optimized based on cluster analysis of 32 peaks. Using the peak characteristics of adenosine and inosine, the result of HCA of the 12 tested samples was very similar to the one derived from 32 peak characteristics. Therefore, the characteristics of peaks, especially adenosine and inosine, from electrophoretic profiles of nucleosides could be used as markers for discrimination and quality control of natural and cultured *Cordyceps* (32). HCA could also be used for differentiation of different locations of same species of herbal medicines (Table 10.3).

10.6. CONCLUSION

Advances in herbal medicines have hastened the need for high-throughput CE methods that can effectively screen and resolve numerous compounds in a short period of time. Chemometric experimental design and optimization techniques will continue to increase as new developments in sample preparation, method optimization, and data processing in CE analysis of herbal medicines occur.

ACKNOWLEDGMENTS

We are grateful to Mr. Qian Zheng-ming, Mr. Yang Cheng, Miss Meng Qiong, Miss Lv Guang-ping, Miss Yang Jing, Mr. Hu De-jun, and Mr. Xu Jun for their help on literature searching. The research was supported by grants from the Macao Science and Technology Development Fund (082/2006/A2).

REFERENCES

1. World Health Organization. (1998) *Guidelines for the Appropriate Use of Herbal Medicines*, Manila, p. 6.
2. World Health Organization. Traditional medicine, http://www.who.int/mediacentre/factsheets/fs134/en/ (accessed July 7, 2009).
3. World Health Organization. (2000) *General Guidelines for Methodologies on Research and Evaluation of Traditional Medicines*, Geneva, p. 1.
4. Pharmacopoeia Commission of PRC (ed.) (2005) *Pharmacopoeia of the People's Republic of China*, Vol. **I**., Chemical Industry Press, Beijing.
5. Jorgenson, J.W. and Lukacs, K.D. (1981) *Anal Chem*, **53**, 1298–1302.
6. Guan, J., Chen, X.J. and Li, S.P. (2008) Recent development on analytical techniques for quality control of Chinese herbs, in *Pharmacological Activity Based Quality Control of Chinese Herbs* (eds. S.P. Li and Y.T. Wang), Nova Science Publishers, Inc., New York, pp. 73–113.
7. Ganzera, M. (2008) *Electrophoresis*, **29**, 3489–3503.
8. Suntornsuk, L. (2007) *J Chromatogr Sci*, **45**, 559–577.
9. García-Cañas, V. and Cifuentes, A. (2008) *Electrophoresis*, **29**, 294–309.
10. Ehlen, J.C., Albers, H.E., and Breyer, E.D. (2005) *J Neurosci Methods*, **147**, 36–47.
11. Gemperline, P.J. (2006) *Introduction to chemometrics, in Practical Guide to Chemometrics*, 2nd ed. (ed. P. Gemperline), Taylor & Francis Group, LLC, Boca Raton, FL, p. 2.
12. Altria, K.D., Clark, B.J., Filbey, S.D., Kelly, M.A., and Rudd, D.R. (1995) *Electrophoresis*, **16**, 2143–2148.
13. Siouffi, A.M. and Phan-Tan-Luu, R. (2000) *J Chromatogr A*, **892**, 75–106.
14. Hanrahan, G., Montes, R., and Gomez, F.A. (2008) *Anal Bioanal Chem*, **390**, 169–179.
15. Bezerra, M.A., Santelli, R.E., Oliveira, E.P., Villar, L.S. and Escaleira, L.A. (2008) *Talanta*, **76**, 965–977.
16. Zhao, R., Xu, G., Yue, B., Liebich, H.M., and Zhang, Y. (1998) *J Chromatogr A*, **828**, 489–496.
17. Sentellas, S., Saurina, J., Hernández-Cassou, S., Galceran, M.T., and Puignou, L. (2003) *J Chromatogr Sci*, **41**, 145–150.
18. Yücel, Y. and Demir, C. (2004) *Talanta*, **63**, 451–459.
19. Weiss, D.J., Austria, E.J., Anderton, C.R., Hompesch, R., and Jander, A. (2006) *J Chromatogr A*, **1117**, 103–108.

20. Li, C., Liu, J.X., Zhao, L., Di, D.L., Meng, M., and Jiang, S.X. (2008) *J Pharm Biomed Anal*, **48**, 749–753.
21. Li, Y., He, X., Qi, S., Gao, W., Chen, X., and Hu, Z. (2006) *J Pharm Biomed Anal*, **41**, 400–407.
22. Liu, X., Zhang, J., and Chen, X. (2007) *J Chromatogr B*, **852**, 325–332.
23. Li, W., Nadig, D., Rasmussen, H.T., Patel, K., and Shah, T. (2005) *J Pharm Biomed Anal*, **37**, 493–498.
24. Brachet, A., Christen, P., and Veuthey, J.L. (2002) *Phytochem Anal*, **13**, 162–169.
25. Li, P., Li, S.P., Lao, S.C., Fu, C.M., Kan, K.K.W., and Wang, Y.T. (2006) *J Pharm Biomed Anal*, **40**, 1073–1079.
26. Furlanetto, S., Pinzauti, S., La Porta, E., Chiarugi, A., Mura, P., and Orlandini, S. (1998) *J Pharm Biomed Anal*, **17**, 1015–1028.
27. Lamas, J.P., Salgado-Petinal, C., García-Jares, C., Llompart, M., Cela, R., and Gómez, M. (2004) *J Chromatogr A*, **1046**, 241–247.
28. Gotti, R., Furlanetto, S., Pinzauti, S., and Cavrini, V. (2006) *J Chromatogr A*, **1112**, 345–352.
29. Spanilá, M., Pazourek, J., Farková, M., and Havel, J. (2005) *J Chromatogr A*, **1084**, 180–185.
30. Sentellas, S. and Saurina, J. (2003) *J Sep Sci*, **26**, 875–885.
31. Ferreira, S.L.C., Bruns, R.E., Ferreira, H.S., Matos, G.D., David, J.M., Brandão, G.C., da Siva, E.G.P., Portugal, L.A., dos Reis, P.S., Souza, A.S., and dos Santos, W.N.L. (2007) *Anal Chim Acta*, **587**, 179–186.
32. Gong, Y.X., Li, S.P., Li, P., Liu, J.J., and Wang, Y.T. (2004) *J Chromatogr A*, **1055**, 215–221.
33. Yang, F.Q., Li, S., Li, P., and Wang, Y.T. (2007) *Electrophoresis*, **28**, 1681–1688.
34. Liu, J.J., Li, S.P., and Wang, Y.T. (2006) *J Chromatogr A*, **1103**, 344–349.
35. Chen, X.J., Tu, P.F., Jiang, Y., Wang, Y.T., and Li, S.P. (2009) *J Sep Sci*, **32**, 275–281.
36. Ferreira, S.L.C., dos Santos, W.N.L., Quintella, C.M., Neto, B.B., and Bosque-Sendra, J.M. (2004) *Talanta*, **63**, 1061–1067.
37. Fakhari, A.R., Breadmore, M.C., Macka, M., and Haddad, P.R. (2006) *Anal Chim Acta*, **580**, 188–193.
38. Hameda, A.B., Elosta, S., and Havel, J. (2005) *J Chromatogr A*, **1084**, 7–12.
39. Havel, J., Peña, E.M., Rojas-Hernández, A., Doucet, J.P., and Panaye, A. (1998) *J Chromatogr A*, **793**, 317–329.
40. Elosta, S., Gajdosová, D., and Havel, J. (2006) *J Sep Sci*, **29**, 1174–1179.
41. Ben Hameda, A., Gajdošová, D., and Havel, J. (2006) *J Sep Sci*, **29**, 1188–1192.
42. Micke, G.A., Fujiya, N.M., Tonin, F.G., de Oliveira Costa, A.C., and Tavares, M.F.M. (2006) *J Pharm Biomed Anal*, **41**, 1625–1632.
43. Liu, H., Song, J., Han, P., Li, Y., Zhang, S., Liu, H., and Wu, Y. (2006) *J Sep Sci*, **29**, 1038–1044.
44. Ji, Y.B., Alaerts, G., Xu, C.J., Hu, Y.Z., and Vander Heyden, Y. (2006) *J Chromatogr A*, **1128**, 273–281.

45. Yu, K., Lin, Z., and Cheng, Y. (2006) *Anal Chim Acta*, **562**, 66–72.
46. Yu, K., Gong, Y., Lin, Z., and Cheng, Y. (2007) *J Pharm Biomed Anal*, **43**, 540–548.
47. Orlandini, S., Giannini, I., Pinzauti, S., and Furlanetto, S. (2008) *Talanta*, **74**, 570–577.
48. Jolliffe, I.T. (2002) *Principal Component Analysis*, Springer-Verlag, NewYork.
49. Brereton, R.G. (2003) *Chemometrics: Data Analysis for the Laboratory and Chemical Plant*, John Wiley & Sons Ltd., Chichester.
50. Adamska, K., Voelkel, A., and Héberger, K. (2007) *J Chromatogr A*, **1171**, 90–97.
51. Djaković-Sekulić, T., Smolinski, A., Perisić-Janjić, N., and Janicka, M. (2008) *J Chemometrics*, **22**, 195–202.
52. Bączek, T. (2006) *J Sep Sci*, **29**, 547–554.
53. Vrakas, D., Giaginis, C., and Tsantili-Kakoulidou, A. (2006) *J Chromatogr A*, **1116**, 158–164.
54. Michel, M., Bączek, T., Studzińska, S., Bodzioch, K., Jonsson, T., Kaliszan, R., and Buszewski, B. (2007) *J Chromatogr A*, **1175**, 49–54.
55. Yang, J., Xu, G., Zheng, Y., Kong, H., Pang, T., Lu, S., and Yang, Q. (2004) *J Chromatogr B*, **813**, 59–65.
56. de Oliveira, L.S., de M Rodrigues, F., de Oliveira, F.S., Mesquita, P.R.R., Leal, D.C., Alcântara, A.C., Souza, B.M., Franke, C.R., de P. Pereira, P.A., and de Andrade, J.B. (2008) *J Chromatogr B*, **875**, 392–398.
57. Jiménez, A., Aguilera, M.P., Beltrán, G., and Uceda, M. (2006) *J Chromatogr A*, **1121**, 140–144.
58. Voon, Y.Y., Sheikh Abdul Hamid, N., Rusul, G., Osman, A., and Quek, S.Y. (2007) *Food Chem*, **103**, 1217–1227.
59. Ballabio, D., Skov, T., Leardi, R., and Bro, R. (2008) *J Chemometrics*, **22**, 457–463.
60. Chen, C.Y., Qi, L.W., Li, H.J., Li, P., Yi, L., Ma, H.L., and Tang, D. (2007) *J Sep Sci*, **30**, 3181–3192.
61. Dan, M., Su, M., Gao, X., Zhao, T., Zhao, A., Xie, G., Qiu, Y., Zhou, M., Liu, Z., and Jia, W. (2008) *Phytochemistry*, **69**, 2237–2244.
62. Qiu, Y., Lu, X., Pang, T., Zhu, S., Kong, H., and Xu, G. (2007) *J Pharm Biomed Anal*, **43**, 1721–1727.
63. Bonetti, A., Marotti, I., Catizone, P., Dinelli, G., Maietti, A., Tedeschi, P., and Brandolini, V. (2004) *J Agric Food Chem*, **52**, 4080–4089.
64. Andersen, K.E., Bjergegaard, C., Møller, P., Sørensen, J.C., and Sørensen, H. (2005) *J Agric Food Chem*, **53**, 5809–5817.
65. Burger, F., Dawson, M., Roux, C., Maynard, P., Doble, P., and Kirkbride, P. (2005) *Talanta*, **67**, 368–376.
66. Vallejo, M., Angulo, S., García-Martínez, D., García, A., and Barbas, C. (2008) *J Chromatogr A*, **1187**, 267–274.
67. García-Pérez, I., Whitfield, P., Bartlett, A., Angulo, S., Legido-Quigley, C., Hanna-Brown, M., and Barbas, C. (2008) *Electrophoresis*, **29**, 3201–3206.

68. Fuguet, E., Ràfols, C., Bosch, E., Abraham, M.H., and Rosés, M. (2006) *Electrophoresis*, **27**, 1900–1914.
69. Sturm, S., Seger, C., and Stuppner, H. (2007) *J Chromatogr A*, **1159**, 42–50.
70. Rauchensteiner, F., Matsumura, Y., Yamamoto, Y., Yamaji, S., and Tani, T. (2005) *J Pharm Biomed Anal*, **38**, 594–600.
71. Reid, R.G., Durham, D.G., Boyle, S.P., Low, A.S., andWangboonskul, J. (2007) *Anal Chim Acta*, **605**, 20–27.
72. McLachlan, G.J. (1992) *Stat Methods Med Res*, **1**, 27–48.
73. Henry, D.B., Tolan, P.H., and Gorman-Smith, D. (2005) *J Fam Psychol*, **19**, 121–132.
74. Kafanov, A.I., Borisovets, E.E., and Volvenko, I.V. (2004) *Zh Obshch Biol*, **65**, 250–265.
75. Gollub, J. and Sherlock, G. (2006) *Methods Enzymol*, **411**, 194–213.
76. Yang, F.Q., Li, S.P., Chen, Y., Lao, S.C., Wang, Y.T., Dong, T.T.X., and Tsim, K.W.K. (2005) *J Pharm Biomed Anal*, **39**, 552–558.
77. Qin, N.Y., Yang, F.Q., Wang, Y.T., and Li, S.P. (2007) *J Pharm Biomed Anal*, **43**, 486–492.
78. Chen, X.J., Guo, B.L., Li, S.P., Zhang, Q.W., Tu, P.F., and Wang, Y.T. (2007) *J Chromatogr A*, **1163**, 96–104.
79. Li, S.P., Song, Z.H., Dong, T.T.X., Ji, Z.N., Lo, C.K., Zhu, S.Q., and Tsim, K.W.K. (2004) *Phytomedicine*, **11**, 684–690.
80. Ding, P.L., Yu, Y.Q., and Chen, D.F. (2005) *Phytochem Anal*, **16**, 257–263.

CHAPTER 11

CLINICAL PATTERN RECOGNITION ANALYSIS APPLYING ARTIFICIAL NEURAL NETWORKS BASED ON PRINCIPAL COMPONENT ANALYSIS INPUT SELECTION

YAXIONG ZHANG[1] and HUA LI[2]
[1]School of Chemistry and Material Science, Shan'xi Normal University, Linfen, China
[2]School of Chemistry and Material Science, Northwest University, Xi'an, China

CONTENTS

11.1. INTRODUCTION

Nucleosides in human urine are often used as biomedical markers for cancer diagnosis and therapy (1–3). It has been studied that nucleosides are excreted

Chemometric Methods in Capillary Electrophoresis. Edited by Grady Hanrahan and Frank A. Gomez

abnormally in the urine of cancer patients (4, 5). Recently, urinary nucleosides have been applied as biochemical markers in the clinical studies of different kinds of cancers (6–11). In the clinical studies of urinary nucleosides, reversed phase-high performance liquid chromatography (5, 6) and immunoassays (12, 13) have been applied as the main analytical techniques. Moreover, capillary electrophoresis (CE) methods have also been proven to be successful in the analysis of nucleosides in clinical urinary samples from healthy persons and cancer patients (7, 10, 11, 8, 14).

To establish a correlation between the concentrations of different kinds of nucleosides in a complex metabolic system and normal or abnormal states of human bodies, computer-aided pattern recognition methods are required (15, 16). Different kinds of pattern recognition methods based on multivariate data analysis such as principal component analysis (PCA) (8), partial least squares (16), stepwise discriminant analysis, and canonical discriminant analysis (10, 11) have been reported. Linear discriminant analysis (17, 18) and cluster analysis were also investigated (19, 20). Artificial neural network (ANN) is a branch of chemometrics that resolves regression or classification problems. The applications of ANN in separation science and chemistry have been reported widely (21–23). For pattern recognition analysis in clinical study, ANN was also proven to be a promising method (8).

The purpose of this study was to employ multilayer perceptron (MLP) ANN based on PCA input selection to perform the pattern recognition analysis of urinary nucleosides as tumor markers. Although ANN based on PCA input selection has been applied for quantification in different analytical methods (24–27), MLP ANN based on PCA input selection applied in pattern recognition analysis for clinical CE data has not yet been reported. As a result, a PCA input selection strategy was employed to MLP ANN for pattern recognition analysis of clinical CE data in this chapter. For the first data set, binary values were used to represent the two groups of samples: "1" for healthy people and "–1" for thyroid cancer patients. For the second data set, "1" and "–1" represent the normal and uterine cervical cancer samples, respectively. Samples from uterine myoma patients were indicated by "0." According to the results of this study, when the input selection strategy based on PCA was applied to MLP ANN, the accuracy rate of pattern recognition analysis for the two data sets was improved to some extent, even with much simpler structures of MLP ANN. In addition, the same accuracy rate can be acquired even by simplified structures of MLP ANN. It was proven that MLP ANN based on PCA input selection was a promising approach for pattern recognition analysis in this work.

11.2. THEORY

11.2.1. ANNs

ANN is a kind of information processing chemometrical technique. It simulates some properties of human brain, and is often applied in the field of regression or classification. The theory of ANN has been described thoroughly in several papers (28–30). Although different training algorithms of MLP ANN have been developed, conjugate gradient descent (CGD) algorithm (31) is one of the most widely used. In this chapter, MLP ANN based on a CGD algorithm was applied to perform clinical pattern recognition analysis. The theory of such an approach is briefly given here. MLP ANN is composed of some logic units and connection weights between the units. MLP ANN is divided into three levels in order to understand the process of information processing. These include the input layer, hidden layer, and output layer, with and each consisting of logic units. The logic units are the basic information-processing unit in MLP ANN. Linear postsynaptic potential (PSP) function and logistic activation function were applied in MLP ANN in this chapter. The sum-squared error function monitoring the training process of MLP ANN was used.

The initial search direction of CGD is given by:

$$d_0 = g_0 \quad \text{(Eq. 11.1)}$$

Subsequently, the search direction is updated using the Polak–Rebiere formula (32):

$$d_{j+1} = g_{j+1} + \beta_j d_j \quad \text{(Eq. 11.2)}$$

$$\beta_j = \frac{g_{j+1}^T \left(g_{j+1} - g_j\right)}{g_j^T g_j} \quad \text{(Eq. 11.3)}$$

11.2.2. PCA

PCA is a statistic technique to extract information from multivariate data sets. To do this, the linear combinations of original variables are constructed, which are termed principal components (PCs). The greatest amount of variability of the original multivariate data set is represented by the first component, and the second component explains the maximum variances of the residual data set. Then, the third one will describe the most important variability of the next residual data set, and so on. According to the theory of least squares, the eigenvectors of all PCs are orthogonal each other in multidimensional data space. Generally speaking, only p PCs are enough to account for the most variance in an m-dimensional data set, where p is the number of important PCs of the data set, and m is the number of all the PCs in the data set. It is obvious that p is less than m. Given this information, PCA is generally regarded as a data reduction technique. That is to say, a multidimensional data set can be projected to a lower dimension data space without the loss of information from the original data set. The work of Statheropoulos et al. (33) and Dong and McAvoy (34) described the algorithm of PCA in greater detail.

The selection of input variables to ANN is necessary to avoid "overfitting" (35) in terms of the multiple input parameters offered. As a linear technique for dimensionality reduction, PCA can transform the input data set from its original form (points in m-dimensional space) to its new form (points in p-dimensional space), where p is less than m. During this process, most of the variability of the original input data set is retained. Using the corrected input data set in a lower dimension, a smaller MLP ANN is applied in the performance of pattern recognition analysis. Since PCA is a linear technique, we achieved transformation of input data set by linear ANN with the same number of input and output nodes.

11.2.3. PCA Input Selection Strategy

In this work, input variables (the corresponding concentrations of urinary nucleosides) were employed in different MLP ANN to perform clinical pattern recognition analysis. For this method, the problem of data analysis may be introduced. If the number of weights exceeds the number of samples for the training of ANN to some extent, "overfitting" may result (35). In the case of a high number of input variables, irrelevant, redundant, and noisy variables might be included in the data set, whereas meaningful variables could likely be hidden (36). For a high number of input variables, the probability of chance correlation increases (37). Moreover, a high number of input variables may prevent ANN from finding optimized models (38). Therefore, PCA input selection is necessary in order to improve the precision of pattern recognition analysis with different MLP ANN.

In this chapter, PCA was performed based on linear ANN. After the performance of PCA preprocessing procedure for the input variables, all the PCs of a training data set can be acquired. The eigenvalues of the corresponding PCs were also given in descending order. The PCs with larger eigenvalues represent the more relative amount of variability of the training data set.

Next, the PCs were applied to the corresponding MLP ANN in sequence, that is, the largest PC was first applied as the input variable of the corresponding MLP ANN, and then the subsequent ones were employed as MLP ANN input data set. The processes continued until all the PCs that represented nearly all the variability of the training data set were included in the input data set of the corresponding MLP ANN. The architecture of the corresponding MLP ANN was experimentally determined by Trajan Automatic Network Designer based on simulated annealing algorithm (39) and CGD approach (31). The structures of the corresponding MLP ANN giving the best pattern recognition results were adopted to perform cluster analysis.

11.3. EXPERIMENTAL

11.3.1. Data

In this work, the first group of clinical data was from Reference 11, and the second one was cited from Reference 10. Both of the two data sets were acquired from clinical urinary sample analysis by a CE method. The concentrations of selected nucleosides not detected by the CE analytical method in Reference 11 were regarded as zero.

11.3.2. Software and Data Analysis

All MLP ANN calculations and the performance of PCA input selections were carried out using Trajan software version 3.0 (Durham, UK) on a Lenovo Pentium IV personal computer.

11.4. RESULTS AND DISCUSSION

11.4.1. Pattern Recognition Analysis for the First Data Set

In this data set, 24 urinary samples were investigated, of which 12 samples were from healthy women and the remaining belonging to female thyroid cancer patients. The healthy samples were indicated by "1," and the malignant tumor groups represented by "–1." Therefore, the calculated values larger than or equal to 0.5 were regarded as healthy samples and those smaller than or equal to –0.5 were treated as cancer samples. If the values were smaller than 0.5 but larger than –0.5, the classifications of the corresponding samples were uncertain. Fourteen varieties of nucleosides were applied to describe each sample. The concentrations of each variety in every sample were quantified by the mentioned CE method. Concentrations of the 14 varieties of nucleosides for each sample were used as input variables to the corresponding MLP ANN. Five samples were randomly selected and used as the verification set, while the others were used as training samples. Hence, the training process of the corresponding MLP ANN could be monitored and controlled. Moreover, the Trajan software performing the calculations in this work was able to search for the best iterative times automatically. Therefore, "overtraining" of the corresponding MLP ANN was conveniently avoided. After 1000 iteration times with a unit penalty 0.01, a 14:1:1 MLP ANN was generated. The automatically designed network was expected to possibly give 100% success recognition rate. Moreover, the initial weights of MLP ANN were set randomly. Therefore, different runs of MLP ANN often result in different calculated results.

In this study, the automatic network designer was utilized for 10 parallel runs incorporating the same performance parameters as above. All of the 10 parallel performances proposed a 14:1:1 architecture. Each of the MLP ANN named "modeling network" in this work performed pattern recognition analysis with a 100% accuracy rate. In order to confirm the pattern recognition ability and the robustness of the proposed MLP ANN model, leave-one-out cross validation (40) was also carried out (i.e., the sample to be classified was deleted from the data set for the training of MLP ANN). The MLP ANNs employed to perform the classification were also designed automatically in 1000 iteration times with a unit penalty 0.01. All the samples in this data set were classified correctly.

In order to perform pattern recognition analysis using a simpler architecture of MLP ANN, PCA input selection was introduced to the data set. According to the PCA procedure, three important PCs accounting for 96.21% of the total variability in the original data set can give 100% classification success rate applying the automatically designed MLP ANN (modeling network) in 1000 iteration times with a unit penalty 0.01. Furthermore, all clinical samples can be classified correctly in 10 parallel runs of the automatically designed MLP ANN by the same performance parameters as above. It was shown that the proposed MLP ANN model was robust and appropriate for the classification of the given data set. However, although applying the three important PCs in the automatically designed MLP ANN can classify the corresponding samples correctly, not all samples can be assigned to their own classifications correctly in the leave-one-out cross-validation strategy. When the five important PCs were applied to design the corresponding networks, 100% success classification rate can be acquired for the samples in the data set and those to be classified in the leave-one-out cross-validation strategy. The reproducibility of the pattern recognition results of the modeling MLP ANN is given in Table 11.1. The PCA input selection process for the modeling network and the leave-one-out cross validation is shown in Figure 11.1a,b, respectively. According to this study, 100% success classification rate can be achieved even using much simpler MLP ANN models.

11.4.2. Pattern Recognition Analysis for the Second Data Set

The second data set investigated in this work consisted of 28 samples. Among them, 10 were from healthy women, eight samples were collected from uterine myoma patients, and the remaining from uterine cervical cancer patients. Five randomly selected samples in the data set were applied as the verification set. In this section, "–1" represented the cancer samples, "0" represented uterine

TABLE 11.1. Reproducibility of the classification results of the modeling MLPANN for the first data set

	Number of Input Variables	Relative Standard Deviation (RSD_{n-1}) (%)
One PC	1	2.89
Two PCs	2	2.23
Three PCs	**3**	**0.00**[a]
Four PCs	4	0.00
Five PCs	5	0.00
Original input variables	14	0.00

[a]The best classification results of the modeling MLP ANN.

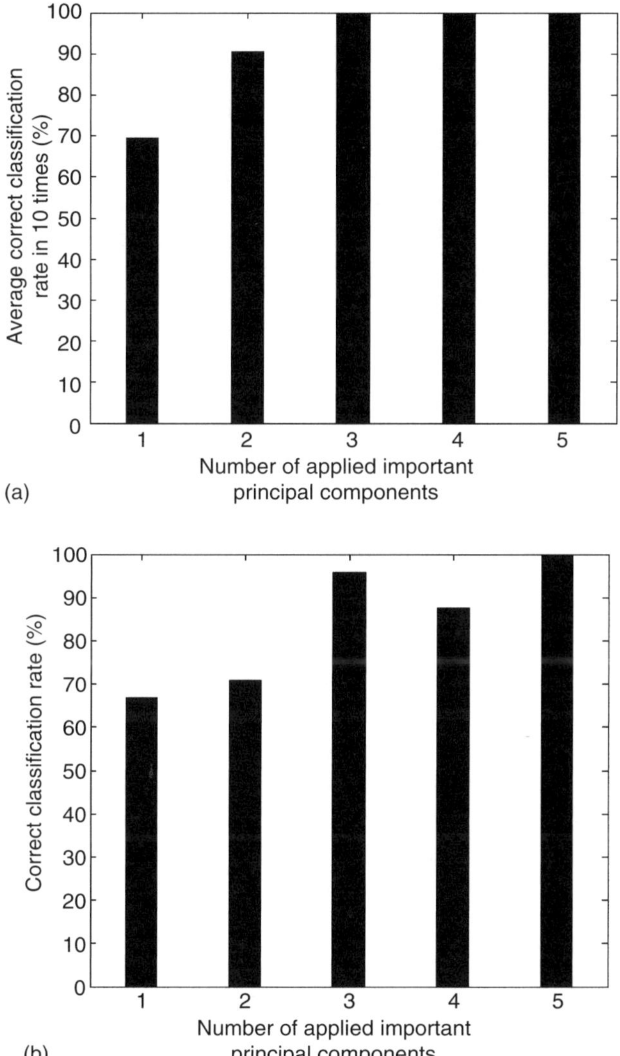

FIGURE 11.1. (a) PCA input selection process for modeling MLP ANN in pattern recognition analysis for the first data set. (b) PCA input selection process for leave-one-out cross validation in pattern recognition analysis for the first data set.

myoma samples, and "1" indicated the healthy samples. If the calculated results of the corresponding MLP ANN for the samples were between –0.5 and 0.5 (larger than –0.5 but smaller than 0.5), the samples were regarded as belonging to uterine myoma patients. Calculated results by MLP ANN for the corresponding samples larger than or equal to 0.5 were considered to be from healthy samples, and those smaller than or equal to –0.5 were judged to be the symbol of uterine cervical cancer samples. Pattern recognition analysis of this data set was divided into four cases.

11.4.2.1. Case I. In this case, the three kinds of clinical samples from 28 women were applied to construct the training data set for the corresponding MLP ANN. From the entire data set, five samples were selected randomly and served as verification samples, with the remaining acting as training samples. For each sample, 14 varieties of nucleosides were employed for pattern recognition analysis. However, the concentrations of two nucleosides varieties (N^2-methylguanosine and N^2,N^2-dimethylguanosine) were summed up for their incomplete resolution under the CE separation conditions described in Reference 10. The automatically designed MLP ANN had 13 input units and performed in 1000 iteration times with a unit penalty 0.01. Pattern recognition analysis of the automatically designed MLP ANN in 10 parallel runs was also performed. The average correct rate of the classification for all the samples was 78.17% with an RSD_{n-1} (relative standard deviation) of 10.07%. In order to investigate the classification ability of the proposed ANN models, the leave-one-out cross-validation procedure was also carried out. The correct classification rate for all the samples was 42.86%. In order to improve the accuracy rate of the pattern recognition analysis, PCA input selection was also investigated. After the PCA input selection, it was shown that when 11 important PCs were applied to design the MLP ANN (1000 iteration times with a unit penalty 0.01), the results of the pattern recognition analysis by the corresponding ANN model were better than those obtained from the neural networks using other numbers of important PCs as input variables.

The input selection process for the modeling network in this section is given in Figure 11.2a. When 11 important PCs were applied in corresponding MLP ANN, 100% success classification rate in 10 parallel runs of MLP ANN was achieved for all the samples. Obviously, after using the PCA input selection strategy, pattern recognition results were improved to some extent with simpler architecture of MLP ANN. The leave-one-out cross-validation approach was also investigated. When six important PCs were used to construct the corresponding MLP ANN, the correct classification rate for all the samples was 67.86%. The input selection process is given in Figure 11.2b. Comparing the leave-one-out cross-validation classification results from MLP ANN of original input variables, it was shown that the PCA input selection strategy can improve the success classification rate even if a much simpler structure of MLP ANN was applied. The reproducibility of the corresponding classification results of the modeling MLP ANN in this section is listed in Table 11.2.

11.4.2.2. Case II. From the calculated results in Case I, it can be seen that the classification ability of the proposed MLP ANN model was poor in the case of the leave-one-out cross-validation procedure. In order to improve the success classification rate, only two kinds of samples were included in the data

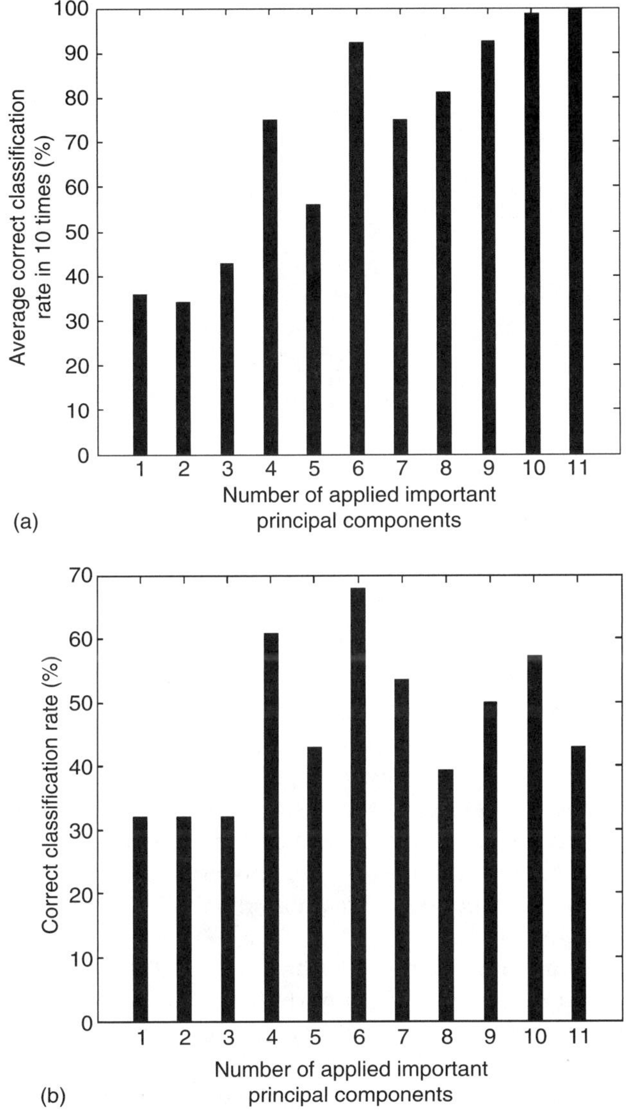

FIGURE 11.2. (a) PCA input selection process for modeling MLP ANN in pattern recognition analysis for the second data set Case I. (b) PCA input selection process for leave-one-out cross validation in pattern recognition analysis for the second data set (Case I).

set (samples from the healthy women and the uterine cervical cancer patients). Each of the two kinds of samples consisted of 10 samples. The healthy samples were also indicated by "1," and those from the uterine cervical cancer patients were symbolized by "–1." The corresponding MLP ANN in this section was

TABLE 11.2. Reproducibility of the classification results of the modeling MLPANN for the second data set in Case I

	Number of Input Variables	Relative Standard Deviation (RSD_{n-1}) (%)
One PC	1	0.00
Two PCs	2	5.55
Three PCs	3	5.56
Four PCs	4	19.5
Five PCs	5	15.62
Six PCs	6	6.00
Seven PCs	7	36.55
Eight PCs	8	5.51
Nine PCs	9	7.15
Ten PCs	10	2.53
Eleven PCs	**11**	**0.00**[a]
Original input variables	13	10.07

[a]The best classification results of the modeling MLP ANN.

designed automatically in 1000 iteration times with a unit penalty 0.01. In this section, three samples selected randomly were included in the verification set. The others were training samples. If the original input variables were employed directly to construct the corresponding MLP ANN, all the samples can be classified correctly in 10 parallel runs of the automatically designed networks. However, for the leave-one-out cross-validation strategy, the success classification rate was only 65.00%. A PCA input selection method was also investigated to improve the results of the pattern recognition analysis in this data set. The 100% success classification results for modeling MLP ANN can be acquired when seven important PCs were applied to design the corresponding MLP ANN. The input selection process based on PCA for modeling networks is described in Figure 11.3a.

For the leave-one-out cross-validation process, 90.00% success classification rate was acquired when five important PCs were applied to design the corresponding MLP ANN automatically. The selection process for the input variables to corresponding ANN applied in leave-one-out cross validation in this section is shown in Figure 11.3b. It can be seen from the figure that the first five important PCs employed as the input variables to the automatically designed MLP ANN could give the best classification results. According to the classification results acquired in this section and those from Case I, two conclusions can be drawn. First, the corresponding MLP ANN can give much better classification results for the two kinds of samples included in the data set than those for the three kinds of samples comprised in the training data set. Second, the proposed PCA input selection strategy can improve the classification results to some extent even when using a simpler architecture of MLP ANN. The reproducibility of the classification results of the modeling MLP ANN in this case is shown in Table 11.3.

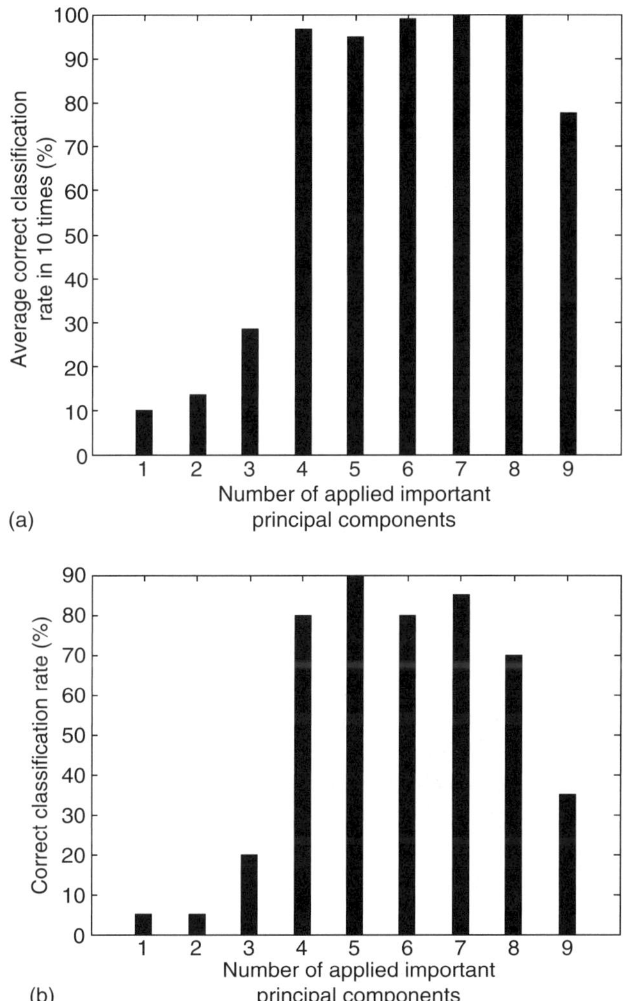

FIGURE 11.3. (a) PCA input selection process for modeling MLP ANN in pattern recognition analysis for the second data set Case II. (b) PCA input selection process for leave-one-out cross validation in pattern recognition analysis for the second data set (Case II).

11.4.2.3. Case III. The data set including the samples from the healthy women and the uterine myoma patients was also studied. The healthy samples were represented by "1," and the uterine myoma samples were denoted by "0." Of all the 18 samples, three of them selected randomly were used as verification set, and the others were training samples. When the 13 original input variables (the concentrations of the 14 kinds of nucleosides in each sample) were adopted to design the corresponding MLP ANN in 1000 iteration times with unit penalty 0.01, all the samples can be classified correctly in the "model-

TABLE 11.3. Reproducibility of the classification results of the modeling MLPANN for the second data set in Case II

	Number of Input Variables	Relative Standard Deviation (RSD_{n-1}) (%)
One PC	1	0.00
Two PCs	2	17.89
Three PCs	3	16.64
Four PCs	4	2.50
Five PCs	5	0.00
Six PCs	6	2.13
Seven PCs	**7**	**0.00**[a]
Eight PCs	8	0.00
Nine PCs	9	20.23
Original input variables	13	0.00

[a]The best classification results of the modeling MLP ANN.

ing" neural networks in 10 runs of the parallel-designed neural networks. Moreover, each sample can also be classified correctly in the leave-one-out cross-validation strategy. The purpose of the study in this section was to perform pattern recognition analysis by a much simpler structure of MLP ANN without any deterioration in the success rate of classification. The original input variables were also projected to a lower dimension data space by PCA input selection strategy. For the "modeling" neural networks, the automatically designed MLP ANN with the four important PCs as input variables classified all the samples to their proper categories.

The MLP ANN suggested by 10 times of parallel running of the network design process could all give the correct classification for all the experimental samples. In the leave-one-out cross-validation procedure, each experimental sample can also be classified correctly when MLP ANN applied in eight important PCs were employed in the pattern recognition analysis. The input selection process for modeling networks is given in Figure 11.4a, and that for the leave-one-out cross validation is shown in Figure 11.4b. According to the classification results in this section, 100% success classification rate can also be acquired by much simpler structure of MLP ANN. The reproducibility of the pattern recognition analysis results of the modeling MLP ANN in this section is given in Table 11.4.

11.4.2.4. Case IV. The pattern recognition analysis for the clinical data from the uterine myoma and the uterine cervical cancer patients was also performed. In this data set, "0" represented uterine myoma samples, and those of uterine cervical cancer patients were indicated by "–1." Eighteen samples were applied for the design of the corresponding MLP ANN. Of all the samples, four of them selected randomly were used as verification set, and the others were training set. The pattern recognition analysis was also performed by MLP ANN method. Without the input selection procedure based on PCA, the mod-

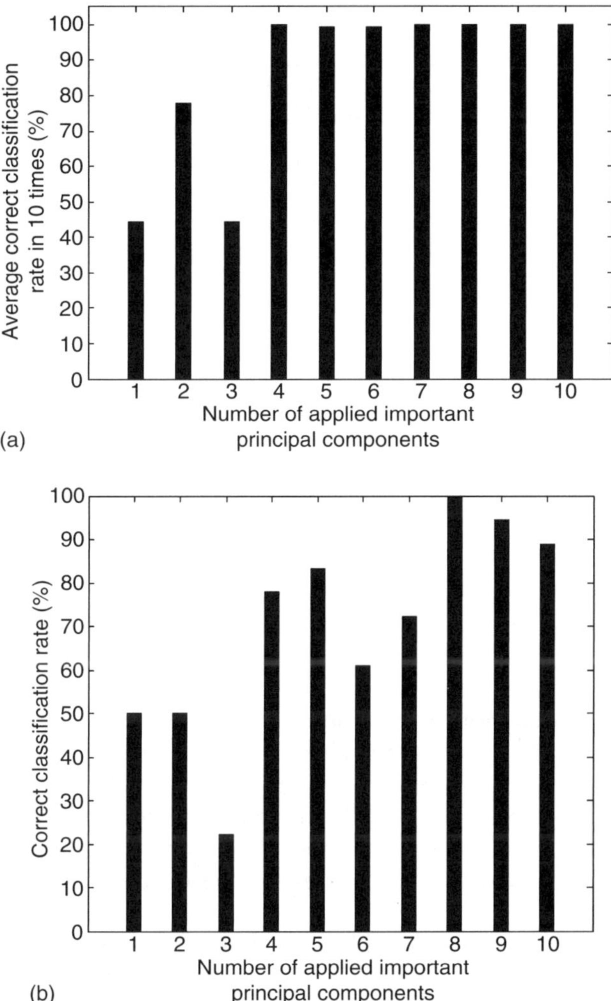

FIGURE 11.4. (a) PCA input selection process for modeling MLP ANN in pattern recognition analysis for the second data set Case III. (b) PCA input selection process for leave-one-out cross validation in pattern recognition analysis for the second data set (Case III).

eling MLP ANN automatically designed in 1000 iteration times with unit penalty 0.01 can give 98.332% average success classification rate in 10 parallel runs with RSD_{n-1} 2.7313%. However, only 50.00% of all the samples were classified correctly in the leave-one-out cross-validation strategy.

In order to acquire better classification results, input selection procedure based on PCA was also carried out. For the modeling neural networks, 94.44% success classification rate can be obtained when five or six important PCs were

TABLE 11.4. Reproducibility of the classification results of the modeling MLPANN for the second data set in Case III

	Number of Input Variables	Relative Standard Deviation (RSD_{n-1}) (%)
One PC	1	0.00
Two PCs	2	0.00
Three PCs	3	0.00
Four PCs	**4**	**0.00**[a]
Five PCs	5	1.77
Six PCs	6	1.77
Seven PCs	7	0.00
Eight PCs	8	0.00
Nine PCs	9	0.00
Ten PCs	10	0.00
Original input variables	13	0.00

[a]The best classification results of the modeling MLP ANN.

applied to design the corresponding MLP ANN. Moreover, each run of the corresponding neural networks can give the same success classification rate. Comparing the classification results of the modeling neural networks with original input variables, the classification ability of the proposed modeling neural networks was more robust despite a little deterioration of its success classification rate. The process to select the input variables for the design of the corresponding MLP ANN is given in Figure 11.5a. The classification ability of the MLP ANN based on PCA input selection for unknown samples was also investigated by leave-one-out cross-validation strategy. When six important PCs were included in the data set for the design of the corresponding MLP ANN, 83.33% success classification rate can be acquired for all the samples. It is shown that the classification ability of the MLP ANN model was also improved to some extent even with a simpler architecture of neural networks by PCA input selection strategy. The input selection process for the neural networks is shown in Figure 11.5b. The reproducibility of the classification results of the corresponding modeling MLP ANN in this section is listed in Table 11.5.

11.5. CONCLUDING REMARKS

The proposed MLP ANN method based on PCA input selection procedure was suitable for the pattern recognition analysis of the clinical urine samples relating to female tumor patients. In the group of the clinical data from Reference 11, when the PCA input selection was introduced to the MLP ANN for pattern recognition analysis, the 100% success classification rate can also be acquired in both the modeling MLP ANN and the leave-one-out cross-

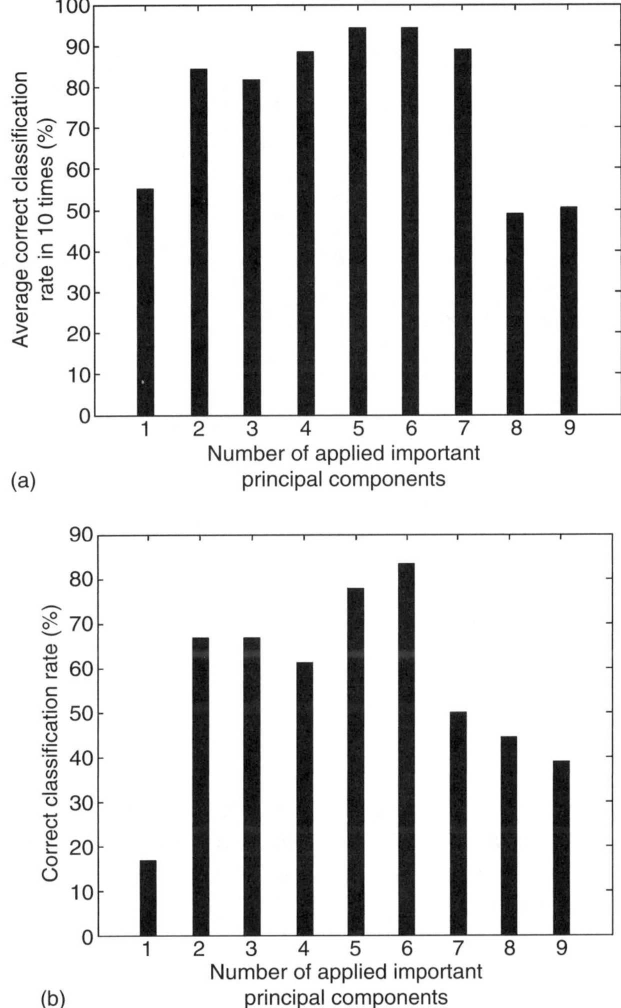

FIGURE 11.5. (a) PCA input selection process for modeling MLP ANN in pattern recognition analysis for the second data set (Case IV). (b) PCA input selection process for leave-one-out cross validation in pattern recognition analysis for the second data set (Case IV).

validation procedure even employing much simpler structures of neural networks. In the second group of the clinical urine samples collected from uterine tumor patients and healthy women cited from Reference 10, the PCA input selection strategy could also be applied in the corresponding MLP ANN to improve the results of pattern recognition analysis by simplified structure of networks in this work. Moreover, the results of the pattern recognition analysis in this study also suggested that the classification results for two kinds of

TABLE 11.5. Reproducibility of the classification results of the modeling MLPANN for the second data set in Case IV

	Number of Input Variables	Relative Standard Deviation (RSD_{n-1}) (%)
One PC	1	3.20
Two PCs	2	4.16
Three PCs	3	4.39
Four PCs	4	6.92
Five PCs	**5**	**0.00**[a]
Six PCs	6	0.00
Seven PCs	7	0.00
Eight PCs	8	15.90
Nine PCs	9	15.93
Original input variables	13	2.73

[a]The best classification results of the modeling MLP ANN.

samples were much better than those for three kinds of samples, that is, MLP ANN method is more suitable for the pattern recognition analysis in dual pattern system. The present study also indicated that the proposed MLP ANN method based on PCA input selection might be a useful clinical tool for the diagnosis or therapeutic monitoring of corresponding tumors.

ACKNOWLEDGMENTS

The authors gratefully acknowledge financial support from the National Natural Science Foundation of China (No. 20675063), the scientific research fund from Shanxi Normal University, China (No. YZ06004), and the Education Department of Shanxi Province, China (No. 2007017).

REFERENCES

1. Apffel, A., Chakel, J.A., Fisher, S., Lichtenwalter, K., and Hancock, W.S. (1997) *Anal Chem*, **69**, 1320–1325.
2. Cohen, A.S., Terabe, S., Smith, J.A., and Karger, B.L. (1987) *Anal Chem*, **59**, 1021–1027.
3. Gaus, H.J., Owens, S.R., Winniman, M., Cooper, S., and Cummins, L.L. (1997) *Anal Chem*, **69**, 313–319.
4. Waalkes, T.P., Abeloff, M.D., Ettinger, D.S., Woo, K.B., Gehrke, C.W., Kuo, K.C., and Borek, E. (1982) *Cancer*, **50**, 2457–2464.
5. Trewyn, R.W., Glaser, R., Kelly, D.R., Jakoson, D.G., Graham,W.P., and Speicher, C.E. (1982) *Cancer*, **49**, 2513–2517.
6. Liebich, H.M., Di Stefano, C., Wixforth, A., and Schmid, H.R. (1997) *J Chromatogr A*, **763**, 193–197.

7. Liebich, H.M., Xu, G., Di Stefano, C., and Lehmann, R.J. (1998) *J Chromatogr A*, **793**, 341–347.
8. Zhao, R., Xu, G., Yue, B., Liebich, H.M., and Zhang, Y. (1998) *J Chromatogr A*, **828**, 489–496.
9. Xu, G., Di Stefano, C., Liebich, H.M., Zhang, Y., and Lu, P. (1999) *J Chromatogr B*, **732**, 307–313.
10. Kim, K.R., La, S., Kim, A., Kim, J.H., and Liebich, H.M. (2001) *J Chromatogr B*, **754**, 97–106.
11. La, S., Cho, J.H., Kim, J.H., and Kim, K.R. (2003) *Anal Chim Acta*, **486**, 171–182.
12. Masuda, M., Nishihira, T., Itoh, K., Mizugak, M., Ishida, N., and Mori, S. (1993) *Cancer*, **72**, 3571–3578.
13. Reynaud, C., Bruno, C., Boullanger, P., Grange, J., Barbesti, S., and Niveleau, A. (1992) *Cancer Lett*, **61**, 255–262.
14. Liebich, H.M., Xu, G., Di Stefano, C., Lehmann, R., Hãring, H.U., Lu, P., and Zhang, Y. (1997) *Chromatographia*, **45**, 396–401.
15. Rhodes, G., Miller, M., McConnell, M.L., and Novotny, M. (1981) *Clin Chem*, **27**, 580–585.
16. Jellum, E., Harboe, M., Bjune, G., and Wold, S. (1991) *J Pharm Biomed Anal*, **9**, 663–669.
17. Chan, K., Lee, T.W., Sample, P.A., Goldbaum, M.H., Weinreb, R.N., and Sejnowski, T.J. (2002) *IEEE Trans Biomed Eng*, **49**, 963–974.
18. Seltzer, S.E., Getty, D.J., Pickett, R.M., Swets, J.A., Sica, G., Brown, J., Saini, S., Mattrey, R.F., Harmon, B., Francis, I.R., Chezmar, J., Schnall, M.O., Siegelman, E.S., Ballerini, R., and Bhat, S. (2002) *Acad Radial*, **9**, 256–269.
19. Marshall, R.J., Turner, R., Yu, H., and Cooper, E.H. (1984) *J Chromatogr A*, **297**, 235–244.
20. Birkenkamp-Demtroder, K., Christensen, L.L., Olesen, S.H., Frederiksen, C.M., Laiho, P., Aaltonen, L.A., Laurberg, S., Sorensen, F.B., Hagemann, R., and Orntoft, T.F. (2002) *Cancer Res*, **62**, 4352–4363.
21. Bocaz-Beneventi, G., Latorre, R., Farková, M., and Havel, J. (2002) *Anal Chim Acta*, **452**, 47–63.
22. Havel, J., Madden, J.E., and Haddad, P.R. (1999) *Chromatographia*, **49**, 481–488.
23. Yannis, L.L. (2000) *J Chromatogr A*, **904**, 119–129.
24. Kompany-Zareh, M., Massoumi, A., and Pezeshk-Zadeh, Sh. (1999) *Talanta*, **48**, 283–292.
25. Khayamian, T., Ensafi, A., and Atabati, M. (2000) *Microchem J*, **65**, 347–351.
26. Wu, W. and Massart D.L. (1996) *Chem Intell Lab Syst*, **35**, 127–135.
27. Brezmes, J., Ferreras, B., Llobet, E., Vilanova, X., and Correig, X. (1997) *Anal Chim Acta*, **348**, 503–509.
28. Zupan, J. and Gasteiger, J. (1991) *Anal Chim Acta*, **248**, 1–30.
29. Sumpter, B.G., Gettino, C., and Noid, D.W. (1994) *Annu Rev Phys Chem*, **45**, 439–481.
30. Sumpter, B.G. and Noid, D.W. (1996) *Annu Rev Mater Sci*, **26**, 223–277.
31. Kinsella, J.A. (1992) *Network*, **3**, 27–35.

32. Polak, E. and Rebiere, G. (1969) *Operationette*, **13**, 35–43.
33. Statheropoulos, M., Pappa, A., Karamertzanis, P., and Meuzelaar, H.L.C. (1999) *Anal Chim Acta*, **401**, 35–43.
34. Dong, D. and McAvoy, T.J. (1996) *Computers Chem Enging*, **20**, 65–78.
35. Tetko, I.V., Luik, A.I., and Poda, G.I. (1993) *J Med Chem*, **36**, 811–814.
36. Seasholtz, M.B. and Kowalski, B. (1993) *Anal Chim Acta*, **277**, 165–177.
37. Livingstone, D.J. and Manallack, D.T. (1993) *J Med Chem*, **36**, 65–70.
38. Broadhurst, D., Goodacre, R., Jones, A., Rowland, J.J., and Kell, B. (1997) *Anal Chim Acta*, **348**, 71–86.
39. Kirkpatrick, S., Gelatt, C.D., and Vecchi, M.P. (1983) *Science*, **220**, 671–680.
40. Courtois, S. and Phan-Tan-Luu, R. (1998) *Analusis*, **26**, 304–309.

CHAPTER 12

CHEMOMETRIC METHODS APPLIED TO GENETIC ANALYSES BY CAPILLARY ELECTROPHORESIS AND ELECTROPHORESIS MICROCHIP TECHNOLOGIES

MARIBEL ELIZABETH FUNES-HUACCA, JULIANA VIEIRA ALBERICE, LUCAS BLANES, and EMANUEL CARRILHO
Grupo de Bioanalítica, Microfabricação, e Separações, Instituto de Química de São Carlos, Universidade de São Paulo, São Carlos, SP, Brazil

CONTENTS

Chemometric Methods in Capillary Electrophoresis. Edited by Grady Hanrahan and Frank A. Gomez

12.1. INTRODUCTION

In this chapter we summarize the complex issues that are involved in the analysis of sizing DNA by capillary electrophoresis (CE), and how chemometric methods can help to optimize a high number of interrelated variables. It is impressive to observe how diverse is the obtainable biological information despite the size of the double-stranded DNA molecule. We also briefly introduce some typical genetic assays that rely on sizing DNA molecules, and how some chemometric approaches are used to correlate sizes of DNA with population and or evolution of species.

12.1.1. Analysis of DNA by CE

The use of CE for genetic analysis has increased exponentially in the last decade, especially with the conclusion of large genome projects such as the Human Genome Project (1, 2). In the genetic analysis field, fast analysis time and high resolution are required for a large range of DNA sizes, and because of that, CE has become a fundamental tool in this area. Today, separation of DNA through polymeric matrices in CE is the dominant technology for high-throughput sequencing, at least until the next-generation sequencing technology becomes widely available (3). CE has become very popular for several reasons: the possibility of full automation, high data storage capability, fast analysis time, and high-resolution analyses using highly sensitive laser-induced fluorescence (LIF) detection (4, 5).

Originally, DNA sequencing was performed on slab gel electrophoresis (SGE), a low-cost technique that still is largely used in biochemistry and molecular biology laboratories. Despite being a very simple technology, SGE is time-consuming and labor-intensive and does not facilitate the use of sensitive detectors. Consequently, it is not a desired technique when fast quantitative analyses are required (4, 6, 7).

Initially, researchers successfully transferred slab-gel technology to CE by filling the capillary with poly(acrylamide), which was cross-linked *in situ*. However, the presence of a permanent matrix in the capillary gel electrophoresis (CGE) was impracticable and failed to yield reproducible results. The problems associated with CGE were resolved by filling the capillary with linear polymer solutions, allowing facile replacement of the polymeric matrix between runs, and enabling the complete automation of DNA sequencing (8).

Several polymers and copolymers can be used to separate DNA in capillary electrophoresis with polymer solutions (CEPS). An ideal matrix should be chemically and physically stable in run conditions, hydrophilic, and relatively low in viscosity. The formation of a robust entangled network matrix providing good sequencing performance is also expected of a good polymer (5).

In order to suppress the electroosmotic flow (EOF) generated at the capillary inner walls, permanent or dynamic coatings can be applied when polymer solutions are used as DNA sieving media. Some examples of polymers that have been used as sieving matrices are linear poly(acrylamide) and poly(dimethylacrylamide), hydroxyethylcellulose, hydroxypropylcellulose, poly(dimethylacrylamide-co-beta-D-glucopyranoside), poly-(ethyleneoxide), and poly(vinylpyrrolidone), to name just a few (9, 10). Specific separation goals can be achieved by tailoring electrophoretic conditions and the matrix composition. For example, ultra-fast separations can be carried out in short capillaries and high electric fields. Alternatively, large range sizes of DNA can be separated by selecting the appropriate mixture of polymers; that is, by manipulating electric field and concentration of the matrix, it is possible to obtain any degree of base pair resolution at any given size of DNA. Heller reviewed the theoretical and empirical mechanisms of electrophoretic migration of DNA in CEPS by means of a systematic study of the separation matrices and of the factors that are relevant for the DNA mobility and its migration mechanism (11).

An important aspect of DNA analysis in CEPS is sample introduction. Most CE applications use hydrodynamic (HD) injection because it is theoretically well established, has a negligible bias, and can be easily managed. However, in the case of nucleic acids, the separation matrix is rather viscous and inhibits the sample injection through pressure application. Therefore, DNA is preferably injected using an electric field (electrokinetic [EK] injection), which shows advantages such as sensitivity enhancement and ease of use. Some disadvantages are also inherent to EK injection. When compared with HD, EK has poorer repeatability for migration time and peak area. EK also suffers from matrix effects (both separation and sample matrix) and biased injection (12). Fortunately, DNA fragments do not suffer from the latter because every DNA fragment in the buffer solution has the same charge/size ratio and thus the same electrophoretic mobility. However, depending on the injection conditions (voltage, ionic strength of sample solution, concentration, and type of the separation matrix), the DNA can assume different conformations leading to different separation mechanisms (13). Since several parameters have a strong influence on DNA separation, the use of chemometric tools are recommended to optimize separation conditions and analysis.

12.1.2. Microchip Electrophoresis Platform

Microchip CE, also known as lab-on-a-chip, is a relatively new method of separation that uses microfabrication technology to produce small electrophoresis devices for high-speed separations. In recent years, these devices have emerged as an effective tool for genetic analysis because this system is a relatively low-cost technology with a high capacity of analysis. The advancement of miniaturized platforms for genetic analyses has become an alternative to labor-intensive SGE and to capillary array devices, which are expensive and complex to utilize. The use of microchips for DNA analysis has some advantages when compared with CEPS. For example, microchips use approximately 10 times less sample (~0.1–1 nL) and run at least four times faster using similar strength fields (100–300 V/cm) (14).

Typical microchips consist of microstructures of glass or a polymer substrate, ranging in design from a single separation channel to a complex system that can include processing steps such as sample input, pre- and post-column reaction chambers, separation columns, and detectors (15). The separations are performed directly in the microchannels constructed in these devices, which begin and end in reservoirs. The typical lengths of these microchannels are several centimeters, with widths of 10–100 μm and depths between 15 and 40 μm. As in CEPS, the channels need to be filled with a polymer matrix to separate DNA. It is important to note that the dynamic coatings used to suppress the EOF in devices based in silica may not be compatible with the chemical surface of polymeric microchips.

Normally, the glass-based microchips have good performance because the surface property is similar to the inner surface of conventional capillaries and high optical transparency. The samples are normally loaded by EK injection and detected with UV or LIF detectors positioned at the end of the channel. In recent years, companies such as Agilent, Hitachi, and Shimadzu have developed equipment based on microchip technology for biochemical analysis, and such equipment is now commercially available (16).

As in conventional CEPS, the composition of the sieving matrix, temperature, electric field strength, injection time, and electric field applied during the injection are important factors to be considered to obtain the best separation and detection sensitivity in microchips. Factors such as injection system and column geometry, electric field distribution along the channels, and heat generated due to the Joule heating also should be considered simultaneously in order to obtain the optimal separation. Due to the high number of variables to optimize, the use of chemometric methods could be useful to determine an optimal microchip design and operation method.

In the last decade, many fundamental studies were carried out in the field of DNA separation using microchips. Their use has been reported in all fields of genetic analysis, from the determination of DNA sizing, analysis of digestion fragments, analysis of nucleotide polymorphisms, analysis of functional genomics, and gene mutation to the diagnosis of diseases via the analysis of polymerase chain reaction (PCR) products. The use of microchips with multiple channels has also been reported as a good option for genomic sequencing and is considered a future alternative to produce personal genomes at lower costs.

Although SGE, CEPS, and microchip analysis are of fundamental importance for genetic analysis, it is important to note that several new promising technologies already exist that allow DNA sequencing without using the classic Sanger biochemistry principle, capillaries, or microchannels (3, 17).

12.1.3. Chemometric Approaches to DNA Analysis

In DNA analysis by CE with polymer solutions, there are many variables that can be optimized. Strength of electric field, concentration of polymer solution, and temperature during analysis are the most relevant variables responsible for fragment resolution and analysis time (18–20). Optimization of such parameters can be obtained simultaneously using chemometric techniques ultimately reaching the optimum working conditions within a few experiments (21). Optimization of several variables can potentially be a costly and a difficult task due to the complicating interactions that exist between variables. The individual optimization of each variable—a univariate approach—usually is a time-consuming process and the results can lead to a local optimum. To overcome such limitations, a multivariate approach can be used to optimize several variables simultaneously, a method particularly well suited to optimize separation conditions in CE (21).

12.1.3.1. Simplex Optimization. The simplex method is probably the most efficient and easily employed procedure to optimize any given system. This method was first developed by Spendly et al. (22) and later improved by Nelder and Mead (23). Simplex is defined as a geometric figure with one more vertex than the number of factors being optimized. Therefore, the optimization of two factors results in a triangle, for three factors a tetrahedron, and so forth. The method is developed through a set of experimental conditions, which represents a vertex, ranked from worst (W) to best (B), based on the output (results). The next experimental condition (R) is determined by reflection of the coordinates from the worst response through the hyper face defined by the other vertices. First, a centroid point (C) is determined and then the reflection is calculated (Eq. 12.1). After that, the worst point is discarded and a new simplex is carried out. The process goes on until eventually reaching an optimum point. The process is better visualized in Figure 12.1, which compares the classical optimization approach for two variables with the chemometric approach.

$$R = C + (C - W) \qquad \text{(Eq. 12.1)}$$

Although the method is not so rigorous mathematically, it is very efficient. It does not use the traditional test of significance and is, therefore, faster and simpler than other methods (24). Simplex optimization has been successfully applied to a wide variety of systems that require optimization (4), and it is explained in greater detail in Chapter 2.

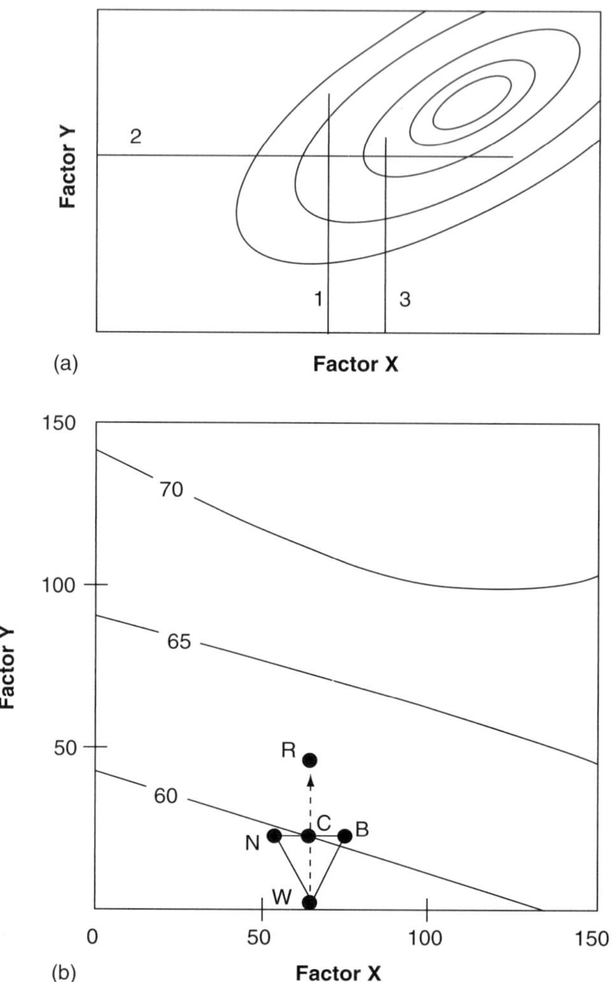

FIGURE 12.1. (a) Representation of a univariate optimization scheme. The concentric circles represent a surface response and the center is the maximum response. (1) The x-variable (or factor) value is fixed and variable y is optimized; (2) y is fixed at best response while x is varied; (3) during optimization of x, a better value is found, thus requiring new experiments varying y. According to this experimental setup, intersection of (2) and (3) would be the best response. (b) Representation of a bidimensional simplex BNW and the reflection R of the worse value W. Reprinted with permission from Reference 4.

12.1.3.2. Application of Simplex to Optimize Injection of the DNA Sample. The total amount of DNA introduced in the column during EK injection depends on several parameters. The main parameters are EOF mobility, the concentration and composition of the sample and polymer, the electric field strength applied to the sample, and the duration of the applied voltage. The EOF can contribute negatively to the amount of sample EK injected in CE; however, this effect is negligible if a coated capillary is used or if the viscosity of the separation media is high enough to suppress the EOF. The total amount of DNA injected is an important factor that dictates the sensitivity and the efficiency of the separation. If too much DNA is introduced, there is a loss in separation efficiency due to the contributions from injection variances, electrophoretic dispersion, and the DNA-polymeric matrix interaction. When separation efficiency is low, the resolution, in terms of base pairs, is limited and the sizing accuracy of the DNA fragments is compromised. On the other hand, when a minimal amount of DNA is introduced, detection is limited by the sensitivity of the detection systems, justifying the need for optimization of the injection process to maximize signal and minimize band broadening. For example, Figure 12.2 illustrates how irreproducible a separation of a DNA ladder is if the conditions of the sample or the conditions of the separation matrix are not under control.

DNA is a very flexible polyelectrolyte molecule and under high electric field strength can undergo severe structural changes (25, 26). Catai and Carrilho conducted several experiments to evaluate which were the main factors affecting the introduction of DNA fragments in CE and how they affected the separation efficiency using simplex optimization. These studies showed that changes in the composition of the ionic concentration of the sample is one of the main factors that affect resolution, signal intensity, and reproducibility. However, the chemical and physical condition of the polymeric solution has also shown a strong influence, mainly in the amount of DNA injected and the mobility of DNA fragments. The replacement of the matrix before each run is important due to the drop in the electric current by electrolyte depletion during electrophoresis, which decreased the polymeric solution conductivity (9). Figure 12.3 shows how the evolution of the simplex for three variables against a response function is designed to account for both signal strength and resolution of the separation—typically, they are inversely correlated.

Figure 12.4 shows the separation of DNA by CE for three of the vertices shown in Figure 12.3 in which is clearly seen the evolution in terms of resolution of the central pair of peaks by just manipulating injection conditions. Note that the separation times are nearly identical for the three separations, which indicates that the separation conditions were preserved.

12.1.3.3. Molecular Sizing of DNA by Sieving. The first step to determine the size of an unknown DNA fragment using capillary gel electrophoresis is to run a molecular size standard first. With the data from the separation, a logarithmic plot of the mobility (μ) versus the fragment size is obtained, and a linear equation for a specific size range is generated. Subsequently, the sample is analyzed under the same conditions, and the mobility of each fragment is interpolated in the previous plot allowing fragment sizes to be determined. For the analysis of a large DNA size range, however, such as fragments

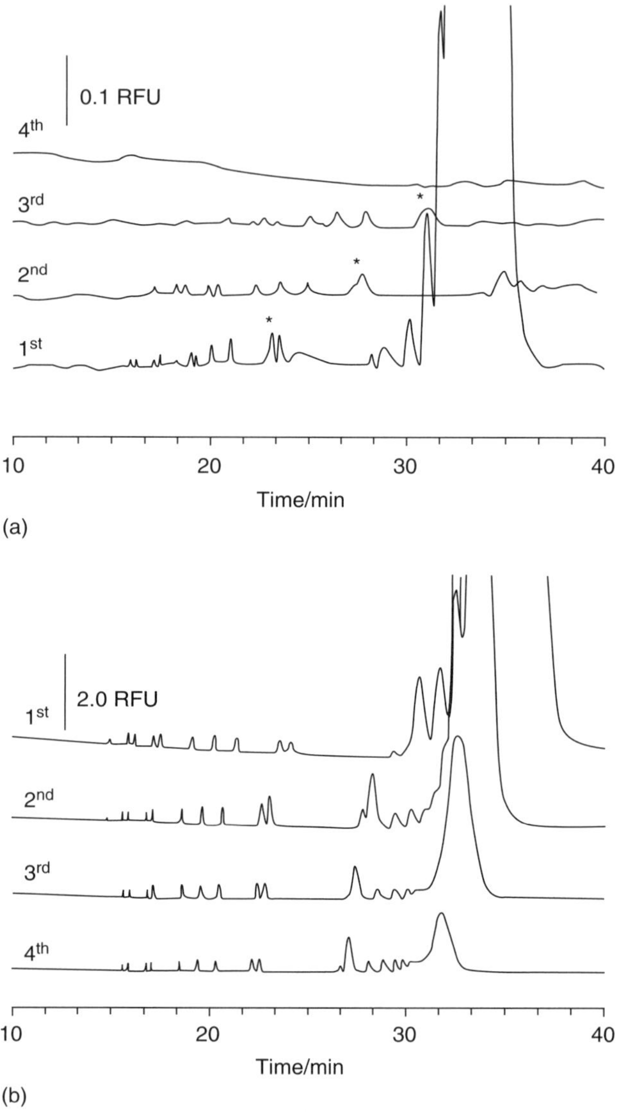

FIGURE 12.2. Separation of 1 kbp DNA ladder in repetitive injections. (a) Four aliquots of sample were injected in the same separation matrix. The 506/517 bp peak pair were labeled with an * for clarity. (b) An aliquot of sample was injected repeatedly in four loads of fresh sieving polymer solution. A 75 μm i.d. poly(vinyl)alcohol (PVA)-coated capillary column 47 cm long (40 cm effective length) was filled with 0.5% hydroxyethyl cellulose (HEC) solution in 100 mmol/L Tris/tris(hydroxymethyl)methyl-3-aminopropanesulfonic acid (TAPS)/ethylenediaminetetraacetic acid (EDTA) buffer, and the separation was carried out with 200 V/cm electric field. The desalted DNA sample (100 μg/mL diluted in deionized water) was intercalated with 10 μmol/L of ethidium bromide for LIF detection (emission at 520 nm) with an Ar-ion laser (excitation at 488 nm), and electrokinetically injected for 20 s under an electric field of 50 V/cm. RFU: relative fluorescence unit. Reprinted with permission from Reference 9.

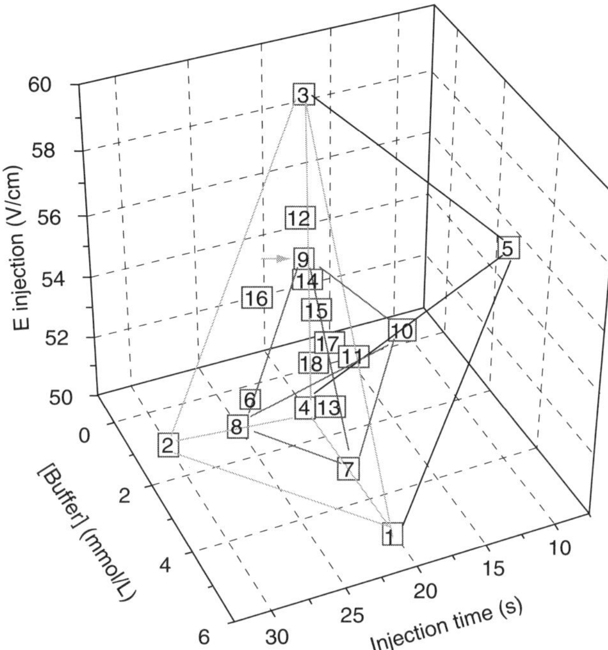

FIGURE 12.3. Spatial evolution of a three-variable simplex. The red lines link the initial conditions (vertices 1–4). The blue lines show the simplex figure after the radical contraction (vertices 4, 7–9) and the first reflection after contraction (vertex 10, blue lines). The arrow points to the best condition. Reprinted with permission from Reference 4. See color insert.

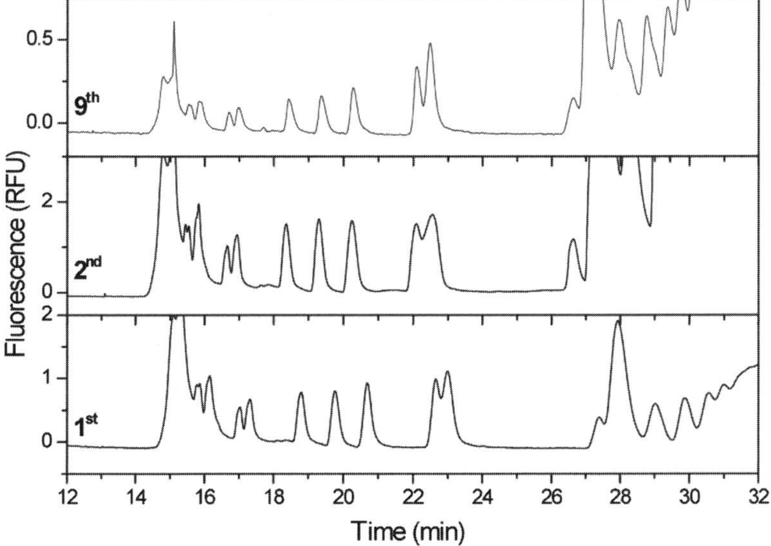

FIGURE 12.4. Electropherograms of the first, second, and of the ninth simplex conditions developed in Figure 12.3. Reprinted with permission from Reference 4.

of randomly amplified polymorphic DNA (RAPD) or PCR by CE, at least three separation mechanisms are observed: (i) Ogston, when the polyelectrolyte forms a random coil that is smaller than the pore size of the sieving polymer; (ii) reptation, when the polyelectrolyte migrates in a reptile-like movement through the pores of the network, and (iii) biased reptation with fluctuation, when all polyelectrolyte molecules migrate with the same mobility (27). As a result of the different mechanisms of DNA separation, the logarithmic plot of μ versus base pair (bp) is a sigmoid, as the one exemplified in Figure 12.5.

Contrarily, to determine the length of large DNA fragments, the separation method should be linear over a wide size range and provide high resolution. In the case of DNA fragments produced by RAPD, in which one of the main objectives is to compare band patterns produced by separation of DNA fragments originated from different individuals or species, the linear range should be approximately between 50 and 4000 bp. Thus, for one to accurately determine the size of a given DNA fragment within this range, the sigmoid must be made linear (Figure 12.6).

In an ideal situation, when r^2 is maximized to its full extent ($r^2 = 1$), all the fragments will migrate according to one separation mechanism, that is, they will have the same migration behavior. In practice, when r^2 increases, a separation mechanism will be favored over others. The correlation coefficient of a log–log curve of μ versus bp, in principle, depends on the analysis conditions. Catai and Carrilho have reported the successful use of simplex in an attempt to optimize the separation of large DNA size range fragments (75–4072 bp),

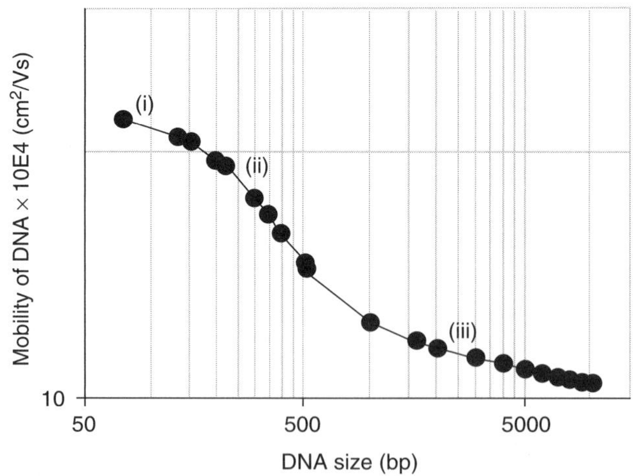

FIGURE 12.5. Different mechanisms of the migration of DNA under an electric field through a sieving matrix (hydroxyethyl cellulose) in capillary electrophoresis. The mechanisms of DNA migration are: (i) Ogston mechanism of sizing; (ii) reptation model; and (iii) reptation with orientation.

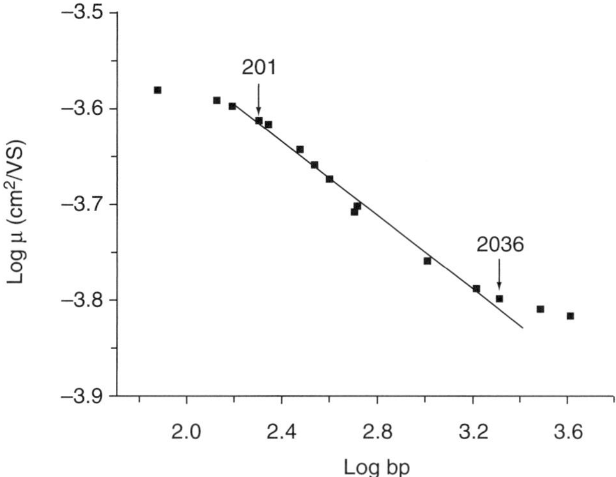

FIGURE 12.6. Plot of mobility of DNA versus DNA size in log scale illustrating that there is a narrow window in which the relationship is linear. Maximization of this linear relationship represents an improvement on the accuracy and precision of the analysis for sizing DNA. Reprinted with permission from Reference 27.

aiming for maximization of the correlation coefficient (r^2) of a logarithmic plot of μ versus bp. In order to obtain high separation resolution and a linear system ($r^2 = 1$), six variables of the CE separation were simultaneously varied by the simplex, eventually reaching an optimum point. In other terms, it was possible to maximize the linearization of the mobility in function of size by extending the reptation mechanism regime.

For the maximization of the correlation coefficient, a simplex with six factors was used. The factors were (i) sample buffer concentration, (ii) injection time, (iii) capillary temperature, (iv) matrix concentration, and electric field strength for (v) injection and for (vi) separation. These variables were chosen because it was demonstrated previously that they were the most relevant in the DNA separation mechanism and performance (resolution) (27). Table 12.1 shows the movements made by the simplex with all variable values and the resulting correlation coefficient (r^2).

The results show three vertices {6, 18, and 21} with $r^2 > 0.98$. Although vertex 6 shows the highest r^2 obtained ($r^2 = 0.98627$), the separation presented low resolution and low signal intensity. This result was probably due to the fact that the CE instrument did not have enough pressure to conduct the proper replacement of a high viscosity matrix inside the capillary at different concentrations. Vertex 18, which presented an $r^2 = 0.98002$, also showed low resolution and the conditions were not completely optimized compared with the results obtained in vertex 21, which was considered the best. The poor result of vertex 18 may be explained by the high salt concentration of the sample.

TABLE 12.1. Movements of the simplex vertices, analysis conditions for the six variables, and subsequent correlation coefficient (r^2)

Vertex	Movement	Sample Buffer Concentration (mmol/L)	Injection Time (s)	$E_{injection}$ (V/cm)	Temp (°C)	Separation Matrix Concentration (%) (v/v)	$E_{separation}$ (V/cm)	r^2
01	–	10.0	20	100.00	20.0	0.5000	400.00	0.96662
02	–	0.0	30	100.00	20.0	0.5000	400.00	0.93963
03	–	0.0	20	150.00	20.0	0.5000	400.00	0.95971
04	–	0.0	20	100.00	30.0	0.5000	400.00	0.96349
05	–	0.0	20	100.00	20.0	1.0000	400.00	0.95474
06	–	0.0	20	100.00	20.0	0.5000	300.00	0.98627
07	–	0.0	20	100.00	20.0	0.5000	400.00	0.96248
08	R (01,03,04,05,06,07)	3.3	10	116.66	23.0	0.6660	366.66	0.95468
09	C_L (01,03,04,05,06,07)	2.5	15	112.50	22.5	0.6250	375.00	0.95823
10	R (01,03,04,06,07,09)	4.2	18	120.83	24.0	0.0417	358.32	0.95154
11	C_{MD} (01,03,04,06,07,09)	1.0	20	105.21	21.0	0.7600	389.57	0.96191
12	R (01,03,04,06,07,11)	1.2	25	105.90	21.2	0.4620	388.19	0.95244
13	C_{MD} (01,03,04,06,07,11)	2.2	17	110.85	22.2	0.5842	378.30	0.96609
14	R (01,04,06,07,11,13)	4.4	19	55.34	24.4	0.6140	355.96	0.96496
15	R (01,04,06,07,13,14)	4.5	19	83.53	24.5	0.3060	355.17	0.97826
16	R (01,04,06,13,14,15)	7.0	19	83.23	27.0	0.5000	329.81	0.97144
17	R (01,06,13,14,15,16)	9.4	18	77.66	16.0	0.5020	306.40	0.96635
18	R (01,06,13,15,16,17)	6.6	19	129.74	18.8	0.3500	333.94	0.98002
19	R (01,06,15,16,17,18)	10.0	21	80.53	19.9	0.3020	296.81	0.97585

20	R (01,06,15,16,18,19)	3.5	21	114.69	27.4	0.3180	365.49	0.97774
21	R (06,15,16,18,19,20)	0.6	19	97.46	25.9	0.2600	260.40	0.98019
22	R (06,15,18,19,20,21)	1.5	21	118.66	18.5	0.1770	307.46	0.97191
23	C_L (06,15,18,19,20,21)	2.9	21	109.80	20.6	0.2580	313.04	0.97318
24	R (06,15,18,19,20,21)	5.6	19	92.00	24.9	0.4200	324.21	0.97222
25	C_{MD} (06,15,18,19,20,21)	3.5	20	105.38	21.7	0.2980	315.85	0.97442
26	R (06,15,18,19,20,21)	4.9	20	96.53	23.8	0.3800	321.43	0.97412
27	C_{MD} (06,15,18,19,20,21)	3.9	20	103.17	22.2	0.3180	317.23	0.97355
28	–	0.3	20	98.62	22.9	0.3800	280.19	0.97264
29	–	5.2	20	90.27	20.0	0.4000	298.40	0.97353
30	–	3.3	19	114.87	19.4	0.4250	316.96	0.97817
31	–	1.7	20	107.34	23.7	0.4088	332.74	0.97626
32	–	2.2	20	91.76	22.2	0.4030	327.57	0.96661
33	–	1.8	20	102.68	20.8	0.4000	307.91	0.97446
34	R (06,28,29,30,31,33)	1.8	20	112.83	20.0	0.4350	284.50	0.97364
35	R (06,29,30,31,33,34)	4.3	20	110.72	18.4	0.4768	333.30	0.96922
36	C_{MD} (06,29,30,31,33,34)	1.3	20	101.64	21.8	0.4039	293.49	0.97941
37	R (06,30,31,33,34,36)	–1.8	20	122.86	22.0	0.4560	313.47	0.01000
38	C_{MD} (06,30,31,33,34,36)	3.4	20	98.4	20.5	0.4150	302.17	0.97094

R = reflection; C_L = simple contraction; C_{MD} = contraction with change of direction; – = initial vertex of the simplex.
Source: Adapted from Reference 27, with permission.

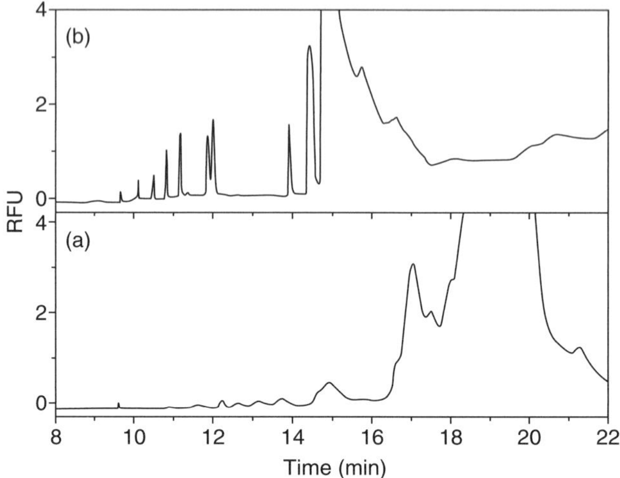

FIGURE 12.7. DNA standard 1 kbp analyzed according to (a) the conditions of vertex 6, and (b) the optimized conditions of vertex 21 (see Table 12.1). Reprinted with permission from Reference 27.

The application of the simplex resulted in finding the best compromise for the separation of small DNA sizes as well for large ones. To illustrate the evolution of the simplex in terms of DNA separation, the electropherograms for vertices 6 and 21 are shown in Figure 12.7.

After optimization by simplex, the plot from Figure 12.6 improved and yielded a linear equation ($\log[\mu] = -3.14 - 0.20 \log[\text{bp}]$, $r^2 = 0.998$) suitable for the analysis of the 201–2036 bp size range. This equation was further used to determine the size of unknown DNA fragments (27). Thus, the simplex method was shown to be an efficient way to optimize an electrophoretic separation of DNA, since several variables could be simultaneously optimized.

12.2. GENETIC MARKERS

12.2.1. PCR Based on Discovery of Genetic Markers

The invention of the PCR technique at the end of the 1980s had a tremendous impact on genomic research and contributed to the development and application of many molecular markers (28, 29). Simplicity of the reaction and high probability of success contributed to the widespread use of this method. The PCR technique and its variations allowed significant advances in all types of DNA analysis, including genetic population and evolution studies, and phylogenetic analysis, without environmental influences or organism development levels (30).

Despite the revolution caused by PCR, some limitations restricted its application. The main problems include high costs and the need to "know the DNA sequence" that will be amplified; such information requires cloning and sequencing of the target region. Aiming to solve this problem, a technique that uses short primers and arbitrary sequences to start the reaction was developed. This method is a variation of PCR protocol, with just one primer used instead of two.

Three research groups independently developed the method, with small differences between them. Williams et al. (31) patented the RAPD technique (Randomly Amplified Polymorphic DNA), which became the most popular. Welsh and McClelland (32) used primers with 20 nucleotides and called the technique Arbitrary Primed-PCR. Finally, Caetano-Anollés et al. (33) described the same technology with name DNA Amplification Fingerprint.

12.2.2. RAPD—Principles and Genetic Basis

As mentioned before, RAPD is a variation of PCR protocol. The use of only one primer with arbitrary sequence and low stringency is the main difference between the methods. The primers used in RAPD have, in general, 10 nucleotides and its G + C content varies from 50% to 70% (34).

The basis of the RAPD technique is the differential amplification of genomic DNA. In the beginning of the reaction, the primer binds itself to the complementary DNA sequence, so the effective amplification takes place between two adjacent priming sites and a DNA polymerase enzyme promotes the extension if the orientation between them overlaps (see Fig. 12.8).

The maximum distance between primer binding sites must be from 3000 to 4000 bp, because the enzyme, Taq polymerase, cannot promote the reaction beyond this size of fragments (29, 34). Low stringency is important in the pairing step as the amplification takes place even if the hybridization occurs without a complete match between primer and binding site. It is known that RAPD segments are amplified even if complementarity is not perfect, that is, a perfect match between the primer sequence and the target DNA sequence. The complementarity is more critical at the 3′ end than at the 5′ end of the primer. The residence time of primer at the priming site is also an important parameter; if this time is short, the fragment cannot be amplified.

Since the reaction takes place by hybridization of the primer and the DNA template, changes at the nucleotide sequence (which can be unique for any given region) result in characteristic patterns allowing the identification and discrimination of different species. Each primer drives the synthesis of several DNA segments at different points in the DNA, generating many bands with distinct sizes. It is important to highlight that the quality of the amplification products and the complexity of the pattern of polymorphisms are directly influenced by the primer.

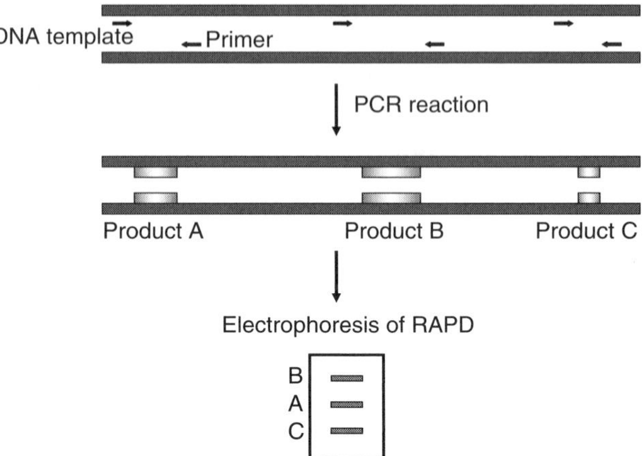

FIGURE 12.8. Simplified diagram of RAPD technique. The PCR products A, B, and C are separated according to their size by gel, CE, or microchip electrophoresis.

Williams et al. (31) reported that change in just one base in the primer-binding site is enough to compromise the amplification. This can generate fragments of different sizes and, as a consequence, modify the pattern of the bands. Other kinds of polymorphisms include site deletions and insertion or deletion between adjacent sites.

RAPD polymorphisms have a binary nature, that is, the polymorphism is present or absent. Different size fragments are indicative of different loci. There is no evidence about RAPD marker grouping in a specific region of the genome; that is, RAPD marker loci are distributed randomly along the DNA strand, from a unique sequence to highly repetitive sequences (35).

As described originally by Williams et al. (31), RAPD uses short primers and 45 PCR cycles with pairing temperature of 36 °C. There are several variations of the original protocol, but all of them share the basic concept of subtyping organisms based on generation of complex patterns of PCR products in a single reaction by using unspecific primers (36).

12.2.2.1. Dominance of RAPD Markers. RAPD markers are called dominants because they cannot discriminate between heterozygous and homozygous genotypes. When a band is visualized in the gel, it is impossible to distinguish if the band is from a homozygous diploid individual (AA) or a heterozygous individual (Aa). Only the recessive homozygous genotype is identified by the absence of a band (37) as shown in Figure 12.9.

12.2.2.2. Competition among Amplification Sites. It is expected that the use of several primers in the same reaction will increase the number of polymorphisms; however, this does not happen due to the competition between initiation sites of reaction. Each site competes for substrates (deoxynucleotides) and enzymes during PCR and, because of this, the fragments tend to be

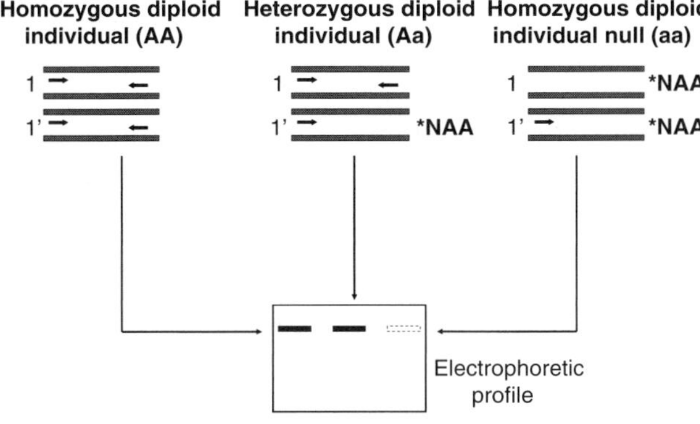

FIGURE 12.9. Dominating behavior of RAPD markers.

amplified with different efficiencies. The more competitive the site is, the better the amplification of that region will be (31).

It is common also to imagine that the number of amplified fragments increases proportionally with the complexity of a genome. Some studies have shown, however, that the number of amplified fragments is independent of the genome complexity. When there is a comparison between organisms of different complexity, the most complex is favored because it has greater complementary to the DNA templates. In general, RAPD reaction favors amplification of fragments with a better match between primers and the site of initiation (31).

12.2.3. Interpretation of Results and Generation of Dendograms

Data obtained in RAPD must be interpreted by a reliable method to quantitatively correlate the results. The interpretation must contain a measure of similarity or disparity for all possible combinations of samples. The aim of comparison between RAPD patterns is to identify similarities or differences between the samples under study (38). The profiles generated are discrete and a computer system can automatically identify the samples. Correlation coefficients between bands are created by statistical software packages that use (i) ordination techniques, such as principal component analysis (PCA) or principal coordinate analysis; (ii) distance matrix or cluster analysis methods, such as neighbor-joining and unweighted pair group method using arithmetic average (UPGMA) algorithms; or (iii) parsimony strategies, such as phylogenetic analysis using parsimony (PAUP). From these coefficients, it is necessary to generate dendograms with software such as PHYLIP (J. Felsenstein, Dept. of Genetics, University of Washington, Seattle, WA, USA) and NTSYS (Exeter Software, Setauket, NY, USA), and parsimony analysis applications in PHYLIP, PAUP (D.L. Swofford, Illinois Natural History Survey, Champaign,

IL, USA), MacClade (Maddison and Maddison), and Hennnigs 86 (J.S. Harris, Port Jefferson Station, NY, USA).

12.2.4. Advantages and Limitations

The RAPD technique is a simple, fast, and low-cost method. The last characteristic allows many laboratories to use it, including low technology laboratories. Because of its simplicity, it does not require a strong understanding of or experience in molecular biology. It needs a minimum quantity of DNA (5–20 ng) and no prior knowledge of the organism to be analyzed.

The use of arbitrary primers has made the technique universal; that is, the same primers can be used for any organisms and no previous work is necessary. The RAPD technique has the ability to generate many markers for genetic mapping and provides information about all genomes, that is, repetitive or unique sequence regions, coding or noncoding regions (34). Each RAPD marker is equivalent to one target site. This simplifies greatly the information transferred in collaborative research programs. Since there are no restrictions on the digestion of PCR products, there is no need to purify samples that can be directly examined by electrophoresis. Another advantage of the RAPD technique is its potential for automation due to the binary nature of the method.

Some restrictions limit the practical application of RAPD analysis. Dominance is the most important limitation of RAPD. Dominant markers are not as discriminating as codominant markers to study population genetics. And as a result, more individuals must be sampled per loci for dominant markers (28).

Reproducibility is another important concern related to RAPD. Small differences in PCR conditions can result in different RAPD profiles. The factors that affect the reaction include concentration of the primer, concentration and quality of DNA template, concentration and source of DNA polymerase, concentration of $MgCl_2$ and the equipment for thermal cycling, and the number of cycles used. Although each one of these parameters can be optimized, the RAPD sensibility for different experimental conditions raises serious doubts about the comparability of results obtained from different laboratories (36). Still, in relation to the limitations of the method, some bands can be ambiguous and thus must be carefully analyzed. The ambiguity may be related to (i) low discrimination of the primer between distinct amplification sites, (ii) competition between different amplification sites, and (iii) variation of amplification conditions (37, 39).

12.3. CHEMOMETRIC APPROACHES TO GENETIC ANALYSIS

12.3.1. Optimization Methods

Several robust methods for multivariate analysis have recently been developed in statistics and chemometrics. Most multivariate methods applied to chemical and genetic data are based on the technique of least squares (LS). For instance, PCA, multiple linear regression, principal component regression[1] (PCR), and partial least squares (PLS) regression are all LS techniques. Multivariate projection techniques, such as PCA and PLS, cover areas such as large-volume high-density data structures obtained within genomics, proteomics, and metabonomic/metabolomic data. PCA and PLS and their extensions derive their usefulness from their ability to analyze data with many, noisy, collinear, and even incomplete variables in both X and Y Cartesian dimensions (40). Furthermore, hierarchical PLS and PCA are two recent modifications of the LS technique, which simplify interpretation in applications involving many variables. In such a situation, plots and lists of loadings, weights, and coefficients tend to become messy and the results are often difficult to overview. Instead of reducing the number of variables, and thus reducing the validity of the modeling, a better alternative is often to divide the variables into conceptually meaningful blocks and apply hierarchical PCA or PLS (41).

12.3.2. Classification Methods

Pattern recognition methods have become well-established tools for the analysis of multivariate chemical data sets. Over the years, these techniques have been applied to problems of classification in a wide variety of fields and the areas of application continue to grow as the methodologies become more broadly recognized and readily available (42). The usefulness of these methods arises not only from their ability to classify unknown samples, but also from their capacity to elucidate how various chemical features contribute to class distinctions, thereby leading to a better understanding of fundamental differences among classes. One area in which the application of multivariate methods has been particularly successful is in the classification of biological species through chemical markers, biological markers, and chemotaxonomy (43).

As an alternative to classification based on morphological or biometric features, chemometrics may be a more reliable or efficient tool in certain cases. Differences in the chemical makeup of biological organisms are a natural consequence of molecular evolution. Chemotaxonomic approaches may examine similarities between genetic sequences in different species directly, or focus on the amino acid sequences in selected proteins or enzymes that are coded from the DNA. Alternatively, changes in the enzymatic environment of the organism may manifest themselves through changes in the distributions of other chemical families (e.g., proteins, free amino acids, and cuticular hydrocarbons). These differences can be more difficult to interpret

[1]The use of "PCR" in this chapter is ambiguous because it is largely used as polymerase chain reaction, in molecular biology, and as principal component regression, in chemometrics. Since this chapter deals with both topics, we will try to be explicit.

since they are less direct, arising from numerous complex biochemical interactions and environmental factors (44).

On the other hand, the application of PCR has helped to classify the DNA sequences of the genes and has proven especially useful in their study and comparison. In particular, RAPD, as discussed, is used to compare and classify different organisms and quantify their overall similarity. RAPD employs short primers of arbitrary sequences to amplify random portions of the sample DNA by PCR. Since each primer is short, it will anneal to many sites throughout the target DNA; a fragment is amplified whenever two of these primers anneal close enough together and in the proper orientation with respect to one another. Individuals that have different sequences will have primers that anneal in different places and therefore produce a different spectrum of fragments from the PCR, that is, a different genetic "fingerprint." Because each primer generates relatively few (5–15) distinct bands when separated on an agarose gel, several reactions must be run, using several different sets of primers, and the results combined to obtain the desired number of markers. Pooled results can then be compared between samples and the percent similarity computed. Using multiple primers also helps ensure that a sufficiently large region of the target DNA is scanned when an estimate of overall variance between samples is desired. Typically, 10–15 primers (~100 bands) are required for statistical comparison of samples using RAPD markers (45, 46).

12.3.3. Genetic Algorithms

Genetic algorithms have been proposed by Holland in the 1960s, but it was possible to apply them with reasonable computing times only since the 1990s, when computers became much faster. General information on genetic algorithms relevant to this topic can be found, for example, in References 47–49. A wealth of information can also be found on the Web sites of various organizations (50–52).

The basic idea is to perform a computer simulation of what occurs in nature, and the first problem to be solved is how to code the information in such a way that the computer can manipulate it. It can therefore be said that the fitness to the environment is a function of the genetic material, in the same way as the result of an experiment is a function of the experimental conditions. Therefore, a correspondence between genetic material and experimental conditions can be established.

At a lower level, we can say that the genetic material is defined by the genes, in the same way as an experimental condition is defined by the values of the variables involved in the experiment. Therefore, corresponding gene variables can be established. On an even lower level, we can see that the information contained in each gene is defined by a sequence of DNA bases: since there are four bases, each gene can be considered as a word of variable length, written in a four-letter alphabet. In the same way, we can use the binary code to transform the value of a variable in a word of variable length, written in binary codes, a two-letter alphabet, 0 and 1.

12.3.3.1. Parameters of the Genetic Algorithms. According to the theory of evolution, the physical adaptation of a species occurs through a very high number of generations because the genetic material of its individuals is constantly changing. This is because those individuals whose physical and thus genetic traits are not complementary to their environment do not survive. Conversely, those who possess the particular traits that allow them to adapt to the surroundings will survive, and in turn have a greater probability of passing on their genetic material to the succeeding generation. Beyond this "logical" development, mutations allow the exploration of new "experimental conditions"; usually, mutations produce undesirable results (e.g., severe pathologies), but it can happen that these random changes of DNA bases end up in a better genome. Several genetic algorithms have been developed; beyond the common basic idea mimicking the evolution of a species, they can have relevant differences.

When describing a genetic algorithm, the details about the different parameters must be given: they can have very different values and can have a very strong effect on the final result. It has to be well understood that an "optimal" form of the genetic algorithm does not exist, and that for each problem the best results can be obtained by a specifically designed genetic algorithm.

All of them have three fundamental steps that can be performed in different ways. These three steps are (i) creation of the original population, (ii) reproduction, and (iii) mutations. The following is a short description of each one of them (53).

12.3.3.1.1. Population Size. The population size stays constant throughout the experiment or analysis. The number of individuals can be quite different, and usually is in the range 20–500 individuals (later in the chapter we will describe the influence of this parameter on the performance of the genetic algorithms). After having decided the population size (p), the genetic material of p individuals is randomly determined. This means that every single bit of each chromosome is randomly set to 0 or 1. If this chromosome corresponds to a possible experimental condition (i.e., inside the experimental domain), its response is evaluated. A population formed by many individuals maintains a great variety among the chromosomes, and therefore exploring at the same time several different regions. With a small population it can happen that all the individuals are extremely similar. Conversely, in the same computing time, a greater population will produce a smaller number of generations than the smaller population. This means that a very good chromosome found in generation n will need much more time in producing its effects, by generating offspring: this will happen only in generation $n + 1$.

In the literature, population sizes ranging between 20 and 500 individuals can be found. To choose the population size, the time required to evaluate the response is also important: if the time is quite short, then a large population can be used, since the time interval between the generations will be short; on the other hand, if it is quite long, then it would be better to work with a reduced genetic variability given an acceptable time interval between generations.

12.3.3.1.2. Reproduction. After having created the original population (or first generation), the individuals start "mating" and "produce offspring." This is the step in which the different genetic algorithms have the greatest variations, although all of them follow the same idea: the probability of the best chromosomes (the ones giving the best responses) producing offspring is higher than that of the worst chromosomes, and the offspring originated by breeding are a recombination of the parents' chromosomes.

The first step is creating the population of the second generation simply by randomly copying p times a chromosome of the first generation. If the drawing would be totally random, then each chromosome would have the same probability of going to the next generation and therefore the average response of the generation $n + 1$ would be statistically the same as that of generation n.

Ideally, each individual has the same opportunities to pass on their genetic information; however, in nature, those that display the most suitable traits for a given environment have a greater probability of surviving and successfully breeding. In the same way, the drawing performed to select the chromosomes that will be copied must take into account the response of the individuals, giving the best ones a higher probability. Hence, a biased drawing is performed, one in which the probability of each individual being selected is a function of its response. To visualize this process in a simple way, consider performing the selection with a roulette wheel in which the slots corresponding to the best individuals are larger than those corresponding to the worst ones.

12.3.3.1.3. Mutation Probability. The mutation is introduced to prevent premature convergence to local optima by randomly sampling new points in the search space. It sets the fraction of bits in the binary strings, which are randomly flipped each generation. The validation procedure is also applied at each step. The selected variables with the lowest prediction error are cross validated and tested on an independent sample. This process is repeated until either the specified number of generations is reached or the solutions converge. In general, the goal of supervised classification is prediction, so a model that is best for prediction of new data should be found (54).

12.4. METHODS IN PHYLOGENETIC RESEARCH

12.4.1. Genetic Distance

When genetic data are available from several populations, it is natural to ask, "how genetically similar are the populations?" In general, genetic distance is considered as related to the time since the population diverged from a single ancestral population. This, in turn, needs a genetic model specifying the process, such as mutation and genetic drift, causing the population divergence. The most widely used measure of genetic distance was proposed by Nei in 1972 (55). This method is based on a statistical process for estimating codon differences and the divergence time between closely related species. The key feature in Nei's genetic distance is that it can compare electrophoretic data from different species. According to this method, the biological unit of measurement is the number of nucleotide or codon differences per unit length of DNA. One important assumption in this method is that the mutation rate is constant over generations (56).

12.4.2. Construction of Phylogenetic Trees

One of the most important achievements in the study of molecular evolution is the understanding of the constancy of the rate of amino acid or nucleotide substitution. The constancy of the rate of amino acid or nucleotide substitution is held only approximately in any given population. However, molecular data show a much more regular pattern of evolutionary change by amino acid or nucleotide substitution compared with changes in morphological and physiological characters. Thus, molecular data provide a clearer picture of the evolutionary relationships existing among organisms than morphological characters do. Also, while it is difficult to give an evolutionary time scale for a morphological tree, it can be done routinely for a molecular tree.

For evolutionary studies, the classification of species also allows the construction of phylogenies, which may shed light on the relationship between observed pattern of speciation and the nature of evolutionary forces. A distinction should be made between "phenetic" and "cladistic" data. The phenetic relationships are similarities based on the degree of similarity, whereas cladistic relationships contain information about ancestry and can be used to study evolutionary pathways. Both of these relationships are best portrayed as phylogenetic trees or dendrograms, respectively (57).

Many different methods are available for reconstructing phylogenetic trees from molecular data. Two of the most popular are the distance matrix method and the maximum parsimony method. In the distance matrix method, evolutionary or genetic distance is computed for all pairs of species or population, and a phylogenetic tree is constructed by considering the relationships among these genetic distance values. In the maximum parsimony method, the nucleotide or amino acid sequences of ancestral species are inferred from those of extant species, and a tree is produced by minimizing the number of evolutionary changes for that given tree. In general, it is difficult to reconstruct the true evolutionary tree through which the extant species or population evolved.

The simplest method for developing a genetic distance matrix is the average distance method or UPGMA (58). This method is used not only to construct a phenogram, but it can also be used to construct a phylogenetic tree. In UPGMA, a measure of evolutionary distance is computed for all pairs of operational taxonomic unit (OTU), that is, species or populations, and the distance values are obtained in a matrix. Clustering of OTU starts from the two OTU with the smallest distance. Then, more distantly related OTU are gradually added to the cluster.

Results obtained with RAPD markers can be used to rapidly obtain information on the genetic diversity of species and can be used for their classifications. The data set and reproducible bands are used to calculate pair-wise similarity coefficients following Jaccard (59). This matrix of similarity coefficients is subjected to UPGMA to generate a dendrogram using average linkage procedure. The standardized data matrix is used to calculate correlations among variables and these correlations are subjected to eigenvector analysis to extract the most informative principal components. These principal components can be plotted in several possible combinations to study the pattern of variations observed among the species.

12.4.3. PCA

PCA is a well-known multivariate technique and detailed descriptions on the subject are available elsewhere (60). The idea of PCA is to take p variables (X_1, X_2, ... X_p) and combinations of those variables to create uncorrelated indices, Z_1, Z_2, ... Z_p, whereby each index measures a different dimension in the data. Further, the indices are also ordered so that Z_1 explains the largest amount of variation. Eigenvalues and eigenvectors are developed as the output of the analysis. The eigenvalue illustrates the percentage of total variation attributable to each component. In other words, the first principal component accounts for the largest amount of variation, the second principal component for the second largest, and so on. On the other hand, the eigenvector provides a coefficient (weight) for each variable, and this results in a new score for each observation. The advantage of PCA is that by observing the first two or three principal components, conclusions can be made about the pattern of variability. The coefficient of the eigenvector also indicates the relative importance of the original variables. Another method of examining the pattern of variation is to plot the scores of the first principal component against the scores of the second principal component, second principal component against the scores of the third principal component, etc.

12.4.4. Hierarchical Analysis

The hierarchical clustering method of multivariate data attempts to find the groups of data sets that have similar characteristics. These groups can then be further analyzed in detail to gain insight from the common characteristics of the data sets in each group. The knowledge of the process acquired from the clustering can be extremely valuable for activities such as process improvement or fault diagnosis, where each new operating condition could be classified as either an existing condition or a new condition.

The clustering methodology is based on calculating the degree of similarity using PCA and distance similarity factors. Many researchers have used PCA with clustering to reduce the dimensionality of the feature space. The number of linearly dependent features is reduced and their scores are calculated. The scores are then used as "new" uncorrelated features that are clustered (61, 62).

12.5. APPLICATIONS

12.5.1. Example of Classification Methods in RAPD Analysis

The main application of RAPD is analysis and determination of genetic diversity in natural populations. This technique has been used in studies of phylogenetic relatedness, differentiation between species, and detection of hybrids and genetically modified organisms. Working with a large number of markers makes it possible to find specific genera, species, subspecies, or breeds, allowing its use to establish taxonomic relationship.

The process requires minimum quantities of DNA to be used in studies of endangered or threatened species. Also, since it does not need prior knowledge of the target organism, it can be used on a large scale in research of organisms that are not well known. Other applications include genetic diversity to assess germplasm databanks, fingerprint production, genetic map construction, and polyploidy studies.

The simplicity of RAPD should not be taken as a triviality. The results obtained must be carefully interpreted. Sometimes, just one band does not give enough information about the genetic nature (37). When a RAPD assay is carried out, some care must be taken. First of all is the optimization of reaction conditions and maintenance of these conditions throughout the assay. After obtaining the results, careful analysis of the data should take into consideration the dominant nature of markers and origin of the samples.

Chemometric applications using PCA and HCA methods were used in the RAPD technique to classify the genetic variability of populations of horn fly from all five geographic regions of Brazil: North, Northeast, Center West, Southeast, and South. In this work, the authors evaluated the genotypical similarity of the different populations studied and obtained an RAPD marker capable of identifying the geographic origin of each of the populations studied (63).

The analyzed DNA was obtained through samples of adult populations of horn flies from five different Brazilian localities: Boa Vista (RR), Mossoró (RN), Seropédica (RJ), Campo Grande (MS), and Rosário do Sul (RS), each one in a different geographical region of the country. These samples were amplified with 60 RAPD primers and their amplified products were assessed for the number and quality of polymorphic loci. Only 16 primers that amplified reproducible polymorphic bands were selected for chemometric analysis.

The RAPD polymorphic bands of each sample were recorded as bands present {1} or bands absent {0} and the data were used to construct a pair-wise similarity matrix between genotypes using the Jaccard coefficient (64). The similarity coeficient is given as $J = a/(a + b + c)$, where a is the number of positive bands shared by both individuals x and y, and b and c are the numbers of fragments present in individuals x and y, respectively. These statistical analyses were carried out using the computer program NTSYS (65) (Exerter Softwares, Setauket, NY, USA). The chemometric analysis was performed using Einsight 3.0 software (Infometrix Inc., Seattle, WA, USA), which uses analysis by hierarchical groups as well as the main components. The construction of a dendogram makes it possible to observe the intercorrelations among the several genotypes, and also the same process estimates the Euclidean distance among the samples (66).

The genomic DNA amplification of horn fly populations using the 16 RAPD selected primers produced 321 fragments. These fragments varied from 1714 to 229 bp. The total number of bands produced by each primer varied from 10 (primer H20) to 28 (primer G4 and G16). From the 16 selected primers, 12 generated 15 bands or more. Regarding the total number of bands produced by population, the most polymorphic was RN (70 bands), followed by RJ (68 bands), MS and RS (62 bands), and RR (59 bands) (Table 12.2).

In this study, PCA and HCA analyses were carried out in order to perform a variable reduction and to identify the most useful variables to discriminate the five geographical regions. The plot of the principal components shows that Roraima (RR) was the farthest population, presenting zero similarity to the others, while the closest populations were Rio Grande do Sul (RS) and Mato Grosso do Sul (MS) with 0.063 similarity, and Rio de Janeiro (RJ) and Rio Grande do Norte (RN), which showed 0.036 similarity among these populations (Fig. 12.10). At the same time through PCA, it was possible to observe that the Brazilian populations of horn fly showed polymorphic loci by which they were able to be characterized genotypically through the OpE9, OpE11, OpE13, OpE15, OpG4, and OpH8 primers.

12.6. CONCLUDING REMARKS

This chapter discussed several aspects of DNA analysis by CE and microchip technologies using polymer solutions as the sieving matrix. Analysis of DNA is a multivariate system by nature and both the separation *and* the result of the separation are suitable to a large number of chemometric tools.

ACKNOWLEDGMENTS

The authors gratefully acknowledge the assistance of the staff at the University of Sao Paulo and the financial support from: Fundação de Amparo à Pesquisa

TABLE 12.2. Random amplified polymorphic DNA primers used and number of fragments generated in five different Brazilian populations of horn fly

Primer	Sequence	Number of RAPDs
OPE1	ccc aag gtc c	21
OPE9	ctt cac ccg a	14
OPE11	gag tct cag g	15
OPE10	cac cag gtg a	13
OPE13	ccc gat tcg g	17
OPE14	tgg cgc tga c	14
OPE15	acg cac aac c	24
OPE18	gga ctg cag a	27
OPG4	agc gtg tct g	28
OPG6	gtg act aac c	27
OPG16	agc gtc ctc c	20
OPG19	gtc agg gca a	15
OPH8	gaa aca ccc c	29
OPH12	acg cgc atg t	26
OPH16	tct cag ctg g	21
OPH20	ggg aga cat c	10
Total		321

Source: Reprinted with permission from Reference 63.

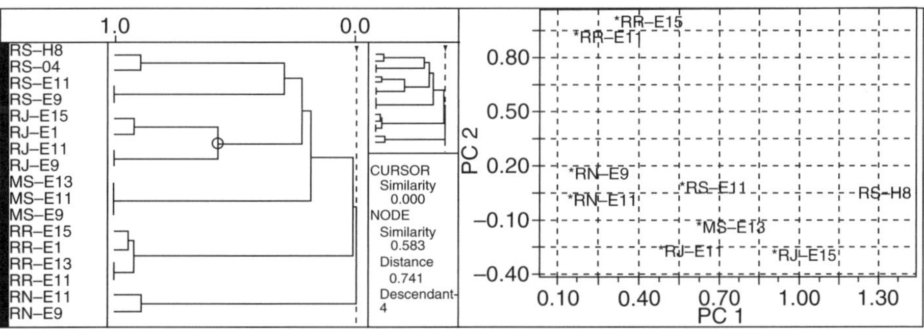

FIGURE 12.10. Relationships among Brazilian populations of *Haematobia irritans* based on principal components after the exclusion of the outliers. Reprinted with permission from Reference 63.

do Estado de São Paulo (FAPESP), Conselho Nacional de Desenvolvimento Científico e Tecnológico (CNPq), and Coordenação de Aperfeiçoamento de Pessoal de Nível Superior (CAPES). The authors would like to thank Ms. Amanda Van Gramberg from the Department of Chemistry, Materials and Forensic Science, University of Technology Sydney, Australia, for language assistance.

REFERENCES

1. Marshall, E. (2000) *Science*, **288**, 2294–2295.
2. Smaglik, P. (2000) *Nature*, **404**, 111.
3. Shendure, J. and Li, H. (2008) *Nature Biotechnol*, **26**, 1135–1145.
4. Catai, J.R. and Carrilho, E. (2003) *Electrophoresis*, **24**, 648–654.
5. Doherty, E.A.S., Kan, C.W., and Barron, A.E. (2003) *Electrophoresis*, **24**, 4170–4180.
6. Chrambach, A. and Rodbard, D. (1971) *Science*, **172**, 440–450.
7. Bishop, D.H., Claybrook, L., and Spiegel, M.S. (1967) *J Mol Biol*, **26**, 373–378.
8. Ruiz-Martinez, M.C., Berka, J., Belenkii, A., Foret, F., Miller, A.W., and Karger, B.L. (1993) *Anal Chem*, **65**, 2851–2858.
9. Catai, J.R. and Carrilho, E. (2004) *J Braz Chem Soc*, **15**, 413–420.
10. Carrilho, E. (2000) *Electrophoresis*, **21**, 55–65.
11. Heller, C. (2001) *Electrophoresis*, **22**, 629–643.
12. Kriváсsy, Z., Gelencser, A., Hlavay, J., Kiss, G., and Sárvári, Z. (1999) *J Chromatogr A*, **834**, 21–44.
13. Heller, C. (1999) *Electrophoresis*, **20**, 1962–1976.
14. Sinville, R. and Soper, S.A. (2007) *J Sep Sci*, **30**, 1714–1728.
15. Easley, C.J., Karlinsey, J.M., Bienvenue, J.M., Legendre, L.A., Roper, M.G., Feldman, S.H., Hughes, M.A., Hewlett, E.L., Merkel, T.J., Ferrance, J.P., and Landers, J.P. (2006) *Proc Natl Acad Sci USA*, **103**, 12272–12277.
16. Zhang, L., Dang, F., and Baba, Y. (2003) *J Pharm Anal*, **30**, 1645–1654.
17. Mukhopadhyay, R. (2009) *Anal Chem*, **81**, 1736–1740.
18. Quesada, M.A. (1997) *Curr Opin Biotechnol*, **8**, 82–93.
19. Mitnik, L., Salomé, L., Viovy, J.L., and Heller, C. (1995) *J Chromatogr A*, **710**, 309–321.
20. Grossman, P.D. and Colburn, J.C. (1992) *Capillary Electrophoresis: Theory and Practice*, Academic Press Inc., San Diego, CA.
21. Siouffi, A.M. and Phan-Tan-Luu, R. (2000) *J Chromatogr A*, **892**, 75–106.
22. Spendley, W., Hesat, G.R., and Himsworth, F.R. (1962) *Technometrics*, **4**, 441–461.
23. Nelder, J.R. and Mead, R. (1965) *Computer J*, **7**, 308–313.
24. Burton, K.W.C. and Nickless, G. (1987) *Chemometrics Intel Lab Sys*, **1**, 135–149.
25. Slater, G., Rousseau, J., Noolandi, J., Turmel, C., and Lalande, M. (1988) *Biopolymers*, **27**, 509–524.
26. Viovy, J.L. and Duke, T. (1993) *Electrophoresis*, **14**, 322–329.
27. Catai, J.R., Formenton-Catai, A.P., and Carrilho, E. (2005) *Electrophoresis*, **26**, 1680–1686.
28. Ali, B.A., Huang, T.H., Qin, D.N., and Wang, X.M. (2004) *Rev Fish Biol Fisheries*, **14**, 443–453.
29. Agarwal, M., Shrivastava, N., and Padh, H. (2008) *Plant Cell Rep*, **27**, 617–631.
30. Newton, A.C., Allnutt, T.R., Gillies, A.C.M., Lowe, A.J., and Ennos, R.A. (1999) *Trends Ecol Evolut*, **14**, 140–146.

31. Williams, J.G.K., Kubelik, A.R., Livak, K.J., Rafalski, J.A., and Tingey, S.V. (1990) *Nucleic Acids Res*, **18**, 6531–6535.
32. Welsh, J. and McClelland, M. (1990) *Nucleic Acids Res*, **21**, 7213–7218.
33. Caetano-Anollés, G., Bassam, B.J., and Gresshoff, P.M. (1991) *Biotechnology*, **9**, 553–556.
34. Fritsch, P. and Rieseberg, L.H. (1996) The use of random amplified polymorphic DNA (RAPD) in conservation genetics, in *Molecular Genetic Approaches in Conservation* (eds. T.B. Smith and R.K. Wayne), Oxford University Press, New York, pp. 54–73.
35. Williams, J.G.K., Hanafey, M.K., Rafalski, J.A., and Tingey, S.V. (1992) *Methods Enzymol*, **218**, 704–740.
36. Swaminathan, B. and Barrett, T.J. (1995) *J Microbiol Methods*, **23**, 129–139.
37. Ferreira, M.E. and Grattapalia, D. (1998) *Introdução ao uso de Marcadores Moleculares em Análise Genética*, EMBRAPA-CENARGEN, Brasília.
38. Dassanayake, R.S. and Samaranayare, L.P. (2003) *Crit Rev Microbiol*, **29**, 1–24.
39. Wang, D., Waye, M.M.Y., Taricani, M., Buckingam, K., and Sandham, H.J. (1993) *Biotechniques*, **14**, 214–218.
40. Eriksson, L., Antti, H., Gottfries, J., Holmes, E., Johansson, E., Lindgren, F., Long, I., Lundstedt, T., Trygg, J., and Wold, S. (2004) *Anal Bioanal Chem*, **380**, 419–429.
41. Eriksson, L., Johansson, E., Kettaneh-Wold, N., and Wold, S. (2001) *Multi- and Megavariate Data Analysis: Principles and Applications*, Umetrics AB, Umea.
42. Massart, D.L., Vandeginste, V.G.M., Deming, S.N., Michotte, Y., and Kaufman, L. (1988) *Chemometrics: A Textbook*, Elsevier, Amsterdam.
43. Stace, C.A. (1989) *Plant Taxonomy and Biosystematics*, 2nd ed., Edward Arnold Publishers, London.
44. White, R.L., Wentzell, P.D., and Beasy, M.A. (1993) *Anal Chim Acta*, **217**, 333–346.
45. Xia, X., Bollinger, J., and Ogram, A. (1993) *Mol Ecol*, **4**, 17–28.
46. Demeke, T. and Adams, R.P. (1994) The use of PCR-RAPD analysis of plant taxonomy and evolution, in *PCR Technology: Current Innovations* (eds. H.G. Griffin and A.M. Griffin), CRC Press, Boca Raton, FL, pp. 179–191.
47. Goldberg, D.E. (1989) *Genetic Algorithms in Search, Optimization and Machine Learning*, Addison-Wesley, Berkeley, CA.
48. Leardi, R. (2003) Nature-inspired methods in chemometrics: Genetic algorithms and artificial neural networks, in *Data Handling in Science and Technology*, Vol. **23** (ed. R. Leardi), Elsevier, Amsterdam.
49. Lucasius, C.B. and Kateman, G. (1994) *Chemometr Intell Lab Syst*, **25**, 99–146.
50. LIPS (Laboratory for Intelligent Process Systems). Purdue University, http://cobweb.ecn.purdue.edu/~lips/ (accessed August 4, 2009).
51. Marczyk, A. (2004) The Talk Origins Archive, Genetic Algorithms and Evolutionary Computation, http://www.talkorigins.org/faqs/genalg/genalg.html (accessed August 4, 2009).
52. Holland, J. (2007) Genetic algorithms. L. Tesfatsion homepage, Department of Economics, Iowa State University, http://www.econ.iastate.edu/tesfatsi/holland.GAIntro.htm (accessed August 4, 2009).

53. Leardi, R. (2007) *J Chromatogr A*, **1158**, 226–233.
54. Ramadan, Z., Song, X.H., Hopke, P.K., Johnson, M.J., and Scow, K.M. (2001) *Anal Chim Acta*, **446**, 233–244.
55. Nei, M. (1972) *Am Nat*, **106**, 283–292.
56. Weir, B.S. (1990) Phylogeny construction, in *Genetic Data Analysis* (ed. B.S. Weir), Sinauer Associates, Inc. Publishers, Sunderland, MA.
57. Nei, M. (1987) *Molecular Evolutionary Genetics*, Columbia University Press, New York.
58. Michener, C.D. and Sokal, R.R. (1957) *Evolution*, **11**, 130–162.
59. Jaccard, P. (1908) *Bull Soc Vaud Sci Nat*, **44**, 223–270.
60. Beebe, K.R., Pell, R.J., and Seasholtz, M.B. (1998) *Chemometrics, a Practical Guide*, Wiley, New York.
61. Sudjianto, A. and Wasserman, G.S. (1996) *IIE Trans*, **28**, 1023–1028.
62. Jun, B.S., Ghosh, T.K., and Loyalka, S.K. (2000) Determination of CHF pattern using principal component analysis and the hierarchical clustering method (critical heat flux in reactors). *Proceedings of the American Nuclear Society 2000 Summer Meeting*, June 4–8, San Diego, CA. In *Trans Am Nucl Soc*, **82**, 250–251 (2000).
63. Brito, L.G., Regitano, L.C.A., Funes-Huacca, M.E., Carrilho, E., and Borja, G.E.M. (2007) *Pesq Vet Bras*, **27**, 1–5.
64. Jaccard, P. (1901) *Bull Soc Vaud Sci Nat*, **37**, 547–579.
65. Rohlf, F.J. (1993) *NTSYS-PC: Numerical Taxonomy and Multivariate Analysis System, Version 1.7*, Aplied Biostatistic, Setauket, NY.
66. *Einsight User's Manual* (1991) Infometrix, Seattle, WA.

CHAPTER 13

EXPLORATORY DATA ANALYSIS AND CLASSIFICATION OF CAPILLARY ELECTROPHORETIC DATA

MELANIE DUMAREY, BIEKE DEJAEGHER, ALEXANDRA DURAND, and YVAN VANDER HEYDEN*
*Department of Analytical Chemistry and Pharmaceutical Technology, Vrije Universiteit Brussel—VUB, Brussels, Belgium

CONTENTS

Chemometric Methods in Capillary Electrophoresis. Edited by Grady Hanrahan and Frank A. Gomez

13.1. INTRODUCTION

Capillary electrophoresis (CE) is well known for its fast separation speed and high efficiency (1–3). Nowadays, the common detector in CE systems registers by default four signals per second. However, although often only the default settings are used, the analyst is free to choose the number of collected signals between 0.5 and 32 per second in the software of the equipment. As a consequence, the resulting electropherogram consists of a huge amount of numbers, which is equally complex as a chromatogram. An electropherogram recorded at four signals per second during 10 min, for example, can easily contain 2400 signals measured at 2400 consecutive scan times. A common practice to handle this type of data is to inspect the electropherograms visually and then select the peaks of interest (with their corresponding data points) to calculate quantitative aspects. For instance, the concentration of the main compound of a mixture can be determined based on its peak area (4). In that case, only few data points from the entire electropherogram are employed.

In order to gain maximal information from the multivariate character of the electrophoretic data, chemometric tools can be applied. They enable the handling of a large amount of output variables, resulting in an easily interpretable result based on the complete electropherograms (5). Before starting the chemometric treatment, the electrophoretic data need to be organized in a matrix, where each row represents one CE profile and each column the signal measured at a specific time (Fig. 13.1). It is important that corresponding information from different electropherograms, for instance, peak maxima, are located in the same column of the matrix. Therefore, warping or peak-aligning techniques can be used (Fig. 13.1). Eventually, chemometric techniques, such as exploratory analysis, classification, peak resolution, or multivariate calibration, can be applied.

In the first mentioned type of application, electrophoretic data are subjected to exploratory analysis techniques, such as principal component analysis (PCA) (5–8), robust PCA (rPCA) (9–13), projection pursuit (PP) (6, 14–18), or cluster analysis (8, 19, 20). They all result in a simple low-dimensional visualization of the multivariate data. As a consequence, it will be easier for the analyst to get insight in the data in order to see whether there is a given

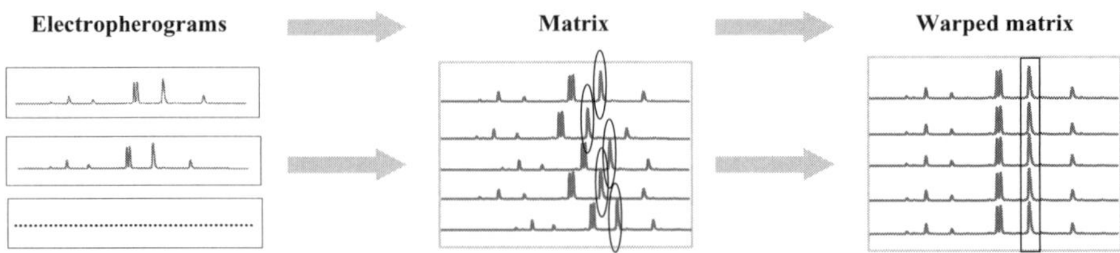

FIGURE 13.1. Schematic view of data pretreatment (peak alignment) prior to chemometric analysis.

structure or clustering tendency in the data set. Practically, these techniques enable the grouping of similar electropherograms and the detection of strongly deviating, that is, outlying, samples. This information cannot always be deduced by comparing the electropherograms visually.

A second interesting application is the classification of pharmaceutical samples based on their electropherograms. Suitable techniques for this purpose are linear discriminant analysis and quadratic discriminant analysis (LDA and QDA) (8, 21–23), k-nearest neighbor (kNN) (8, 24, 25), classification and regression tree (CART) (26–28), partial least squares discriminant analysis (PLSDA) (8), soft independent modeling of class analogy (SIMCA) (29–31), and support vector machines (SVMs) (32–34). These are all supervised pattern recognition methods, which means that they calculate classification rules based on a training set of samples belonging to a priori known classes (8).

Third, chemometric tools can be useful to resolve overlapping peaks in an electropherogram. Orthogonal projection approach (8, 35, 36), evolving factor analysis (8, 36), or window factor analysis (8, 36) are commonly used for liquid chromatographic data obtained with diode array detection (high performance liquid chromatography–diode array detector [HPLC–DAD]). These techniques should also be applicable in curve resolution of CE data. Finally, multivariate calibration can be applied on electrophoretic data in order to model and predict a property of interest of the samples, for example, the antioxidant, antimicrobial, or cytotoxic activity. Commonly used techniques in this field are principal components regression and partial least squares (PLS) regression (8).

In this chapter, different techniques for exploratory analysis and classification of CE data will be discussed and supplemented with some theoretical background. Examples of the application of each technique in the CE field will also be provided, if available. If not, the technique will be illustrated with a chromatographic or spectroscopic case study, because mathematically, they deliver an output similar to electropherograms.

13.2. DATA PRETREATMENT

Before exploratory analysis or classification is started, the electrophoretic data need to be organized in an $m \times n$ matrix $\mathbf{X}$. Each row (1 until m) of this matrix represents the electropherogram of a sample (with m the number of electropherograms), while each column (1 until n) represents a given time at which the signal was measured (with n the number of scan points), and the content of the matrix contains the measured signals (intensity, absorption).

It is generally known that the repeatability of CE analyses is not optimal due to irreproducible flow rates (37). Therefore, it is recommended to align the corresponding peaks in the different electropherograms before chemometric data analysis (exploration or classification) is started. This alignment results in a data matrix, where the signals of the corresponding peaks of the different samples are located in the same column of the matrix (Fig. 13.1). Different warping techniques, such as correlation optimized warping (COW) (38), dynamic time warping (39), and parametric time warping (40), are applicable for this purpose. COW, one of the most popular warping techniques, aligns two electropherograms by maximizing the correlation between both signals by piecewise stretching and compression. However, warping only allows correcting for peak shifts, but does not correct for other disadvantageous consequences of the irreproducible flow rates, like inconsistent injection volumes and irreproducible detector responses. This implies that electrophoretic data are less suited for chemometric analysis than HPLC data.

Besides the warping or peak-aligning techniques, other often-applied preprocessing techniques are column-centering, normalization, baseline correction, and multiplicative signal correction (MSC) (8, 41, 42). Column-centering, which removes the column mean from each corresponding column, is frequently applied because it is a basic and essential part of many techniques, such as, for example, PCA and PLS (8). Normalization scales the rows to a constant total, and can, for instance, be achieved by dividing each row by its corresponding norm or by the sum of the data of each row (~electropherogram) (8). This preprocessing technique can, for example, be useful to remove uncontrolled variations of the general signal intensity, such as differences caused by varying amounts of injected samples. In CE, this is even more important than in HPLC, because of the lower reproducibility of the injection volume. When the baseline is drifting, a baseline correction is recommended. This correction is already included in many instrumental data treatment software. If not, chemometric pretreatment techniques can be applied for this purpose. Although originally developed for spectroscopic data pretreatment, MSC is a preprocessing technique that can also be applied to correct for irreproducible detector responses in HPLC or CE data (41, 42). The irreproducible responses can, for instance, originate from measurements coming from different companies, analysts, instruments, and times. The CE response correction leads to data where all electropherograms have the same zero component response, that is, the same average zero level. In contrast to baseline correction, where the baseline of each electropherogram is used to remove the shift in that given electropherogram, MSC uses an average baseline from different electropherograms to remove the shift in all electropherograms.

13.3. EXPLORATORY DATA ANALYSIS

13.3.1. PCA

13.3.1.1. Theory. PCA is a frequently used variable reduction technique, which can be used to visualize the objects of a multivariate data set in a lower-dimensional space. This technique calculates new latent variables, called principal components (PCs), which are linear combinations of the original manifest variables, describing the maximal variance of the data. The PCs are mutually orthogonal and the first contains the maximal variance (Fig. 13.2). This reduction of the number of variables finds a compromise between two conflicting objectives: choosing a lower dimensional feature space and keeping maximally the information. Scores are determined by projecting the samples from the original data space on the PCs. Finally, a 2-dimensional (occasionally 3-dimensional) visualization giving information about the samples can be obtained by plotting the scores on two PCs versus each other. The largest amount of variation will then be shown in the PC1–PC2 score plot, providing information related to the (dis)similarity of the samples. For example, in Figure 13.3a, the PC1–PC2 score plot is given for a data set, where for 10 different vegetables, subjected to different cooking styles, the concentrations of seven elements, that is, calcium (Ca), copper (Cu), iron (Fe), zinc (Zn), potassium (K), sodium (Na), and magnesium (Mg), were measured. The score plot, obtained after autoscaling the variables to zero mean and unit variance, clearly allows distinguishing the celery samples from all other samples along PC2. Also the white cabbage, the carrot, the red cabbage, the onion, and the chicory samples are situated in rather distinct clusters on the plot. On the other hand, the cauliflower, the leek, the French bean, and the sprout samples are not clearly separated in distinct groups. PCA also allows calculating the contribution of each original variable to the scores of the objects on a PC, that is, the loading. The relative importance of the variables can then easily be determined by plotting the loadings on two PCs versus each other (5–7). For example, in Figure 13.3b, the PC1–PC2 loading plot, obtained after autoscaling the variables, is given for the above vegetable data set. From this plot, it can be concluded that along PC2, mainly the sodium concentration is responsible for the clear separation of the celery samples from all other vegetables.

The scores and loadings of a data set can be determined by the singular value decomposition method (8), which decomposes the $m \times n$ matrix $\mathbf{X}$ according to the following relationship:

$$\mathbf{X} = \mathbf{U} \cdot \mathbf{\Lambda} \cdot \mathbf{V}^{\mathrm{T}} \qquad \text{(Eq. 13.1)}$$

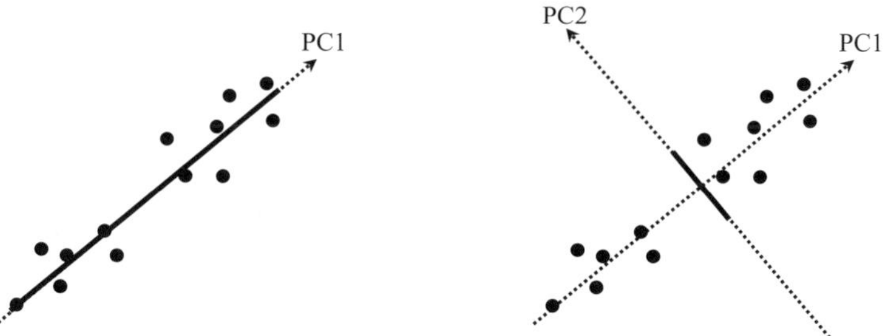

FIGURE 13.2. Principal component analysis: definition of PC1 and PC2 for a two-dimensional data set.

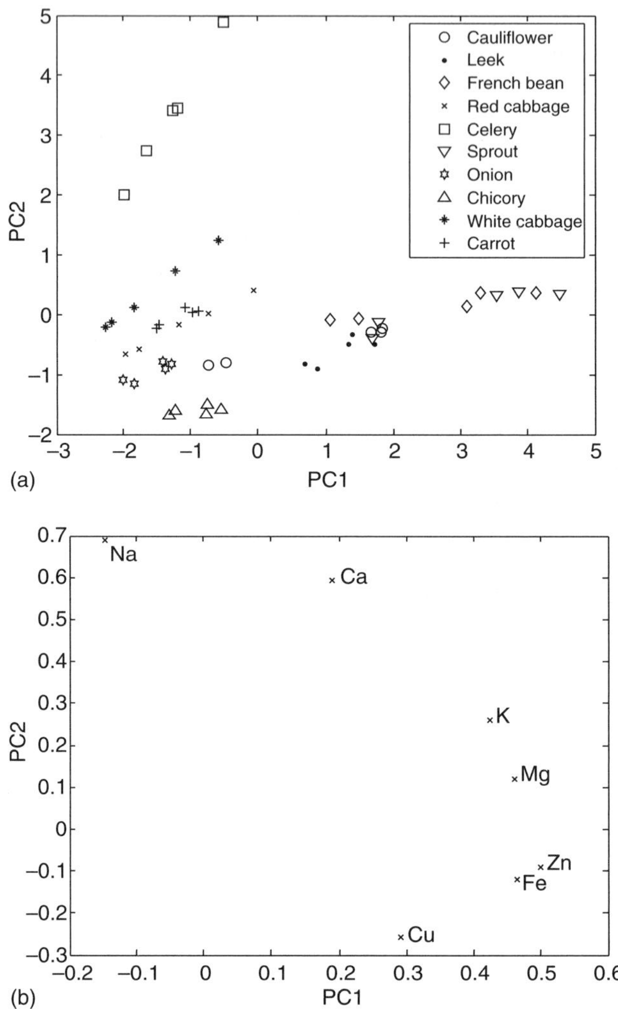

FIGURE 13.3. (a) A PC1–PC2 score plot representing different types of vegetables, which were subjected to different cooking styles. The plot is based on the concentrations of several elements. (b) The corresponding loading plot.

where the $m \times r$ matrix $\mathbf{U}$ is related to the scores of the objects, the $n \times r$ matrix $\mathbf{V}$ is related to the loadings of the manifest variables, and the $r \times r$ matrix $\mathbf{\Lambda}$ is the singular values matrix. The dimension r can at the most be equal to the smaller of the dimensions m or n. The diagonal of $\mathbf{\Lambda}$ contains the square roots of the so-called eigenvalues and gives information about the variation explained by the successive PCs. The PC associated with the highest eigenvalue determines the direction of the maximal variance.

After singular value decomposition, the $m \times r$ score matrix $\mathbf{S}$ can be calculated with the following equation:

$$\mathbf{S} = \mathbf{U} \cdot \mathbf{\Lambda}^{\alpha} \quad \text{(Eq. 13.2)}$$

and the loading matrix **L** can be calculated according to:

$$\mathbf{L} = \mathbf{V} \cdot \mathbf{\Lambda}^{\beta} \quad \text{(Eq. 13.3)}$$

where α and β are factor scaling coefficients, usually assigned with values 0, 0.5, or 1. Depending on the choice of α and β, different features of the data in the factor space can be reconstructed (8). When $\alpha = 1$, the cross products between the rows of the data can be reproduced, while when $\beta = 1$, those cross products between the columns can be reproduced. If the data in **X** should be reconstructed, the requirement $\alpha + \beta = 1$ should be fulfilled. Therefore, frequently $\alpha = 1$ and $\beta = 0$ are selected (5–8).

13.3.1.2. Applications. Reid et al. (43) developed micellar capillary electrophoresis separations of several opium extracts in order to differentiate samples from four different locations. The resulting electropherograms were subjected to PCA, resulting in score plots. The PC1–PC3 score plot (Fig. 13.4) clearly reveals four groups corresponding to the regions of origin of the samples.

In another application, Sturm et al. (44) evaluated CE–mass spectrometric (capillary electrophoresis–mass spectrometry [CE–MS]) data from different *Corydalis* species, that is, *Corydalis cava* from two different regions (CM and CB), *Corydalis pumila* (PU), *Corydalis intermedia* (IN), and *Corydalis solida* from two different regions (SG and SE). Exploring the CE–MS data with PCA succeeded in distinguishing the six *Corydalis* species samples. In this example, only a comprehensive list of peak areas of analytes was chosen as initial data

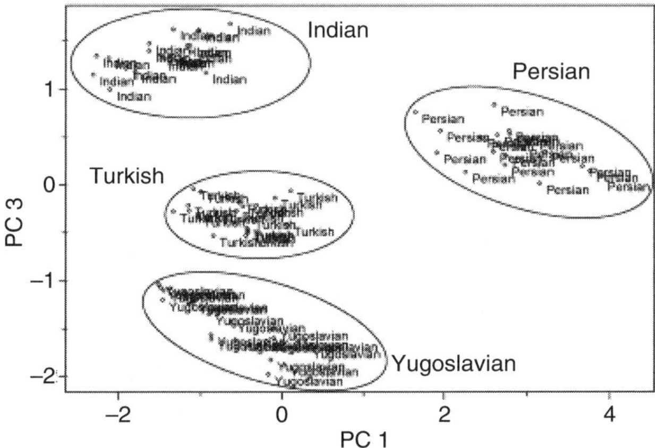

FIGURE 13.4. PC1–PC3 score plot of the electropherograms for opium samples from four different locations. Reproduced with permission from Reid et al. (43).

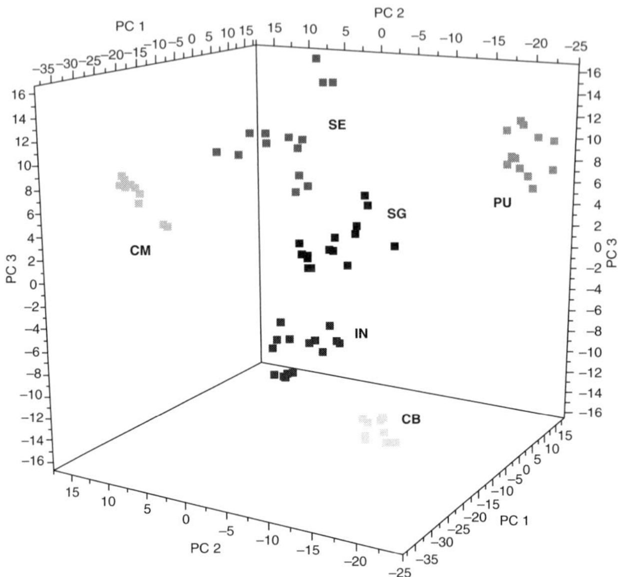

FIGURE 13.5. Three-dimensional score plot (PC1–PC2–PC3) of different *Corydalis* species electropherograms. Reproduced with permission from Sturm et al. (44). See color insert.

matrix instead of the complete electropherogram. A 3-dimensional score plot (Fig. 13.5) was drawn in order to discriminate visually the different clusters.

In Szymańska et al. (45), PCA was performed on electrophoretic data of urinary nucleoside profiles, in order to distinguish profiles of healthy controls from cancer patients. Prior to PCA, the data were preprocessed using baseline correction, COW, and normalization according to creatinine concentration. After adequate preprocessing, PCA allowed us to reveal data structure and to evaluate differences between the healthy controls and the cancer patient profiles.

13.3.2. rPCA

13.3.2.1. Theory. The variance criterion (i.e., maximizing the variance in the data) of classical PCA is very sensitive to outlying samples. As a consequence, the real structure of the data cannot always be revealed. To overcome this problem, rPCA (9–13) was introduced, which aims to obtain PCs that are less influenced by outliers. Additionally, robust methods should be able to detect the outlying observations. These goals are achieved by applying a more robust parameter (than variance) as projection index.

Several algorithms are already developed to perform rPCA. Since the algorithm proposed by Croux and Ruiz-Gazen (9,10) is generally applied and is the basis for some other rPCA methods, it will be explained in more detail.

The first step of Croux and Ruiz-Gazen making PCA more robust is centering the data with a robust criterion, the L1-median, that is, the point which minimizes the sum of Euclidean distances to all points of the data. In a next step, directions in the data space, which are not influenced by outliers, are determined by maximizing a robust parameter, the Q_n estimator. To calculate this estimator, first all objects are projected onto normalized vectors passing through each point and the L1-median center. Then for each projection, the Q_n, that is, the first quartile of all pairwise differences, is calculated as follows:

$$Q_n = 2.2219 * c_n * \{|z_i - z_j|; i < j\}_{(k)} \qquad \text{(Eq. 13.4)}$$

where $k = \binom{h}{2} \approx \binom{m}{2}\Big/4$, $h = [m/2] + 1$, $(z_1, \ldots, z_n)$ is the univariate data set and c_n is a correction factor, which tends to 1 when the number of objects, m, increases. Then the vector with the maximal value for this projection index, that is, a robust PC, is selected. When the data are projected onto the orthogonal complement of the robust PC, the above procedure can be repeated. In that way, new robust PCs are determined until a certain number of vectors is calculated. As mentioned before, other algorithms for rPCA are also available. Hubert et al., for instance, developed the RAPCA algorithm (11), as well as the ROBCA algorithm (12), modified versions of the approach of Croux and Ruiz-Gazen.

The rPCA technique allows one to construct a score diagnostic plot (Fig. 13.6) to detect outlying samples. In such a graph, the distance of an object from the data majority (robust distance) versus its distance from the rPCA

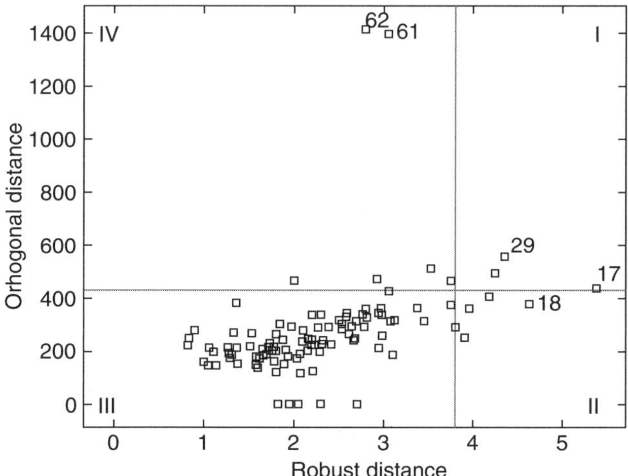

FIGURE 13.6. The score diagnostic plot of 110 green tea fingerprints. The orthogonal distance is plotted versus the robust distance. The cutoff values are determined in the space of five rPCs. Reproduced with permission from van Nederkassel et al. (46).

model space (orthogonal distance) is plotted. Samples found in quadrant III are considered ordinary samples. When a sample exceeds the cutoff value for the orthogonal distance (quadrants I and IV), then it will influence the model building in a negative way, and these samples are considered outliers in the PCs' space. On the other hand, a sample is considered an outlier in the robust PC space when the cutoff value for the robust distance is exceeded (quadrants I and II). This implies that such an outlier will not always influence the model negatively, but only when its orthogonal distance is also high (quadrant I). Samples in quadrants I, II, and IV are considered bad leverages, good leverages, and orthogonal outliers, respectively. Nevertheless, before removing any outlying samples from quadrants I, II, and/or IV, it should be evaluated whether it is necessary to eliminate them for further analysis. Although outliers might exhibit some extreme characteristics, in some situations, it can be considered unnecessary to remove them.

13.3.2.2. Applications. Since no applications were found in CE data handling, a chromatographic example of van Nederkassel et al. (46) is provided. They predicted the total antioxidant capacity of green tea from chromatographic fingerprints. rPCA was applied to detect the outliers, resulting in the score diagnostic plot of Figure 13.6. Prior to the multivariate calibration, the orthogonal outliers (quadrant IV), indicated with 61 and 62, were removed, in order to obtain the best possible predictions for future samples. Samples 61 and 62 are replicates of one tea sample and they contain an exceptionally high peak, which in other samples is at least five times smaller. Therefore, in this data set, they can be considered as atypical samples. For the other samples that exceed the cutoff value(s), with samples 17, 18, and 29 as the most extreme, the distance to the majority of objects is not high compared with that of the two extreme outlying objects (61 and 62), and therefore, they were not removed.

13.3.3. PP

13.3.3.1. Theory. PP is also a variable reduction method, very similar to PCA. In fact, PP can be considered a generalization of classical PCA (6, 14–18). While in PCA the PCs are determined by maximizing variance, in PP, the latent variables, called the projection pursuit features (PPFs), are obtained by optimizing a given projection index that describes the inhomogeneity of the data, instead of its variance (6, 18). In the literature, many PP indices have been described.

To determine the possible directions in the data space, the algorithm proposed by Croux and Ruiz-Gazen (9) (see also rPCA), for instance, can be used. First, the data are preprocessed, called sphering or whitening, leading to a zero mean (first central moment) and a unit variance (second central moment) for each variable. Then all objects are projected onto all possible normalized directions going through the objects and the data origin, in contrast to PCA where the directions are not required to contain objects. The projection index for all projections is then estimated, and consecutively, that direction with the highest index is selected. The next direction with the highest index is found in the residual data space, that is, the space remaining after removing one projection from the former space. The procedure continues iteratively, until the desired number of orthogonal directions is obtained. Finally, all objects are projected onto the found directions, resulting in the PPFs.

By applying specially designed projection indices, the visual detection of clusters and outliers should be more evident than by using PCA. One of the most popular indices is entropy, which is a measure for the structure in the data. It can be calculated as follows:

$$h(x) = \int f(x)\log(f(x))\,\mathrm{d}x \qquad \text{(Eq. 13.5)}$$

where $f(x)$ is a density estimate of the projected data. Maximizing this index will lead to nonuniform distributions of the projections, and, as a consequence, possibly present clusters will be revealed (6, 18). The entropy can also be approximated by higher-order cumulants, for instance by the kurtosis index,

$$kurt\,(x) = \frac{x(4)}{(x(2))^4} = \frac{x(4)}{\sigma^4} \qquad \text{(Eq. 13.6)}$$

where $x(4)$ is the fourth central moment, and $x(2)$ the second central moment or the standard deviation σ (17, 18). The kurtosis equals zero for a normally distributed projection. Such projection is noninteresting from the PP point of view, because PP searches for inhomogeneities. Both a positive and a negative kurtosis value represent a measure of deviation of a projection from the normal distribution. When the data contain clusters, the distribution becomes multimodal and negative. For instance, kurtosis goes through a minimum for two clusters containing the same number of objects. The larger (positive) the entropy value is, the larger the data inhomogeneity is. Thus, extreme observations, that is, possible outliers, are highlighted.

Another well-known measure is the Yenyukov index, which is the ratio, Q, of the mean of all inter-object distances, D, and the average nearest neighbor distance, d. When objects are located in the same cluster clearly separated from the other data, the average nearest neighbor distance will be small and the average inter-objects distance large. As a consequence, Q will be large when clusters are present in the data. Clusters in data can thus be revealed by maximizing the Yenyukov index (6).

13.3.3.2. Applications. Schoonjans and Massart (47) combined mass spectrometric (MS) and infrared (IR) spectra of compounds in order to characterize the (dis)similarity of their chemical structures by means of chemometric exploration. The application of PP on the log-transformed combined spectra resulted in a separation of the steroids from the amino acids and the β-blockers along the PPF1 direction. Along the PPF2 direction, the β-blockers were found in the lower part of the plot, while the groups of amino acids and steroids fell apart. PP showed a separation of groups of compounds, and also allowed detection of inhomogeneities in the data, that is, two outliers were indicated. These latter were much more difficult to distinguish on the PCA plots.

13.3.4. Cluster Analysis

13.3.4.1. Theory. The goal of cluster analysis is to group objects based on their values for a set of variables (8, 19, 20). The clustering techniques can be divided in hierarchical and nonhierarchical methods. The latter contain, among others, partition methods, density methods, and heuristic methods (19, 20). However, hierarchical clustering (8, 19, 20) dominates the applications, and therefore, only these methods will be further explained. In divisive hierarchical clustering, the data set is divided into smaller partitions, optimizing a given criterion (correlation or distance). When the most similar objects are sequentially merged in clusters (based on criteria as correlation or distance) until one big cluster is obtained, the hierarchical clustering technique is called agglomerative. Based on the (dis)similarity criteria used, different agglomerative hierarchical clustering techniques can be distinguished. In linkage clustering, the distance between two groups is optimized. However, this distance can be calculated in different ways. In single linkage, the distance considered between two groups is the smallest distance between two objects of both groups, while in complete linkage it is the largest distance between two objects of two groups. In (weighted) average linkage, the (weighted) average of the distances between all objects of both groups is applied. When the hierarchical clustering is based on the centroid criterion, the squared Euclidean distances between the centroids of two groups are maximized. The Ward method uses a heterogeneity criterion, which is defined as the sum of the squared distances of each member of a cluster to the centroid of the cluster. When objects and/or clusters are merged, the increase of the sum of heterogeneities should be as small as possible. The above-discussed criteria are most common; additional ones can be found in (19, 20). In all these agglomerative clustering techniques, the distances between two objects i and i' are commonly calculated with the equation for the Euclidean distance (8):

$$D_{ii'} = \sqrt{\sum_{j=1}^{n} (x_{ij} - x_{i'j})^2} \qquad \text{(Eq. 13.7)}$$

where n is the number of variables. Other possible measures for (dis)similarity can be found in References 19 and 20.

The result of hierarchical clustering methods can be visualized in a dendrogram (Fig. 13.7). The higher two objects are connected in the dendrogram, the more dissimilar they are. The hierarchical analyses do not naturally recover

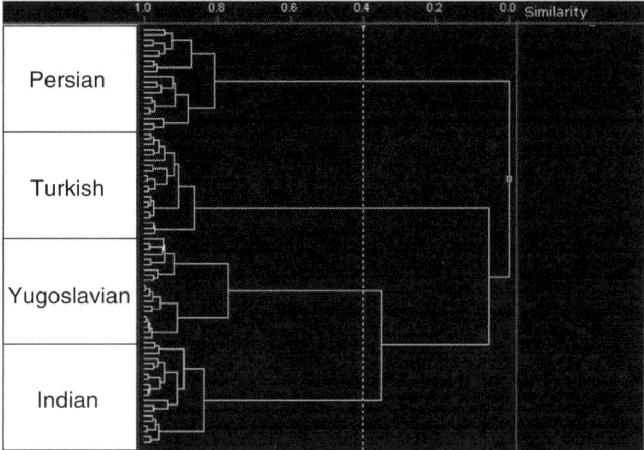

FIGURE 13.7. Dendrogram for opium samples from four different locations. Reproduced with permission from Reid et al. (43).

distinct clusters, but this can be accomplished by cutting the dendrogram at an appropriate point, which is determined arbitrarily by the analyst.

13.3.4.2. Applications. Reid et al. (43) performed hierarchical clustering on the earlier described micellar CE data, which resulted, as with PCA, in the distinction of opium samples from four different locations. The authors did not mention the similarity criterion used for the clustering. In the dendrogram (Fig. 13.7), a similarity value of 0.4 was set as cutoff value to distinguish the different groups.

A spectrometric application was performed by López-Sánchez et al. (48), who applied hierarchical clustering with the Ward algorithm on attenuated total reflection Fourier transform infrared spectra of toothpastes in order to establish different groups in the sample population.

13.4. CLASSIFICATION

The aim of supervised classification is to create rules based on a set of training samples belonging to a priori known classes. Then the resulting rules are used to classify new samples in none, one, or several of the classes. Supervised pattern recognition methods can be classified as parametric or nonparametric and linear or nonlinear. The term parametric means that the method makes an assumption about the distribution of the data, for instance, a Gaussian distribution. Frequently used parametric methods are LDA, QDA, PLSDA, and SIMCA. On the contrary, kNN and CART make no assumption about the distribution of the data, so these procedures are considered as nonparametric. Another distinction between the classification techniques concerns the linearity or nonlinearity of the method, that is, the nature of the function used to discriminate the different classes. Examples of linear methods are LDA, QDA, PLSDA, and SIMCA. Among the nonlinear procedures, artificial neural networks (ANNs) and SVM are frequently applied when the data set presents some nonlinear variability.

In general, supervised learning techniques, such as multivariate calibration or classification methods, use a calibration or training set, respectively, in order to build the model or to obtain the classification. In case of classification methods, the classes to which the objects of the training set belong are a priori known. This knowledge is then used to obtain the classification by means of a given technique, hence the name supervised learning. To validate the predictive ability of the calibration model or the classification, either a cross-validation (CV) procedure, or an independent test or prediction set is used. Preferably, the latter approach is applied. Then, the predictive character of the model is evaluated by means of its root mean square error of prediction. However, in cases where the number of objects is small, a division of the data into a calibration/training set and an independent test set is not possible. In these cases, often the CV approach is used, where the root mean square error of CV will be evaluated in order to validate the model.

13.4.1. LDA and QDA

13.4.1.1. Theory. LDA, a popular method for supervised classification, was introduced by Fisher in 1936 (21). The goal of this method is to classify the samples, establishing a linear function based on the variables x_i (i ranges from 1 to n, the number of considered variables), which separates the classes existing in the training set (Fig. 13.8). Classification is based on the interclass discrimination (22). It is a parametric method because the method assumes that the distribution of the samples in the classes is Gaussian.

Similar to PCA, LDA is a feature reduction method. For this purpose, a 1-dimensional space, that is, a line, on which the objects will be projected from

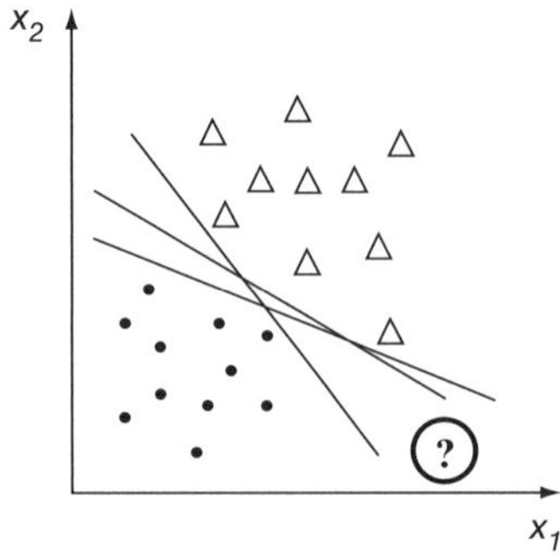

FIGURE 13.8. Principle of classification with LDA.

higher-dimensional space, is determined. Whereas PCA selects the first PC direction according to maximal data variance, LDA selects the direction that achieves maximal separation among the classes. The thus obtained latent variable is a linear combination of the original variables, and this function is called a canonical variate. When k classes are present, $k - 1$ canonical variates can be determined. Similar to PC1–PC2 score plots, the data can be visualized 2-dimensionally by plotting one canonical variate against another (Fig. 13.9).

The discriminant power of the variables will be high when the centroids of the two classes of samples are sufficiently distant from each other and when the samples in the classes are dense. This means that the variance between classes is higher than the variances in the classes. LDA will search a linear function, D, of the variables, which maximizes the ratio between the variances of two classes K and L (8). The discriminant function for n variables is given by the following equation:

$$D = \mathbf{w}^T \mathbf{x} + w_0 \qquad \text{(Eq. 13.8)}$$

where the weight vector **w** and the weight w_0 are adapted to the characteristics of the data to allow the discrimination, and **x** is the vector containing the variables (8).

QDA is identical to LDA, but this method is based on a quadratic classification curve instead of a straight line. The data must be normally distributed as for the LDA method. QDA is thus a linear parametric method.

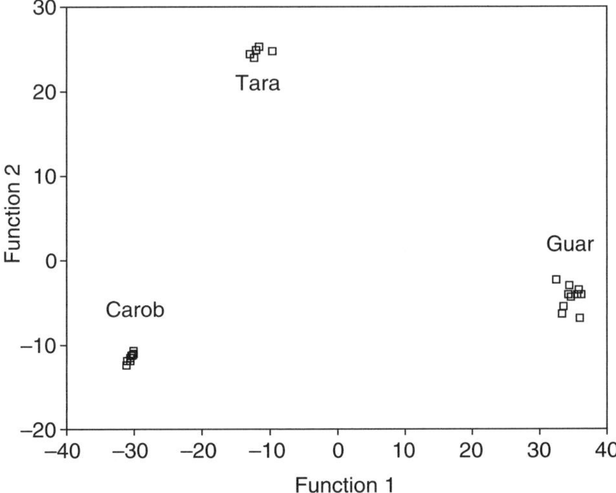

FIGURE 13.9. Projection of carob, tara, and guar gum samples on the plane of the two discriminant vectors showing the resolution between the three categories. Reproduced with permission from Ruiz-Ángel et al. (49).

LDA cannot be used if the number of variables (n) is higher than the total number of samples (m), while QDA requires that the number of variables (n) must be lower than the number of objects in the smallest class ($m_{smallest}$) (8). These problems can be overcome by reducing the number of variables with PCA prior to LDA or QDA (23).

13.4.1.2. Applications. Ruiz-Ángel et al. (49) separated the proteins of *Leguminosae* gums with capillary zone electrophoresis. The characteristic peaks of the resulting protein profiles were subjected to LDA, which was capable of correctly classifying all samples in both the calibration and prediction set in three classes, which were different types of *Leguminosae* gums, that is, carob, guar, and tara gum (Fig. 13.9).

Beltrán et al. (50) succeeded in classifying 172 Chilean wines according to the type of grapes (cabernet sauvignon, merlot, and carménère). First, phenolic compound chromatograms were developed with HPLC–DAD. Second, features were extracted from the chromatographic data with different feature extraction techniques, like discrete Fourier transform and Wavelet transform. Finally, next to other different classification techniques, LDA and QDA were applied. From CV, both methods were found to result in acceptable correct classification rates without statistically significant difference between both rates.

13.4.2. kNN

13.4.2.1. Theory. kNN is a nonparametric method based on the distance measurements between an unknown object and all others objects present in the training set (8). First, a small number (k) of nearest neighbors, that is, objects of the training set with the smallest distances to the unknown sample, is selected. Usually, distance measures such as the Euclidean or the Mahalanobis distances are employed for this purpose. However, for strongly correlated variables, a correlation-based measure as the correlation coefficient will be preferred. The k-value, preferably a small number (e.g., 3 or 5), is determined by optimizing the predictive ability of the kNN method by testing several k-values. Finally, a majority rule is applied, which classifies the unknown in the group to which the majority of the kNN belong (8, 24, 25).

The kNN method is illustrated in Figure 13.10, where the data obviously contain two clusters K and L. In the first case (Fig. 13.10a), the unknown sample (▲) is situated in between the samples of class L and the kNN method classifies the sample correctly in that class. When the unknown object (■) is located at the border of, for instance, class L, but also close to the other class (Fig. 13.10b), kNN will allocate the object to the class with the majority of the k nearest objects, in this case class L. In the third case (Fig. 13.10c), the unknown (◆) is situated at the border of class K and far from class L. Since all kNNs are belonging to class K, the object will be classified in that

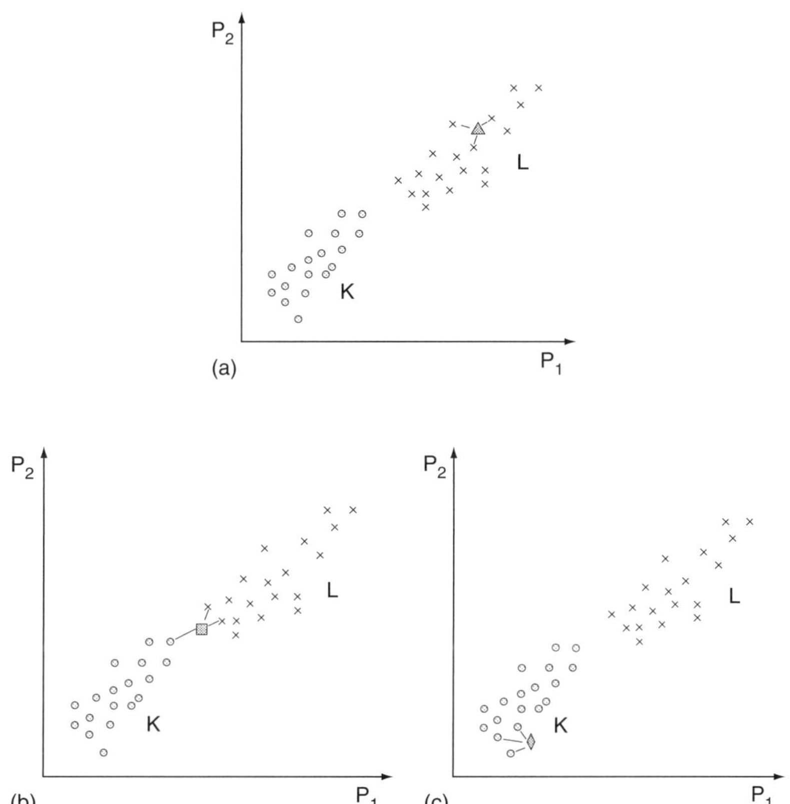

FIGURE 13.10. Three examples (a, b, and c) of the classification of a new sample with kNN, when two clusters K and L are present in the data.

class. Thus, kNN is a mathematically simple method, but has the disadvantage that it is sensitive to large inequalities in the number of objects between classes (8).

13.4.2.2. Applications. Schirm et al. (51) applied several chemometric methods to CE fingerprints in order to detect quality differences between different pentosan polysulfate sodium samples. First, the electropherograms were elaborately aligned and differing baseline shapes were removed with wavelet transformation. Then kNN, among other techniques, was applied on the electropherograms, to detect quality differences between the different samples. This computational easy method succeeded in revealing all relevant changes in the profile of pentosan polysulfate sodium. However, it should be noticed that the algorithm classifies each object, even if it is far away from the training set samples. In this study kNN was not able to detect small differences between samples and, as a consequence, 10% of samples were wrongly classified. Nevertheless, Schirm et al. (51) concluded that the automated classification is clearly superior to a visual inspection, especially when exploring data sets with small variations.

Beltrán et al. (50) also tested kNN to classify the Chilean wines according to their grape type. Again different feature extraction techniques were tested to reduce the dimensionality of the chromatographic data, describing the phenolic compounds. In most cases, kNN resulted in a slightly lower average correct classification rate than LDA and QDA.

13.4.3. CART

13.4.3.1. Theory. CART was introduced by Breiman et al. in 1984 (26) to explain and/or predict both categorical and continuous responses with CART, respectively. The goal of exploration with CART is to produce subsets of the initial data set, which are as homogeneous as possible with respect to the response variable. When this variable is the class to which the object belongs, CART is used as a nonparametric classification technique. CART is applicable for both exploration and classification.

In the CART method, a classification tree is built by binary recursive partitioning. Practically, a classification tree starts with a root node containing all objects. This node is then divided by a binary split based on the value for an explanatory variable, for example, logP = 2.47 in Figure 13.11, resulting in a node containing objects with smaller and one with larger logP values. Each value of each explanatory variable is once considered as a possible split and the split, for which the highest reduction in impurity is achieved, is selected as the optimal. This impurity can be defined with, for example, the information index, which minimizes the within-group diversity (27). Other possible measures for impurity are the so-called gini index, the twoing index, and the deviance index (27). For regression trees with continuous responses, the total sum of squares of the response values about the mean of the node is the most popular impurity measure (26).

After the split, each child node is individually treated as a parent node and the procedure described above is repeated until all terminal nodes are small (containing only one or a predefined number of objects) or pure (all objects in the node have the same response variables) (26). This learning procedure is represented in Figure 13.11. In fact, a regression tree is shown, but the principle is the same for a classification tree. The use of this tree is explained further (see section 13.4.3.2). It is in fact finally transformed to a kind of classification tree.

The resulting over-large maximal tree (Fig. 13.11a) is then gradually shrunk in a next step by pruning away branches, in order to obtain a smaller tree with a better predictive ability without losing much accuracy (26). For all smaller subtrees, a cost-complexity measure is calculated, which depends on the resubstitution error, the size of the subtree, and the complexity parameter ($0 \leq \alpha \leq 1$). For a given α value, there is then only one tree among all subtrees

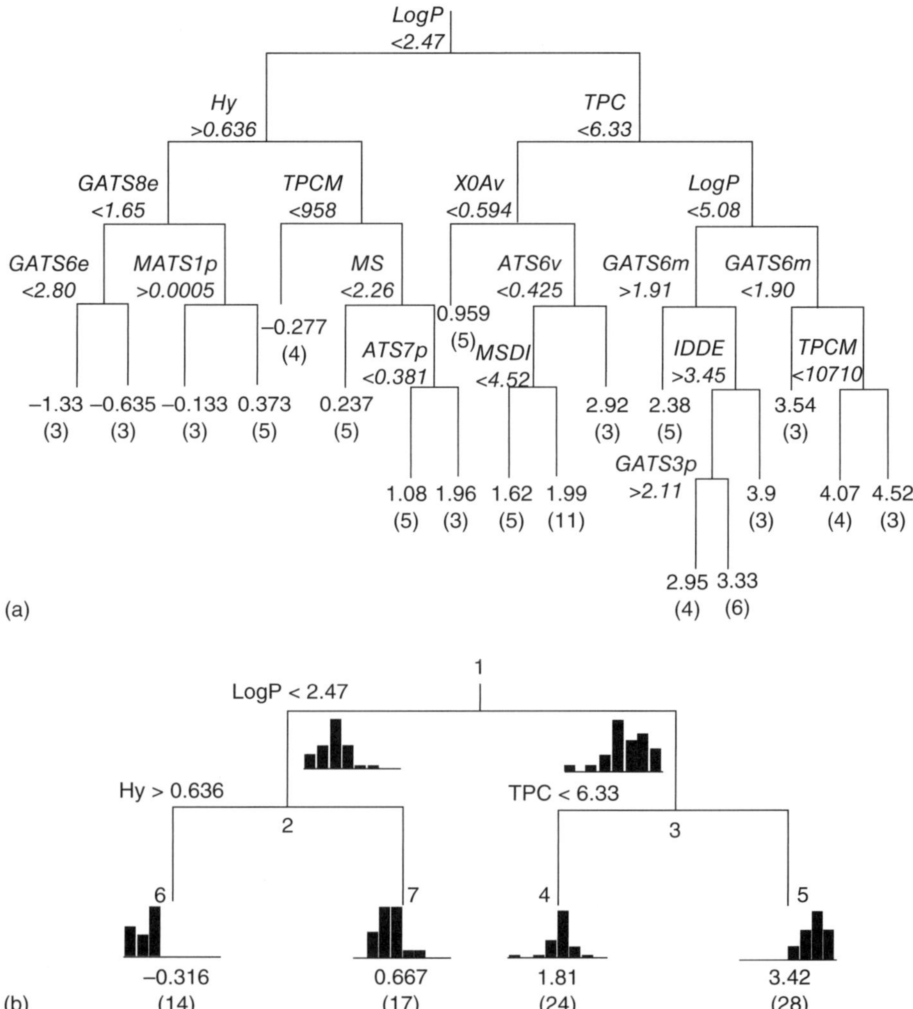

FIGURE 13.11. Classification and regression tree: (a) maximal tree and (b) optimal tree. Reproduced with permission from Put et al. (52).

of the same size that minimizes the cost-complexity measure. This procedure thus leads to a sequence of nested trees with decreasing size.

Finally, all these pruned subtrees will be subject to CV, in order to select the optimal tree size. The optimal tree (Fig. 13.11b) is selected as the simplest among those that have a CV error within one standard error deviation of the minimal CV error (26, 28). Another approach to determine the optimal tree size, preferred when a large number of training samples is available, is the use of an independent test set (26). After obtaining the final model, new samples can be classified by using the rules (split criteria) given by the model.

13.4.3.2. Applications. CART is not generally established yet, and as a consequence, not many applications for electrophoretic or similar data in the pharmaceutical field are found. Put et al. (52) applied CART in a quantitative structure–retention relationship context on a retention data set of 83 structurally diverse drugs, in order to predict chromatographic retention. There were 266 molecular descriptors calculated and used as explanatory variables (**X** matrix). The considered response (**y**) was the retention factor of the compounds, predicted for a pure aqueous mobile phase. The total sum of squares of the response values about the mean of the node was applied as impurity measure. From all descriptors, three were selected to describe and predict the retention, and four terminal nodes were obtained (Fig. 13.11b). Arbitrarily, the drugs were then divided into five retention classes. Each terminal node was then labeled with either one or two class names. The regression tree thus becomes a classification tree. From CV, it was concluded that only 9% serious misclassifications were observed.

Deconinck et al. (53) used CART in a quantitative structure–activity relationship context on an intestinal absorption data set of 141 drug-like molecules. Many theoretical molecular descriptors were calculated and used as explanatory variables (**X** matrix). The considered response (**y**) was the percentage human intestinal absorption of the compounds. The total sum of squares of the response values about the mean of the node was applied as impurity measure. From all descriptors, only two were chosen to describe and predict the intestinal absorption, and this resulted in three terminal nodes. However, the tree thus obtained did not allow defining classes with a limited absorption range, and therefore more complex trees were evaluated. Finally, a tree with 11 terminal nodes was selected. The absorption of the molecules was divided into five (absorption) classes. Each terminal node was labeled with one or two class symbols. From an external test set, three out of 27 molecules were wrongly classified (11.1%).

Caetano et al. (54) applied CART on Fourier-transform infrared spectra of olive oils to discriminate samples from Italian and non-Italian origin. Several earlier mentioned indexes were considered as split criteria, but finally the gini index was used in the final tree. Data were split into a calibration and test set with the duplex algorithm. The signal measured at wavenumber 1035.8/cm was selected as optimal split and divided the original data into two terminal nodes containing Italian and non-Italian samples, respectively. However, the terminal nodes were not pure. Of the Italian samples of the test set, 9.3% were misclassified, representing a relatively good sensitivity, that is, the percentage of correct classification of the Italian samples (90.7%). On the other hand, 86.7% of the non-Italian samples were misclassified, resulting in a poor selectivity, that is, percentage of non-Italian samples correctly classified (13.3%).

The above also shows that when using a spectrum, a chromatogram, or an electropherogram as explanatory variables, the splits in the tree are caused by the measurements at only one wavelength or wavenumber from the spectrum, or at one time point from the electropherogram or chromatogram.

13.4.4. PLSDA

13.4.4.1. Theory. When PLSDA (8) is used to allocate new samples in different classes, first, a classical PLS model is built for a calibration set of samples. In classical PLS, first, the number of explanatory variables is reduced by creating new latent variables (factors), which maximize the covariance between the explanatory and response variables. The obtained factors are then used to build a linear regression model.

Contrary to classical PLS, in PLSDA, the response variables used for the model construction are qualitative and discrete. They are coded in a vector with one number, 0 or 1, per class. The value 1 is attributed to the class to which a sample belongs and 0 to all other classes. For example, when simple PLS is applied with only one response variable, samples can be classified in two classes. The response variable has then the values 1 or 0. When more than two classes need to be distinguished, PLS2, that is, a modified version of PLS which is able to handle multiple response variables, is required. For instance, when three groups are present in the data, each sample is then associated with one of the three following vectors {1,0,0}, {0,1,0}, {0,0,1}, representing the classes 1, 2, or 3, respectively. For an unknown sample, the predicted value obtained with the PLSDA model is normally distributed around 0 or 1. A value close to zero indicates that the new sample does not belong to the considered class and a value close to one that the new sample belongs to the considered class. To determine the limit from which a sample is considered to be in the class or not, a threshold between zero and one is determined. When a value above the threshold is obtained, a sample is considered to belong to the class, while a value below the threshold indicates that the sample does not belong to the class.

13.4.4.2. Applications. Vallejo et al. (55) succeeded in revealing the effects of an antioxidant treatment on diabetic animals, which were not seen in the control group of nondiabetic animals, in a rapid and simple way without identifying a single marker. In a first step, to obtain a better representation of the sample, two metabolic fingerprints of urine were sequentially developed with two capillary electrophoresis methods: one with cyclodextrin modified micellar electrokinetic chromatography and one with capillary zone electrophoresis. The resulting electrophoretic profiles were then baseline corrected, aligned using COW, normalized and variable scaled, in order to prepare the data for classification. Finally, PLSDA was applied on the combined data from the two CE methods. For each class, a threshold, that is, an upper bound, is determined. However, the approach to determine the considered threshold value was not specified in Reference 55. Possibly, the thresholds are calculated as (95%) confidence limits determined from the distribution of the calibration sample predictions, which would explain the ellipsoidal boundaries of the classes (5, 8). PLSDA resulted in an almost 100% correct classification, allowing a clear distinction between treated and nontreated diabetic animals (Fig. 13.12). The effect of the treatment was not observed in the control groups of

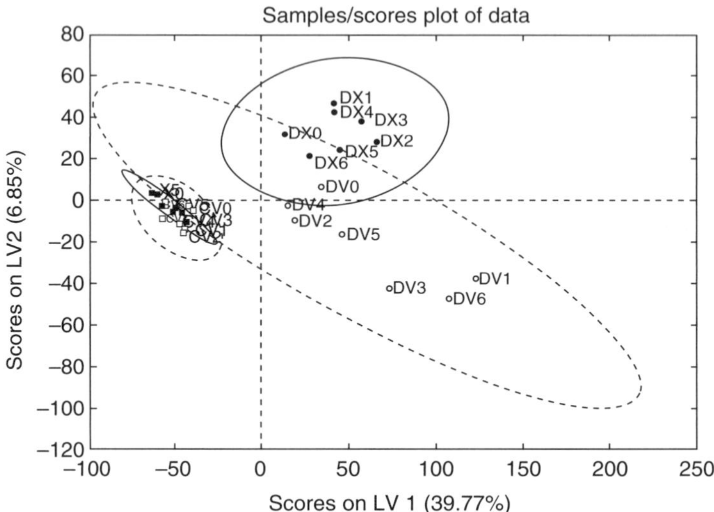

FIGURE 13.12. PLSDA data derived from urine fingerprints with □ representing the control group, ■ the control group treated with antioxidants, ○ the diabetic group, and • the diabetic group treated with antioxidants. LV = latent variable. Adapted from Vallejo et al. (55).

the nondiabetic animals, since no clear distinction is observed between treated and nontreated nondiabetic animals.

A chromatographic example is described by Yi et al. (56), who applied PLSDA successfully on HPLC fingerprints for class separation between authentic Pericarpium *Citri reticulatae* and authentic Pericarpium *Citri reticulatae Viride*. In this application, only the peak areas of 18 characteristic compounds were used as input data.

13.4.5. Soft Independent Modeling of Class Analogy

13.4.5.1. Theory. SIMCA is a parametric classification method introduced by Wold (29), which supposes that the objects of a given class are normally distributed. The particularity of this PCA-based method is that one model is built for each class separately, that is, disjoint class modeling is performed. The algorithm starts by determining the optimal number of PCs for each individual model with CV. The resulting PCs are then used to define a hypervolume for each class. The boundary around one group of objects is then the confidence limit for the residuals of all objects determined by a statistical *F*-test (30, 31). The direction of the PCs and the limits established for these PCs define the model of a class (Fig. 13.13).

A new unknown sample is then compared with the class models and assigned to classes according to its analogy to the training samples. Mathematically, the new sample is projected to the set of latent variables of

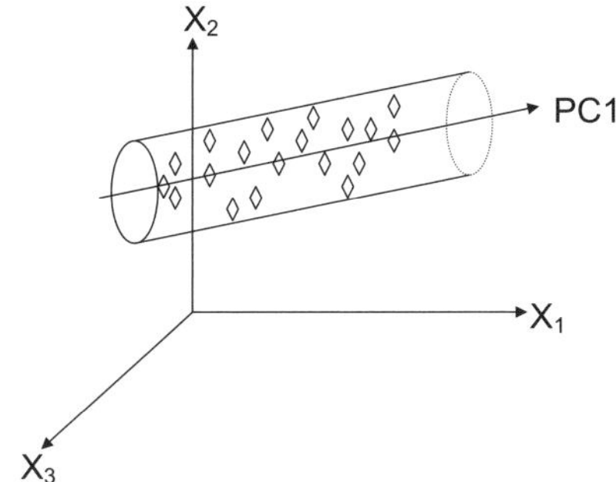

FIGURE 13.13. Principle of building a hypervolume for one class with SIMCA.

each class model. Then, for each model, the residual standard deviation for the new object is calculated and compared to the residuals of all objects from the group. A new object is located within the boundary, if its residuals are smaller than those of the objects in the group. This makes it possible that one object can be assigned to more than one group or to none of the groups. For this reason, SIMCA is called a soft classification technique (30, 31). This is in contrast to hard classification techniques, for example, LDA, QDA, or PLSDA, which will assign each new sample to exactly one class.

Instead of using the residuals to determine the boundary of a class, distance measures can also be applied for classification purposes (31). In this approach, for each class, the Mahalonobis distances (MD) for the objects in the score space, describing the distances to the center of the PCA model, and the orthogonal distances (OD) from the PCA model, describing the deviations to the model or the residuals, are calculated for all objects in the model set and are used to determine their cutoff values. These cutoff values are then used to decide whether or not a new sample belongs to a group. Similarly to the rPCA score diagnostic plot, four situations are possible when plotting the OD as a function of the MD for each object (Fig. 13.14). The samples in quadrant IV are considered ordinary objects, belonging to a certain class. All objects with MD and/or OD larger than the corresponding cutoff value(s) are considered outliers to that class. Moreover, objects situated in quadrant I are called high residual objects or vertical outliers (high residuals from PCA model, high OD), those in quadrant III are good leverage objects (far from majority of data, fit PCA model, high MD), and those in quadrant II are bad leverage objects (both high MD and OD). For each class, such plot can be drawn.

A new unknown sample is then compared with the models and plots of all classes. To verify whether a new object belongs to a given class, it is projected in the space defined by the selected factors (PCs) of the corresponding class model. Then the MD and OD are calculated for this sample, and the sample is plotted on the above Figure 13.14 for the given class. When the sample is located in quadrant IV of the plot, it is considered to belong to this given class.

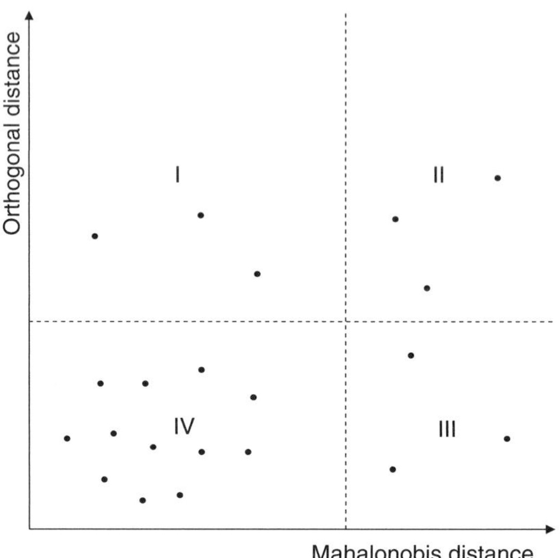

FIGURE 13.14. SIMCA: Types of outlying objects when plotting the orthogonal distance (OD) as a function of the Mahalonobis distance (MD). The ----- lines on this plot represent their cutoff values. I = high residual objects, II = bad leverage objects, III = good leverage objects, IV = ordinary objects.

This procedure is repeated for each class. Finally, it can be concluded whether the new unknown sample belongs to none, one, or several classes.

13.4.5.2. Applications. In Reid et al. (43), the electrophoretic data of the opium samples originating from four different locations were subjected to SIMCA, in order to use the models to determine the origin of new opium samples. When the four established SIMCA models were applied to an external test set, only one sample out of 40 was misclassified.

Next to kNN, Schirm et al. (51) also applied SIMCA to detect quality differences between different pentosan polysulfate sodium samples. SIMCA succeeded in discriminating samples from two different manufacturers. A 5% incorrect classifications occurred, which is, however, lower than the 10% misclassification obtained from kNN.

13.4.6. SVMs

The SVM method, introduced by Vapnik (32) in 1995, is applicable for both classification and regression problems. In case of classification, SVM are used to determine a boundary, a hyperplane, which separates classes independently of the probabilistic distributions of samples in the data set and maximizes the distance between these classes. The decision boundary is determined calculating a function $f(x) = y(x)$ (32–34). The technique is gaining popularity fast in the analytical sciences, because of its ability to model complex nonlinear relationships. The principle of this method in the latter cases is the use of a suitable kernel function, which transforms the input space to a higher-dimensional feature space in which the data can be discriminated using a linear function (34).

13.4.6.1. Linear SVM Classifiers. When the data set is *linearly separable*, the decision function $f(x) = y(x)$ to separate the classes is given by:

$$y(x) = \langle \mathbf{x}, \mathbf{w} \rangle + b \qquad \text{(Eq. 13.9)}$$

where $\mathbf{w} \in \Re^d$ is the weight vector, $b \in \Re$ is the bias, and $\mathbf{x} \in \Re^d$ is a set of input vectors with corresponding labels $y_i \in \{-1, +1\}(i = 1, \ldots, n)$, where -1 and $+1$ indicate the two classes (K and L) (34).

When the data of the two classes are separable, it can be said:

$$\begin{cases} \langle \mathbf{x}, \mathbf{w} \rangle + b \geq +1; \ \forall y = +1 \\ \langle \mathbf{x}, \mathbf{w} \rangle + b \leq -1; \ \forall y = -1 \end{cases} \qquad \text{(Eq. 13.10)}$$

These two sets of inequalities in Equation 13.10 can be combined into one single inequality as follows:

$$y_i(\langle \mathbf{x}_i, \mathbf{w} \rangle + b) \geq 1; \ i = 1, \ldots, n; \ y_i \in \{-1, +1\} \qquad \text{(Eq. 13.11)}$$

However, there are many linear classifiers that might satisfy this property (Fig. 13.15a). The concept of margin (M), shown graphically in Figure 13.15b, is used to quantify the fact that among all solutions to the classification problem,

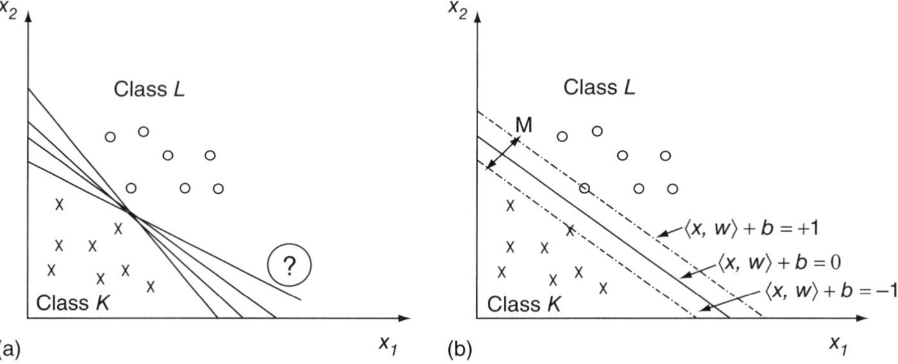

FIGURE 13.15. Example of two linearly separable classes that can be separated with (a) several hyperplanes, but for which SVM defines (b) a unique separating hyperplane. The margin (M) is the distance between the dashed lines through the support vectors.

a hyperplane exists, which is the "optimal" boundary. To calculate the margin, two parallel hyperplanes (dashed lines on Fig. 13.15b) are constructed, one on each side of the decision plane passing through support vectors, that is, samples of the training set closest to the decision plane. The pair of hyperplanes leading to the maximum margin is found by minimizing $\frac{\|\mathbf{w}\|^2}{2}$. The "optimal" boundary is defined as the hyperplane for which the distances to the support vectors of class L (+1) and of class K (−1) are maximized (34).

However, in real life, many *nonseparable* (linear or nonlinear) classification problems occur, which practically means that distributions between two classes are overlapping. This implies that misclassifications should be tolerated. Therefore, a set of slack variables ($\xi_i \geq 0$) is introduced in the margin minimization approach used for the linearly separable case, allowing some samples inside the margin. For this purpose, Equation 13.11 is replaced by Equation 13.12.

$$y_i(\langle \mathbf{x}_i, \mathbf{w} \rangle + b) \geq 1 - \xi_i;\ i = 1, \ldots, n;\ y_i \in \{-1, +1\} \qquad 13.12$$

In the nonseparable case, also a regularization parameter C is added tuning the trade-off between the number of accepted errors and the maximization of the margin. In this situation, one seeks to minimize $\frac{\|\mathbf{w}\|^2}{2} + C\left(\sum_i \xi_i\right)^k$ (with k being a positive integer) instead of $\frac{\|\mathbf{w}\|^2}{2}$. When the value of C is high, the number of samples misclassified is minimized without maximizing the margin. On the contrary, when C is close to zero, the margin is maximized without taking into account the number of samples misclassified. In this case, the model can give aberrant predictions. The choice of the optimal C value can be performed by CV (32–34).

13.4.6.2. NonLinear SVM Classifiers. For nonlinear classification problems, the SVM basic idea is to project samples of the data set, initially defined in $\Re^d$ dimensional space, into another space $\Re^e$ with a higher dimension ($d < e$), where samples then are separated by a linear separation (Fig. 13.16) (34).

This transformation into the higher-dimensional space is realized with a kernel function. The best function used depends on the initial data. In the SVM literature, typical kernel functions applied for classification are linear and polynomial kernels, or radial basis functions. Depending on the applied kernel function, some parameters must be optimized, for instance, the degree of the polynomial function (33, 34). Once the data are transformed to another dimensional space by the kernel function, linear SVM can be applied. The main parameter to optimize with the SVM algorithm for nonseparable cases, as described in the previous section, is the regularization parameter, C.

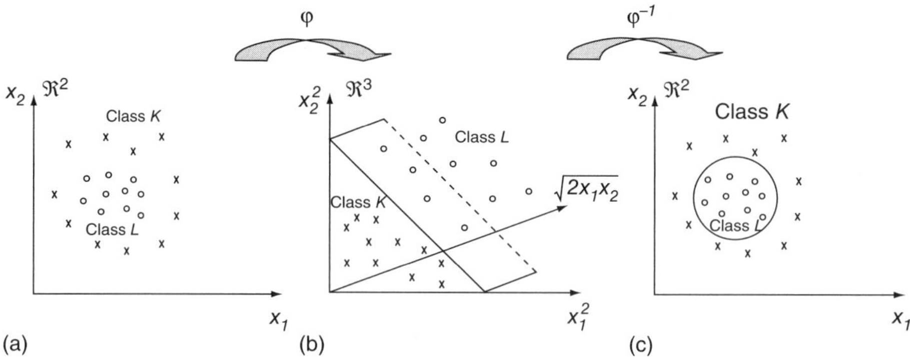

FIGURE 13.16. Principle of classification with nonlinear SVM. For nonlinear classification problems, the SVM basic idea is to project samples of the data set, (a) initially defined in $\Re^d$ dimensional space, (b) into another space R^e with a higher dimension ($d < e$), where samples are separated linearly. The latter separation can then be projected again (c) in the original data space. The transformation into the higher-dimensional space is realized with a kernel function.

13.4.6.3. Applications. Zomer et al. (57) propose a pattern recognition procedure for determining the type of cadmium dosage (chronic or acute) administrated to laboratory rats based on the urinary profiles developed by CE. The classification is not possible by a visual inspection. First, the electropherograms were baseline corrected. The most common peaks in the electropherograms were indicated using the first derivative of the signal. Then, the data matrix was produced, in which the rows referred to the samples and the columns to the peak areas of the most common components of the electropherograms. Only those peaks were retained that occurred in more than half of the samples, and samples where less than half the peaks were detected were removed. Finally, the data matrix was normalized, prior to applying pattern recognition techniques. The above preprocessing techniques reduced the lack of reproducibility and enhanced the contribution of low-level metabolites. The resulting matrix was then subjected to PCA, cluster analysis, discriminant analysis (DA), and SVM. PCA and hierarchical clustering with average linkage allowed distinguishing rats with acute or chronic cadmium intoxication, but no subgroups according to dosage levels could be observed. The latter was achieved with DA and SVM. The advantage of SVM was that no prior variable reduction was required. The training set was described correctly by the SVM model (100% correct classification). However, DA could better classify the samples of an external test set (97% correct classification) than SVM (76% correct classification). This might indicate that the SVM model was overfitting the calibration data.

Another example, but for spectroscopic data, is found in Caetano et al. (54), who applied SVM on the Fourier-transform infrared spectra of olive oils in order to classify them according to geographical region. SVM performed superior to CART for predictive purposes. However, a disadvantage of SVM is that the obtained models cannot be interpreted from a physicochemical point of view.

13.5. CONCLUSIONS

Chemometric techniques can be valuable tools for the exploration of CE data as well as for the classification of samples based on electrophoretic data. The techniques maximally exploit the multivariate character of the data. In several applications, it was demonstrated that chemometric approaches can extract more information from electropherograms than only a visual inspection can. It is very important, especially when using entire electropherograms, that the CE data are preprocessed (e.g., aligned) in an appropriate way prior to other chemometric calculations, because CE analyses generally exhibit a rather poor reproducibility.

13.6. SUMMARY

In modern CE analysis, the detector can easily register between 0.5 and 32 signals per second. By default, four signals per second are registered. As a consequence, the resulting data are highly multivariate and not always easily visually comparable, especially not when a high number of samples is investigated.

Chemometric techniques, which can easily cope with this type of data by the use of matrices, will maximize the benefit of the multivariate character. These calculation techniques require that corresponding data points (for instance the top of a peak) in different electropherograms are located in the same column of the matrix. As a consequence, preprocessing the CE data is recommended. Peak shifts are commonly corrected with warping techniques, for example, COW, while column centering, normalization, baseline correction, and MSC are also frequently performed preprocessing techniques.

Once the data are prepared, they can be explored chemometrically with techniques as PCA, rPCA, PP, and clustering. These enable visualization of the structure of the data set; more specifically, they detect outliers and group similar samples. For several applications, it was confirmed that this approach outperforms the visual comparison of electropherograms. Chemometric techniques can also be applied to classify samples based on their CE profile. When the classes in the data set are a priori known, supervised classification techniques as LDA, QDA, kNN, CART, PLSDA, SIMCA, and SVM can be used. The choice of techniques will often depend on the preference of the analyst and the complexity of the data. However, when nonlinear classification problems occur, a more complex technique as, for instance, SVM, will be outperforming others. In practice, often several techniques are tested before selecting the method with the best predictive classification results.

ACKNOWLEDGMENTS

Melanie Dumarey acknowledges the Institute for the Promotion of Innovation through Science and Technology in Flanders (IWT-Vlaanderen) for the funding of her PhD project.

Bieke Dejaegher is a postdoctoral fellow of the Fund for Scientific Research (FWO), Vlaanderen, Belgium.

REFERENCES

1. Li, S.F.Y. (1992) *Capillary Electrophoresis—Principles, Practice and Applications*, Elsevier, Amsterdam.
2. Jimidar, I., De Smet, M., Sneyers, R., Van Ael, W., Janssens, W., Redlich, D., and Cockaerts, P.J. (2003) *J Cap Elec Microchip Tech*, **8**, 45–52.
3. Visky, D., Jimidar, I., Van Ael, W., Vennekens, T., Redlich, D., and De Smet, M. (2005) *Electrophoresis*, **26**, 1541–1549.
4. Altria, K.D. (1998) *Analysis of Pharmaceuticals by Capillary Electrophoresis*, Vieweg, Braunschweig/Wiesbaden.
5. Massart, D.L., Vandeginste, B.G.M., Buydens, L.M.C., De Jong, S., Lewi, P.J., and Smeyers-Verbeke, J. (1997) *Handbook of Chemometrics and Qualimetrics: Part A*, Elsevier, Amsterdam.
6. Daszykowski, M., Walczak, B., and Massart, D.L. (2003) *Chemometr Intell Lab Syst*, **65**, 97–112.
7. Massart, D.L. and Vander Heyden, Y. (2004) *LC-GC Eur*, **17**, 586–591.
8. Vandeginste, B.G.M., Massart, D.L., Buydens, L.M.C., De Jong, D., Lewi, P.J., and Smeyers-Verbeke, J. (1998) *Handbook of Chemometrics and Qualimetrics: Part B*, Elsevier, Amsterdam.
9. Croux, C. and Ruiz-Gazen, A. (1996) *COMPSTAT: Proceedings in Computational Statistics*, Physica-Verlag, Heidelberg.
10. Croux, C. and Ruiz-Gazen, A. (2005) *J Multivariate Anal*, **95**, 206–226.
11. Hubert, M., Rousseeuw, P.J., and Verboven, S. (2002) *Chemometr Intell Lab Syst*, **60**, 101–111.
12. Hubert, M., Rousseeuw, P.J., and Vanden Branden, K. (2005) *Technometrics*, **47**, 64–79.
13. Stanimirova, I., Walczak, B., Massart, D.L., and Simeonov, V. (2004) *Chemometr Intell Lab Syst*, **71**, 83–95.
14. Friedman, J.H. and Stuetzle, W. (1981) *J Am Stat Assoc*, **76**, 817–823.
15. Huber, P.J. (1985) *Ann Stat*, **13**, 435–475.
16. Friedman, J.H. (1987) *J Am Stat Assoc*, **82**, 817–823.

17. Stanimirova, I., Daszykowski, M., Van Gyseghem, E., Bensaid, F.F., Lees, M., Smeyers-Verbeke, J., Massart, D.L., and Vander Heyden, Y. (2005) *Anal Chim Acta*, **552**, 1–12.

18. Daszykowski, M., Stanimirova, I., Walczak, B., and Coomans, D. (2005) *Chemometr Intell Lab Syst*, **78**, 19–29.

19. Massart, D.L. and Kaufman, L. (1983) *The Interpretation of Analytical Chemical Data by the Use of Cluster Analysis*, John Wiley & Sons, Brisbane.

20. Vogt, M., Nagel, D., and Sator, H. (1987) *Cluster Analysis in Clinical Chemistry: A Model*, John Wiley & Sons, Essex.

21. Fisher, R.A. (1936) *Annal Eugenics*, **7**, 179–188.

22. Gemperline, P.J., Laurie, D., Webber, F., and Cox, O. (1989) *Anal Chem*, **61**, 138–144.

23. Wu, W., Mallet, Y., Walczak, B., Penninckx, W., Massart, D.L., Heuerding, S., and Erni, F. (1996) *Anal Chim Acta*, **329**, 257–265.

24. Coomans, D. and Massart, D.L. (1982) *Anal Chim Acta*, **138**, 153–165.

25. Tominaga, Y. (1999) *Chemometr Intell Lab Syst*, **49**, 105–115.

26. Breiman, L., Friedman, J.H., Olshen, R.A., and Stone, C.J. (1984) *Classification and Regression Trees*, Wadsworth, Monterey, CA.

27. De'ath, G. and Fabricius, K.E. (2000) *Ecology*, **81**, 3178–3192.

28. Questier, F., Put, R., Coomans, D., Walczak, B., and Vander Heyden, Y. (2005) *Chemometr Intell Lab Syst*, **76**, 45–54.

29. Wold, S. (1976) *Pattern Recogn*, **8**, 127–139.

30. Brereton, R.G. (1992) *Multivariate Pattern Recognition in Chemometrics*, Elsevier, Amsterdam.

31. Daszykowski, M., Kaczmarek, K., Stanimirova, I., Vander Heyden, Y., and Walczak, B. (2007) *Chemometr Intell Lab Syst*, **87**, 95–103.

32. Vapnik, V. (1995) *The Nature of Statistical Learning Theory*, Springer, New York.

33. Burges, C.J.C. (1998) *Data Min Knowl Discov*, **2**, 121–167.

34. Suykens, J.A.K., Van Gestel, T., De Brabanter, J., De Moor, B., and Vandewalle, J. (2002) *Least Squares Support Vector Machines*, World Scientific, Singapore.

35. Cuesta Sànchez, F., Toft, J., Van den Bogaert, B., and Massart, D.L. (1996) *Anal Chem*, **68**, 79–85.

36. Cuesta Sànchez, F., Rutan, S.C., Gil García, M.D., and Massart, D.L. (1997) *Chemometr Intell Lab Syst*, **36**, 153–164.

37. Schaeper, J.P. and Sepaniak, M.J. (2000) *Electrophoresis*, **21**, 1421–1429.

38. Nielsen, N.P.V., Carstensen, J.M., and Smedsgaard, J. (1998) *J Chromatogr A*, **805**, 17–35.

39. Pravdova, V., Walczak, B., and Massart, D.L. (2002) *Anal Chim Acta*, **456**, 77–92.

40. Eilers, P.H.C. (2004) *Anal Chem*, **76**, 404–411.

41. Martens, H. and Naes, T. (1989) *Multivariate Calibration*, Wiley, Chichester.

42. Xu, C.J., Liang, Y.Z., Chau, F.T., and Vander Heyden, Y. (2006) *J Chromatogr A*, **1134**, 253–259.

43. Reid, R.G., Durham, D.G., Boyle, S., Low, A.S., and Wangboonskul, J. (2007) *Anal Chim Acta*, **60**, 520–527.
44. Sturm, S., Seger, C., and Stuppner, H. (2007) *J Chromatogr A*, **1159**, 42–50.
45. Szymańska, E., Markuszewski, M.J., Capron, C., van Nederkassel, A.M., Vander Heyden, Y., Markuszewski, M., Krajka, K., and Kaliszan, R. (2007) *J Pharm Biomed Anal*, **43**, 413–420.
46. van Nederkassel, A.M., Daszykowski, M., Massart, D.L., and Vander Heyden, Y. (2005) *J Cromatogr A*, **1096**, 177–186.
47. Schoonjans, V. and Massart, D.L. (2001) *J Pharm Biomed Anal*, **26**, 225–239.
48. López-Sánchez, M., Domínguez-Vidal, A., Ayora-Canada, M.J., and Molina-Díaz, A. (2008) *Anal Chim Acta*, **620**, 113–119.
49. Ruiz-Ángel, M., Simó-Alfonso, E.F., Mongay-Fernández, C., and Ramis-Ramos, G. (2002) *Electrophoresis*, **23**, 1709–1715.
50. Beltrán, N.H., Duarte-Mermoud, M.A., Bustos, M.A., Salah, S.A., Loyala, E.A., Peña-Neira, A.I., and Jalocha, J.W. (2006) *J Food Eng*, **75**, 1–10.
51. Schirm, B., Benend, H., and Wätzig, H. (2001) *Electrophoresis*, **22**, 1150–1162.
52. Put, R., Perrin, C., Questier, F., Coomans, D., Massart, D.L., and Vander Heyden, Y. (2003) *J Chromatogr A*, **988**, 261–276.
53. Deconinck, E., Hancock, T., Coomans, D., Massart, D.L., and Vander Heyden, Y. (2005) *J Pharm Biomed Anal*, **39**, 91–103.
54. Caetano, S., Üstün, B., Hennessy, S., Smeyers-Verbeke, J., Melssen, W., Downey, G., Buydens, L., and Vander Heyden, Y. (2007) *J Chemometr*, **21**, 324–334.
55. Vallejo, M., Angulo, S., García-Martínez, D., García, A., and Barbas, C. (2008) *J Chromatogr A*, **1187**, 267–274.
56. Yi, L., Yuan, D., Liang, Y., Xie, P., and Zhao, Y. (2007) *Anal Chim Acta*, **588**, 207–215.
57. Zomer, S., Guillo, C., Brereton, R.G., and Hanna-Brown, M. (2004) *Anal Bioanal Chem*, **378**, 2008–2020.

PART III

QUANTITATIVE STRUCTURE RELATIONSHIPS

CHAPTER 14

CHEMOMETRICAL MODELING OF ELECTROPHORETIC MOBILITIES IN CAPILLARY ELECTROPHORESIS

MEHDI JALALI-HERAVI
Department of Chemistry, Sharif University of Technology, Tehran, Iran

CONTENTS

14.1. GENERAL OBJECTIVES AND CONCEPTS

In capillary electrophoresis (CE), analytes are separated due to their different velocities under the influence of an electric field. The analytes reach a steady-state velocity that can be expressed independently of the field strength as the electrophoretic mobility (μ_e).

The electrophoretic mobility (μ_e) of an analyte at a given ionic strength can be determined using Equation 14.1.

$$\mu_e = \frac{L_t L_d}{V}\left(\frac{1}{t_r} - \frac{1}{t_o}\right) \qquad \text{(Eq. 14.1)}$$

where L_t is the total length of the capillary, L_d is the separation length (from the upstream end of the capillary to the detection window), V is the applied

Chemometric Methods in Capillary Electrophoresis. Edited by Grady Hanrahan and Frank A. Gomez

voltage, t_r is the analyte retention time, and t_o is the retention time of the EOF marker, such as mesityl oxide.

The search for optimal separation conditions is sometimes time-consuming and tedious. The key parameter for separation of analytes is their electrophoretic mobilities. Therefore, development of theoretical models for estimating the μ_e seems to be useful.

Quantitative Structure–Mobility Relationships (QSMR) is an area of computational research, which is able to build a mathematical model relating the mobility of a series of compounds to physicochemical and structural parameters. One of the most important factors governing the quality of QSMR model is the quantification of structural features. Numerous descriptors developed in commercial special software can be used to build linear and nonlinear models. Therefore, developing a successful QSMR model that is robust with a high predictive ability requires a successful combination of feature selection and feature mapping tools. We have recently developed some hybrid methods in our laboratory consisting of feature selections such as multiple linear regression (MLR) and classification and regression tree (CART) techniques and artificial neural network (ANN) and adaptive neuro-fuzzy inference system (ANFIS) methods as mapping tools. In our laboratory, the application of these hybrid methods is focused on two areas: (1) peptide mobility and peptide mapping and (2) modeling of electrophoretic mobilities of organic acids.

14.2. PEPTIDE MOBILITY AND PEPTIDE MAPPING

Peptide mapping involves digestion of a protein through enzymatic or chemical means and subsequent separation and detection of the resultant peptide mixture. This method is widely used for characterization of protein structure. These maps can be applied to rapid protein identification and the detection of posttranslational modifications. In fact, the peptide maps play the role of "fingerprints" for the proteins. One of the most commonly used techniques for peptide mapping is the gas liquid chromatography–tandem mass spectrometry (GLC–MS/MS) method. While this method provides excellent resolution, it is time-consuming and generally requires relatively large quantities of peptides. CE has received considerable attention as a peptide mapping technique because of its high efficiency, speed, small sample size, automation, and high throughput capability (1).

Analysis and identification of a large number of peptides from complex samples is challenging and time-consuming. Model-based approaches can improve the separation quality and shorten the time normally needed. The calculated electrophoretic mobility can be converted to migration time and a CE electropherogram can be simulated using a Gaussian function. This means that calculation/prediction of this parameter is useful in peptide mapping studies.

Numerous empirical predictive models, based on Stoke's law, have been developed for the prediction of μ_e from the charge-to-size ratio (2–8). However, these models are not robust for accurately predicting this parameter for all categories of peptides by relying on two parameters of charge and size alone (9–12). Two strategies were applied in our laboratory to address this problem. First, an MLR procedure was used for choosing additional peptide descriptors to Offord's charge-over mass term of $Q/M^{2/3}$. Second, ANN as a nonlinear modeling method was introduced to gain more accurate and robust models (9, 11, 13–15). All these methods were derived based on the assumption that the peptide electrophoretic mobility should substantially depend on amino acid compositions.

The methodology used to develop linear multivariable models is described in the following section.

14.2.1. Development of Linear Multivariable Models

Generally, it has been shown that the electrophoretic mobility is proportional to the charge Q and inversely proportional to the molecular mass M as:

$$\mu_e = a\frac{Q}{M^b} \qquad \text{(Eq. 14.2)}$$

where a and b are constants. The main difference between various reported models is the value of b that depends upon the assumption involved in the derivation of the models and the conditions under which the assumptions are valid (3, 9, 10, 16–19).

Compton (4) has shown that the mobilities of small molecules in low-ionic-strength buffer are more closely correlated with $1/M^{1/3}$ while large molecules in high-ionic-strength buffer correlated with $1/M^{2/3}$. Molecules of intermediate size and in moderate-ionic-strength buffers show dependence on $1/M^{1/2}$ (3). Janini et al. (10), based on a data set of 58 peptides, concluded that except for the highly charged and the hydrophobic peptides, the Offord model is superior to the other models.

Recently, Jalali-Heravi et al. (9, 11) have developed a multivariable model in order to improve the predictive ability of the Offord model and understand the effects of further structural descriptors on electrophoretic mobility in capillary zone electrophoresis (CZE), in addition to charge and size. They generated a diverse data set based on a 125-peptide study, which ranges in size from 2 to 14 amino acids and charges of 0.743–5.843. The μ_e of the peptides were measured in bare fused-silica capillaries in CZE mode using 50 mM sodium phosphate buffer at pH 2.5. The detection wavelength was 214 nm and the separation temperature was 37 °C.

As a first step in developing an MLR model, one has to choose the most suitable descriptors contributing to the motion of a peptide in an electric field. Several physicochemical parameters were used by Jalai-Heravi et al. for model generation (9). The best model was achieved by a step-wise MLR method that combined the Offord term with various peptide descriptors and on the basis of r^2, F statistics, and standard error (SE). The Offord charge-to-mass parameter was chosen as the first input for the software package of Minitab (20) to generate the MLR model. Then, the stepwise addition method was used for choosing the other descriptors contributing to the electrophoretic mobilities of model peptides. The following equation was obtained:

$$\mu_e = \rho \frac{Q}{M^{2/3}} + e\sum E_{s,c} + m\sum \text{MR} \qquad \text{(Eq. 14.3)}$$

where $E_{s,c}$ is the corrected steric substituent constant and MR is the molar refractivity. The $E_{s,c}$ has been defined by Taft as log (k/k_o), where k and k_o are the rate constants for the acidic hydrolysis of a substituted ester and of a reference ester, respectively (21). This parameter represents the steric interactions. The molar refractivity is a constitutive-additive property that is calculated by the Lorenz–Lorentz formula (22). MR is strongly related to the volume of the molecules (i.e., molecular bulkiness).

The specifications for the best MLR model are shown in Table 14.1. Also the mean effect for each descriptor is included in this table.

The Offord model shows the largest mean effect among the descriptors appearing in the model. This indicates that the net charge of the peptide and its size play the major roles in the migration mechanism of the peptides in an electric field. The contribution of $E_{s,c}$ and MR to electrophoretic mobility is almost the same, but in an opposite direction. It is shown that the larger peptides show a higher steric constant and, therefore, have a smaller mobility in a CZE system.

Figure 14.1 shows the plot of the MLR-calculated electrophoretic mobility against the experimental values for the validation and test sets. This plot showed an improved correlation of $r^2 = 0.895$ in the predictive ability of the model over the use of the simple Offord relationship ($r^2 = 0.878$). However, some MLR-calculated electrophoretic mobilities showed a large deviation from the experimental values (9). The MLR model overestimated the electrophoretic mobility of peptides containing arginine (R), histidine (H), and lysine amino acids. These amino acids contribute a charge +1 to the peptide. Jalali-

TABLE 14.1. Specifications of the best selected MLR model

Descriptors	Notation	Coefficient	Mean Effect[a]
Charge-to-size ratio	QM	1347.04 (±31.51)	28.102
Corrected steric substituent constant	$E_{s,c}$	1.4476 (±0.4161)	−3.252
Molecular refractivity	MR	0.04979 (±0.01466)	4.266
Constant		0.0	

[a]The mean effect of a descriptor is the product of its mean and the regression coefficient in the MLR model.

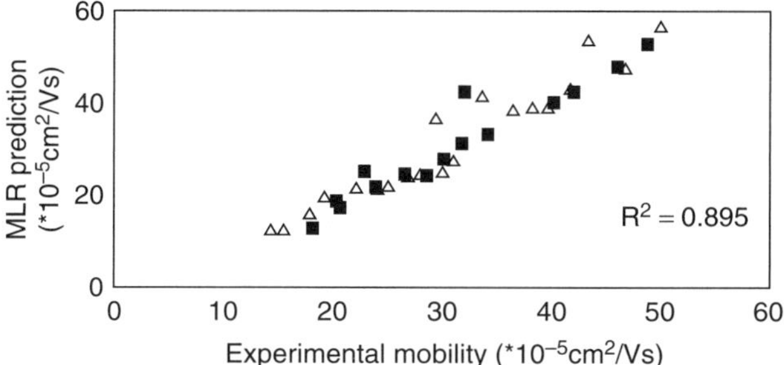

FIGURE 14.1. Plot of the MLR-calculated electrophoretic mobilities against the experimental values for the test and validation sets. (▵) test; (▪) validation.

Heravi et al., therefore, concluded that the linear models are not able to predict the mobility of the peptides with high charges (9). The limited ability of linear models in predicting the electrophoretic mobility of a more diverse set of peptides persuaded some researchers to apply machine learning (ML) techniques, which are more generic, nonlinear modeling tools.

14.2.2. ML as a Tool to Develop QSMR Models

ML is a subfield of artificial intelligence that is concerned with the design and development of algorithms that allow computers (machines) to improve their performance over time (or learn) based on data. A major focus of ML research is to automatically produce models. Many researchers quote Herbert Simon in describing ML (23):

> Learning denotes changes in the system that are adaptive in the sense that they enable the system to do the same task or tasks drawn from the same population more efficiently and more effectively the next time.

However, chemometricians are more interested in ML algorithms and their performance. In their eyes, ML is:

> The process (algorithm) of estimating a model that's true to the real-world problem with a certain probability from a data set (or sample) generated by finite observations in a noisy environment.

However, because the complexities of real-world data make a general learning algorithm impossible, the quality of the data and background knowledge could be the key to ML's success.

ML techniques are well situated for the analysis of molecular sequence data. These methods have been applied successfully to a variety problem, ranging from gene identification to protein structure prediction and sequence classification (24, 25). These techniques have become an important topic for developing QSMR models in CZE (9, 11, 13, 14). This is due to self-learning ability and the potential of these techniques to describe complex data sets without the need for detailed understanding of the underlying phenomena. The ANNs and support vector machine (SVM) are the two most common techniques in exploring the linear/nonlinear characteristics of the electrophoretic mobility of peptides. This article focuses on the principles of ANNs together with their application in QSMR modeling.

14.2.2.1. ANNs. The ANN, or simply neural network (NN), is an ML method that evolved from the idea of simulating the human brain. An ANN consists of simple neurons operating in parallel and organized in layers. The connections between the layers and the transfer functions being used determine the function of the network. The ANNs learn from the given samples by modifying the weights and biases. After training, the networks can accomplish a given task. It means that, for example, they can predict the electrophoretic mobility of peptides.

Figure 14.2 shows the biological representation of a three-layer network used to predict the electrophoretic mobility of peptides. In general, there are three stages for developing each NN approach (see Fig. 14.2).

14.2.2.1.1. Stage 1. In this stage, the samples or their representations (molecular descriptors) are selected, which are to be used as inputs to the NN. This step is taking care of the input layer of the biological representation. For example, Jalali-Heravi et al. selected a diverse data set based on a 125-peptide study, with ranges in size between 2 and 14 amino acids as samples (9). Also, in another attempt, these researchers, to evaluate the robustness of their ANN model, chose a data set of 102 peptides that consisted of larger, more hydrophobic and highly charged peptides compared with the previous data set (19). However, the best features representing these peptides should be used as inputs for developing the network. Choosing the most suitable structural fea-

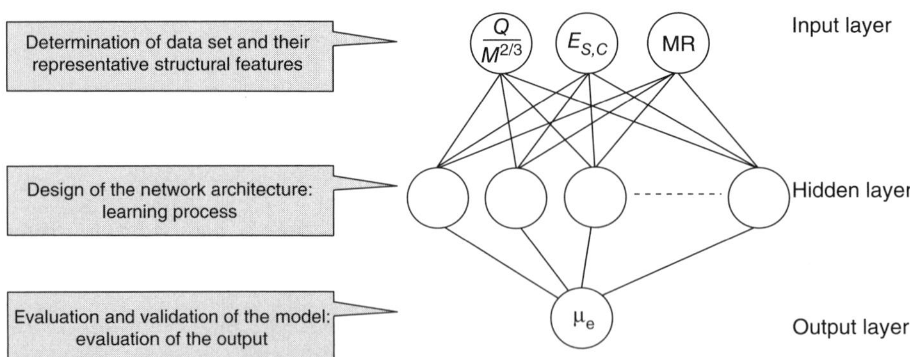

FIGURE 14.2. The biological representation of a three-layer network for prediction of electrophoretic mobility of peptides.

tures is an important factor governing the quality of the ANN model. Different rational methods have been used to design a network. For example, the genetic neural network (GA–ANN) uses a genetic algorithm to select the input features for the neural network.

Jalali-Heravi and coworkers used the three descriptors of their multivariable model, that is, Offord charge-to-mass parameter, corrected steric constant, and molar refractivity, as the input parameters for generating the network (Fig. 14.2). In fact, they proposed an MLR–ANN model for the prediction of the electrophoretic mobility of peptides. The purpose for choosing the MLR parameters as inputs for the ANN mode was to compare the abilities of linear and nonlinear models in predicting the electrophoretic mobilities of peptides (9).

14.2.2.1.2. Stage 2. In this stage, the network architecture is designed. The back-propagation (BP) algorithm seems to be the most attractive choice from the variety of NN architectures. This algorithm is ideally suited for many different applications because of its clear application of supervised learning. BP–ANN is progressively regarded as a standard for chemical pattern recognition due to its inherent superiorities in modeling complex and nonlinear data spaces.

Figure 14.2 shows that three features of $Q/M^{2/3}$, $E_{s,c}$, and MR are used as input parameters for generation of the network. The signals from the output layer represent the electrophoretic mobility of peptides. Therefore, the ANN may be designed as a 3-n_h-1 net in which the number of nodes in the hidden layer should be optimized. The ANN uses a learning process to train the network. During the training, weights are adjusted to desired values.

Hidden neurons communicate only with other neurons. They are part of the large internal pattern that determines a solution to the problem. The information that is passed from one processing element to another is continued within a set of weights. Some of the interconnections are strengthened and some are weakened, so that a neural network will output a more corrected answer. The activation of a neuron is defined as the sum of the weighted input signals to that neuron:

$$Net_j = \sum_i W_{ij} X_i + bias_j \qquad \text{(Eq. 14.4)}$$

where W_{ij} is the weight-connection to neuron j in the actual layer from neuron i in the previous layer and $bias_j$ is the bias of neuron j. The Net_j of the weighted inputs is transformed with a transfer function, which is used to get to the output level. Several functions can be used for this purpose, but the "sigmoid function" is mostly applied. This function is as follows:

$$y_j = \frac{1}{1+e^{-Net_j}} \qquad \text{(Eq. 14.5)}$$

where y_j is output of the neuron j. In order to train the network using the BP algorithm, the differences between the ANN output and its desired value are calculated after each iteration. The changes in the values of the weights can be obtained using the equation:

$$\Delta w_{ij}(n) = \eta \delta_i O_j + \alpha \Delta w_{ij}(n-1) \quad \text{(Eq. 14.6)}$$

where Δw_{ij} is the change in the weight factor for each network node, δ_i is the actual error of node i, and O_j is the output of node j. The coefficients η and α are the learning rate and the momentum factor, respectively. These coefficients control the velocity and the efficiency of the learning process. These parameters would be optimized before training the network. The goal of training a network is to change the weights between the layers in a direction that minimizes the error, E:

$$E = \frac{1}{2} \sum_p \sum_k (y_{pk} - t_{pk})^2 \quad \text{(Eq. 14.7)}$$

The error E of a network is defined as the squared differences between the target value t and the output y of the output neurons summed over p training patterns and k output nodes. In BP learning, the error in prediction is fed backward through the network to adjust the weights and minimize the error, thus preventing the same error from happening again. This process is continued with multiple training set until the error is minimized across many sets.

Jalali-Heravi et al. developed a 3-4-1 BP–ANN model for the set of 125 peptides ranging from 2 to 14 amino acids (9). However, the topology of BP–ANN model developed by these researchers for the set of 102 peptides was 3-3-1 (11). Therefore, designing the network topology involves determining the number of nodes at each layer, the number of layers in the network, and the path of the connections among the nodes.

14.2.2.1.3. Stage 3. In this stage, the generated model is evaluated and validated. There are two common methods to assess the robustness of the developed model: internal validation and external validation. When data size is not too large, one commonly prefers using the internal validation method. Among different methods for this purpose, two methods of cross validation and leave-one-out (Jackknife) are the most common ones. All internal and external techniques use the same principle, that the validation data must not involve any process of model parameter estimation. This means that the data set must be divided into two parts. One is for model development, which is commonly referred to as *training set*. The other is for model evaluation, which is referred as *validation set*.

In cross validation, the data set is randomly divided into m folds. Each fold contains distinctive data points. Every time, one has to select one fold as the validation set and the remaining $m - 1$ folds as the training set for model development. This process is repeated for m times, until each fold has been used for validation once. It means that there are m validation models.

When data size is not too large, one commonly prefers using the leave-one-out cross validation (Jackknife) method. This means that one data point is picked up for validation and the remaining data points are used for training. This process is repeated until each data point has been validated once. In other words, for a data consisted of n points, n validation models should be performed.

External validation is applicable when either a large data set is available or a new data set has become available after generation of the model. In the former case, called the resampling method, we normally randomly sample a certain percentage of data for training and the rest for validation. Such a process can be repeated many times. It is noteworthy that the molecules included in the validation set have no role in model parameter estimation.

Evidently, in QSMR studies, the primary concern should be to build neural network models that are general and robust. *Generalization* means the ability of neural networks to predict the observed response variable for patterns not included in the training set. In contrast, *memorization* means the ability to reproduce the values of the response variable for patterns taken from the training set. If for a fixed data set, we gradually increase the complexity of the neural network, which is defined as the number of connection weights and biases, by adding additional hidden neurons, the generalization error initially decreases, but after reaching optimal network size, starts to increase, although the memorization error decreases all the time. Figure 14.3 shows a typical learning plot.

The phenomenon in which the resulting neural network has bad generalization and good memorization ability is called *overfitting*. The model developers should exert every effort to prevent overfitting. Usually, to prevent overfitting, one must keep the ratio of the number of data points to the number of con-

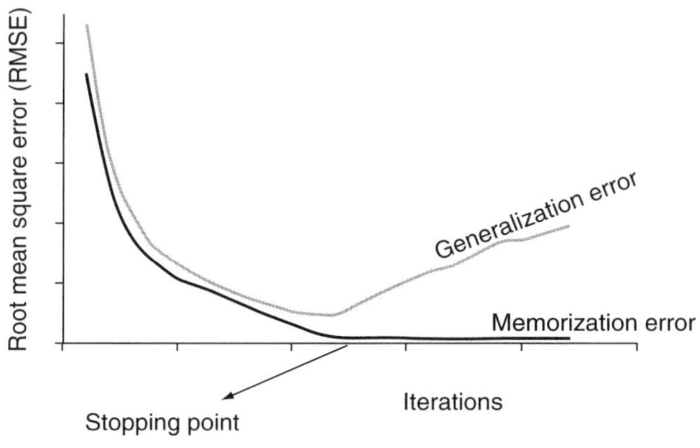

FIGURE 14.3. A typical learning plot.

nections higher than some threshold. In other words, overtraining can be avoided by means of "early stopping" of training after reaching the lowest generalization error. This means that an additional validation data set is required for monitoring the overfitting.

Jalali-Heravi et al. reported an ANN model for the first time to explore the linear/nonlinear characteristics of the electrophoretic mobilities of peptides (9). As described in detail above, they also developed a multivariable QSMR using Offord's charge-over-mass variable, combined with the corrected steric substituent constants and molar refractivity (9). These researchers used these features as inputs for a 3-4-1 BP–ANN model. Inspection of the ANN-calculated mobilities revealed significant improvements in predictive ability ($r^2 = 0.930$, SE = ~2.5) over the MLR-based treatment ($r^2 = 0.895$, SE = ~3.3). This was especially noticeable for highly charged peptides, containing amino acids such as arginine, histidine, and lysine. Before developing this model, it was argued that deviations in the prediction of mobilities for highly charged peptides are due to inaccurate charge calculations (10, 26). This improved correlation by BP–ANN analysis suggested that apparently nonlinear characteristics of the mobility–charge relationships are responsible for such a deviation.

In another attempt, Jalali-Heravi et al. (11), to assess the generalization of their ANN model, developed a 3-3-1 BP–ANN model based on a data set of 102 peptides (19). This data set consisted of peptides ranging in size from 2 to 42 amino acids. In contrast to their previous data set, the peptides of this set were larger, more hydrophobic, and highly charged. The better prediction ability of the BP–ANN model ($r^2 = 0.970$) over the MLR-based model ($r^2 = 0.930$) confirmed the nonlinear characteristics of the electrophoretic mobility. The robustness of ML models was approved by predicting CZE mobilities of a diverse sample set under different experimental conditions. Also, in endoproteinase digest separation simulation of melittin, glucogagon, and horse cytochrome C, the BP–ANN model exhibited good peptide-map prediction (11). The long-range outlook for modeling efforts looks promising, and the ability to predict CE mobilities of peptides precisely and possibly construct a peptide-map database holds the promise of helping current efforts in proteomics.

14.3. PREDICTION OF ELECTROPHORETIC MOBILITY OF ORGANIC ACIDS

CE has been applied in the analysis of organic acids (27). The key parameter in these analyses is electrophoretic mobility, which depends on both molecular structure and separation conditions. Therefore, developing chemometrical models to predict the mobilities of ions will relieve analysts of a large number of costly and time-consuming experiments. Two principal methods based on the quantitative relationship between molecular structures and electrophoretic mobilities are reported in the literature: mechanistic and statistical methods.

Recently, Cheng and Yuan (28), proposed a mechanistic model for predicting the electrophoretic mobility of carboxylic and sulfonic acids. Their model is very simple and is based on Equation 14.8:

$$\mu_e = \frac{a\text{NG}}{b\text{NA}^k + c\text{MW}^k} \qquad \text{(Eq. 14.8)}$$

where NG is the number of acid groups, NA is the number of atoms of organic acid, and MW is its molecular weight. The value of k ranges from 1/3 to 2/3 depending on the magnitude of molecular weight. Although simple molecular structure descriptors are employed in this method, it suffers from a high root mean square error (RMSE) and absolute average relative deviation (AAR). Recently, Jalali-Heravi and Shahbazikhah have developed a statistical model to improve the predictive ability and interpretability of the mechanistic model (29). Their new approach in QSMR studies represents the successful combination of CART as feature selection method and ANFIS as a feature mapping tool. The methodology of the CART–ANFIS model is described briefly in the following section and its results are compared with the mechanistic model. Detailed descriptions of CART and ANFIS can be found elsewhere (30, 31).

14.3.1. CART

CART is widely used for regression and classification in several areas such as medical diagnosis, classification of drugs, and retention prediction (32, 33). This method was introduced for the first time by Breiman et al. in 1984 (30). The aim of this statistical method was to explain the variation of a dependent variable, using a set of independent predictors, via a binary partitioning procedure. CART works by splitting the data into mutually exclusive subgroups, called child nodes, within which the objects have similar values for the response variable. The process starts from the parent node, which contains all objects of the data set. Then, binary splitting is repeated in which the parent node is split in two child nodes, and followed by treating each child node as a parent node, and so on. Each split is defined by a simple rule based on a single explanatory variable. For numerical variables, a cut point (splitting value) is selected to form two groups, which contain objects with values smaller and larger, respectively, than the selected cut point. Trees are grown by selecting the splits in such a way that the impurity of the response variable within each node is minimized. Among all possible splits, the best split is chosen by evaluation of the impurity of the formed nodes, according to some statistical criteria. The final tree is called the *maximal tree*, in which no further split can be performed, that is, all child nodes are homogeneous or contain one or a user-defined minimal number of observations. The terminal nodes of maximal tree, the so-called leaves, represent the final groups formed by the tree. However, the maximal tree is not always the best one, because a large number of leaves of this tree may overfit the learning data set, which will cause a poor predictive ability for new samples (30, 33). Therefore, one optimal tree should be selected by a good compromise between model fit and predictive properties. In general, CART analysis consists of three steps: (i) developing the maximal-tree; (ii) the tree "pruning," that is, generating a sequence of simpler trees by the cutting off of nodes; and (iii) selecting the optimal tree.

14.3.1.1. Maximal-Tree Building. To build the maximal tree, one needs to choose the best splitter to divide each root node into two child nodes. The measure of a good split is the impurity decrease between the parent node and its children:

$$\Delta i(s, t_p) = i_p(t_p) - p_L i(t_L) - p_R i(t_R) \qquad \text{(Eq. 14.9)}$$

where s is the candidate split and p_L and p_R are the fractions of observations of the parent node t that go into the child nodes t_L and t_R, respectively. The best splitter is the one that will maximize $\Delta i(s, t_p)$. The most popular criteria to measure the impurity is the total sum of squares of the response values about the mean of the node:

$$i(t) = \sum_{X_n \in r} (y_n - \bar{y}(t))^2 \qquad \text{(Eq. 14.10)}$$

where $i(t)$ is the impurity of node t, y is the response value of observation x belonging to node t, and $\bar{y}(t)$ is the mean of all observations in node t.

14.3.1.2. Tree Pruning. Usually, the maximal trees are oversized and describe the training set perfectly. This means that the model has been overfitted (34). The predictive ability of such trees is poor, because they tend to also fit the noise in the data. Pruning is a process that takes care of this problem by generating a sequence of smaller trees. These trees can be obtained by removing successively branches of the maximal tree. Since several trees of the same size can be generated from the maximal tree, both accuracy and complexity of the tree are considered to choose the best tree. This is done by a cost-complexity measure, R(T), defined for each subtree, T, as

$$\mathrm{R}_\alpha(\mathrm{T}) = \mathrm{R}(\mathrm{T}) + \alpha|\tilde{\mathrm{T}}| \qquad \text{(Eq. 14.11)}$$

where R(T) is the average within-node sum of squares and $|\tilde{\mathrm{T}}|$ is the tree complexity, which is equal to the total number of nodes of subtree. In this equation α is the complexity parameter, which is a penalty for each additional terminal node, and during the pruning procedure its value will gradually be increased from 0 to 1. This means that, by gradually increasing α, one generates a sequence of pruned subtrees starting from the largest one.

14.3.1.3. Selection of Optimal Tree. The optimal tree (most accurate tree) is the one having the highest predictive ability. Therefore, one has to evaluate the predictive error of the subtrees and choose the optimal one among them. The most common technique for estimating the predictive error is the cross-validation method, especially when the data set is small. The procedure of performing a cross validation is described earlier (see section 14.2.2.1). In practice, the optimal tree is chosen as the simplest tree with a predictive error estimate within one standard error of minimum. It means that the chosen tree is the simplest with an error estimate comparable to that of the most accurate one.

14.3.2. ANFIS

The architecture of an ANFIS model is shown in Figure 14.4. As can be seen, the proposed neuro-fuzzy model in ANFIS is a multilayer neural network-based fuzzy system, which has a total of five layers. The input (layer 1) and output (layer 5) nodes represent the descriptors and the response, respectively. Layer 2 is the *fuzzification* layer in which each node represents a membership. In the hidden layers, there are nodes functioning as membership functions (MFs) and rules. This eliminates the disadvantage of a normal NN, which is difficult for an observer to understand or to modify. The detailed description of ANFIS architecture is given elsewhere (31).

Recently, Jalali-Heravi and Shahbazikhah have developed a CART–ANFIS model for predicting the electrophoretic mobility of carboxylic and sulfonic acids (29). Their work consists of the following steps:

1 Selection of the data set: A total of 115 carboxylic and sulfonic acids were taken from the article published by Wronski (35). These acids are shown in Table 14.2. The data set has been divided into three sets; a training, a prediction, and a test set consisting of 73, 23, and 19 molecules, respectively. The test set was randomly selected from the training set for controlling the construction of the ANFIS model. The prediction set was used for the evaluation of the generated models.

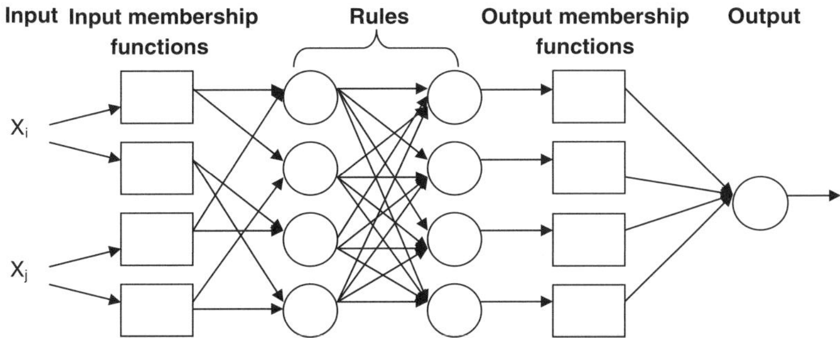

FIGURE 14.4. The architecture of ANFIS.

TABLE 14.2. Experimental and calculated electrophoretic mobilities ($\times 10^{-5}$ cm^2/s/V) of carboxylic and sulfonic acids using mechanistic and CART–ANFIS methods

No.	Compound[a]	Exp.	$Cal._{mechanistic}$	$Cal._{CART-ANFIS}$
1	Fluoroacetic acid	43.9	42.2	42
2	TFA	42.5	40.8	42.5
3	Chloroacetic acid	41.9	41.5	40
4	Dichloroacetic acid	39.4	40.3	39
5	TCA[p]	36.2	39.2	36.1
6	3-Chloropropionic acid	36.8	36	35
7	2-Chlorobutyric acid	32.8	32.2	32.5
8	5-Chlorovaleric acid	30.8	29.4	31.7
9	Bromoacetic acid	38.8	39.9	41.3
10	2-Bromopropionic acid[p]	33.4	34.9	31.6
11	2-Bromobutyric acid	30.8	31.4	30.2
12	4-Bromobutyric acid	32.8	31.4	35
13	5-Bromovaleric acid	30.8	28.7	29.4
14	2,3-Dibromopropionic acid	32.3	33.3	31.4
15	Tribromoacetic acid[p]	34.9	36.3	38.9
16	Iodoacetic acid	40.2	38.6	38.1
17	3-Iodopropionic acid	34.9	33.9	32
18	4-Iodobutyric acid	32.9	30.6	36.5
19	5-Iodovaleric acid[t]	30.8	28.1	29.7
20	Dibromofluoroacetic acid[p]	36.9	37.5	36.2
21	Chlorodibromoacetic acid	34.9	37.2	35.8
22	Glycolic acid[t]	42.3	40.3	38
23	Lactic acid[t]	36.5	35.2	36.3
24	2-Hydroxybutyric acid	34.2	31.7	30.6
25	Glyceric acid[p]	36.3	33.6	37.2
26	Glucuronic acid	26.6	25.2	27
27	Gluconic acid	27.2	24.3	26.4
28	2-Chloro-3-hydroxybutyric acid[t]	32.9	31	31.9
29	Glyoxalic acid	37.8	44.8	37.7
30	Pyruvic acid[p]	40.4	38.1	35.8
31	Trichlorolactic acid[t]	34.2	32.9	32.1
32	Maleic acid	62	69	62
33	Fumaric acid	61.2	69	60.9
34	Tartaric acid	60.5	59.9	61.2
35	Citric acid[p]	70.8	78.6	74.1
36	2-Ketoglutaric acid	59	60.1	59.3
37	Malic acid[t]	59	62.1	59
38	Thiomalic acid	58.5	61.5	59.6
39	2,3-Dimercaptopropanesulfonic acid	34.4	28.6	33.8
40	2-Hydroxyethanesulfonic acid[p]	39.6	33.1	42.2
41	Cyclobutane-1,1-dicarboxylic acid	51.1	57.2	52.8
42	Cyclopentane-1,1-dicarboxylic acid	50	53.2	49.8
43	Cyclohexane-1,1-dicarboxylic acid	48	50	48.3
44	Methylmalonic acid	58.5	64.6	56.8
45	Methylethylmalonic acid[p]	50	54.7	51

TABLE 14.2. *Continued*

No.	Compound[a]	Exp.	$Cal._{mechanistic}$	$Cal._{CART-ANFIS}$
46	Propylmalonic acid[t]	52	54.7	50.4
47	Diethylmalonic acid[t]	49.5	51.2	50.4
48	Ethylpropylmalonic acid	47	48.3	48.2
49	Dipropylmalonic acid[t]	46	45.8	50.1
50	Oxaloacetic acid[p]	56	66	60.6
51	3-Propylglutaric acid	47	48.3	47.4
52	Benzoic acid	34.4	31.3	34
53	Benzenesulfonic acid[t]	38.7	29.8	44.5
54	*p*-Toluenesulfonic acid	31.1	27.6	30.8
55	*o*-Aminobenzoic acid[p]	31.6	29.4	28.9
56	Sulfanilic acid	33.7	28.2	33.9
57	*p*-Fluorobenzoic acid	33.4	31	35
58	*p*-Chlorobenzoic acid	33.4	30.7	32.5
59	*m*-Iodobenzoic acid	33.4	29.3	31.6
60	*p*-Bromobenzoic acid[p]	31.5	30	33.5
61	*p*-Nitrobenzoic acid	32.1	28.9	35.8
62	3,5-Dinitrobenzoic acid	29.5	27.1	31.3
63	*p*-Toluic acid	29.1	28.7	31
64	*p*-Ethylbenzoic acid[t]	26.5	26.7	26.3
65	2,3-Dimethylbenzoic acid[p]	27.1	26.7	27.1
66	*o*-Isopropylbenzoic acid	24.7	25.1	27
67	2,4,6-Trimethybenzoic acid	24.7	25.1	24.4
68	*p*-tert-Butylbenzoic acid[t]	23.2	23.7	24.2
69	*p*-Hydroxybenzoic acid	34	30.2	30.6
70	Salicylic acid[p]	35.4	30.2	36.2
71	2,4-Dihydroxybenzoic acid[t]	32	29.1	31.7
72	3,4-Dihydroxybenzoic acid	34.4	29.1	34.8
73	Gallic acid	34.4	28.2	32.1
74	*p*-Methoxybenzoic acid	28.3	27.8	32.4
75	*p*-Ethoxybenzoic acid[p]	26.6	26	25
76	2-Nitro-3-bromobenzoic acid	28.2	27.9	30.1
77	2-Nitro-3-chlorobenzoic acid	31.3	28.5	30
78	Phenol[t]	34.4	33.9	36.5
79	*p*-Nitrophenol	33.4	31	32.5
80	2,4-Dinitrophenol[p]	31.3	28.7	33.1
81	Pieric acid	31.5	26.9	31.5
82	*p*-Chlorophenol[t]	33.4	33.1	34.7
83	2,4-Dichlorophenol	31.3	32.4	33.9
84	Vanillic acid	27.1	27	27.3
85	Cinnamic acid[p]	28.3	27.9	29.4
86	Phenylacetic acid	31.7	28.7	32.6
87	Phenoxyacetic acid	27.8	27.8	31
88	Nicotinic acid	34.6	32.2	35.7
89	2-Naphthalenesulfonic acid[t]	31.3	25.5	30.2
90	Acetic acid[p]	42.4	43.1	45.5

TABLE 14.2. *Continued*

No.	Compound[a]	Exp.	$Cal._{mechanistic}$	$Cal._{CART-ANFIS}$
91	Propionic acid	36.9	37.1	37.2
92	Butyric acid	33.7	33	33.1
93	Valeric acid	31.6	30.1	29
94	Hexanoic acid	30.2	27.8	28.4
95	Heptanoic acid[p]	28.4	26	28.1
96	Octanoic acid[t]	27.4	24.5	28.4
97	Nonanoic acid	26.7	23.2	28.8
98	Oxalic acid	74.6	83.4	74.4
99	Malonic acid[t]	66	72.2	63
100	Succinic acid[p]	60.3	64.6	61.6
101	Glutaric acid	55.6	59	55.5
102	Adipic acid	52.4	54.7	51.2
103	Pimelic acid	49.9	51.2	47.1
104	Suberic acid	47.2	48.3	44.8
105	Azelaic acid[p]	45.9	45.8	42.5
106	Sebacic acid	44.9	43.7	43.9
107	Methanesulfonic acid	50.5	39.5	49.5
108	Ethanesulfonic acid	42.7	34.7	44
109	Propanesulfonic acid	37.5	31.3	37.3
110	Butanesulfonic acid[p]	33.9	28.7	32.8
111	Pentanesulfonic acid	31.4	26.7	32.5
112	Hexanesulfonic acid[t]	29.4	25.1	28.3
113	Octanesulfonic acid	26.2	22.5	25.8
114	Nonanesulfonic acid	25.1	21.5	24.7
115	Dodecanesulfonic acid[p]	22.3	19.1	21.8

[a]Compounds marked with p and t are included in the prediction and test sets, respectively; remaining molecules are included in the training set.

2 Generation of the descriptors: A total of 1497 0-, 1-, 2-, and 3-D variables were generated using Dragon v 3.0 software (Via V. Pisani, Milan, Italy). These parameters were constitutional, topological, and molecular walk, and path counts, 2-D autocorrelation, aromatic indices, Randic molecular profiles, geometrical, RDF, 3D-MoRSE, WHIM descriptors, GETAWAY, functional group counts, atom-centered fragments, charge, and empirical and molecular properties.

3 Selection of features: Descriptors with the same values for all objects were eliminated and one of the descriptors with correlations higher than 0.98 was removed. The total number of descriptors before and after the screening was 1497 and 1193, respectively. All 1193 descriptors were used for regression tree analysis. One of the advantages of CART is the lack of requirement for preprocessing. As a first step, a maximal tree was built. This tree exhibited a maximum of 7 levels and 21 leaves (terminal nodes).

4 Tree pruning: A 10-fold cross validation was applied for reducing the number of variables and obtaining the best predictive tree. A plot of the number of terminal nodes versus *COST* function showed that the trees with more than four terminal nodes have good predictability. Keeping in mind the least number of variables, Jalali-Heravi et al. examined the descriptors of trees with only four, five, and six leaves for variable selection. The best model was constructed by applying ANFIS using a tree with five terminal nodes or four descriptors (Fig. 14.5). The definition of the parameters selected by optimal tree is included in Table 14.3.

5 Generation of ANFIS model: The variables chosen in step 4 were used as inputs for ANFIS. The number and type of the MFs needed for developing the ANFIS model were optimized using RMSE for the test set. Finally, the optimized models were applied to all data sets, and the results are shown in Table 14.2.

For the sake of comparison, the results of the mechanistic model are also given in Table 14.2. To assess the robustness of the models, a 10-fold cross-validation method was used on all data sets (29). The consistency of the results of cross validation for all groups proved the stability and robustness of the models. Figure 14.6 demonstrates the plot of the CART–ANFIS calculated values for the acid mobilities against the experimental values. The high value of $R^2 = 0.970$ for this plot indicates that the CART–ANFIS model can be considered as a powerful tool for the prediction of the electrophoretic mobility of organic and sulfonic

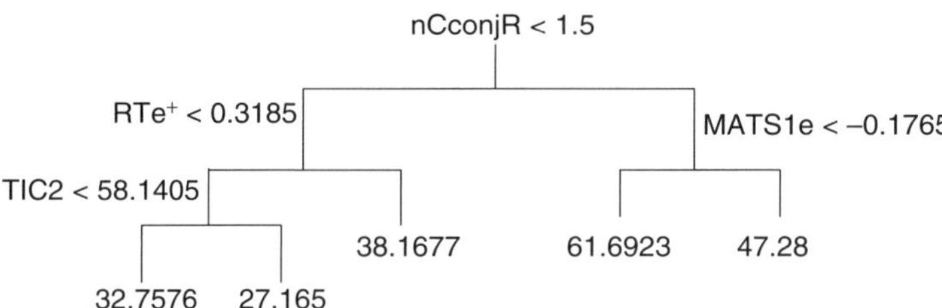

FIGURE 14.5. Selected tree with low RMSE for variable selection.

TABLE 14.3. Important molecular descriptors selected by CART

Descriptor	Definition	Class of Descriptor
nCconjR	Number of exo-conjugated C (sp2)	Functional groups
RTe1	R maximal index/weighted by atomic Sanderson electronegativity	GETAWAY
MATS1e	Moran autocorrelation –lag 1/weighted by atomic Sanderson electronegativities	2-D autocorrelation
TIC2	Total information content index (neighborhood symmetry of second-order)	Topological

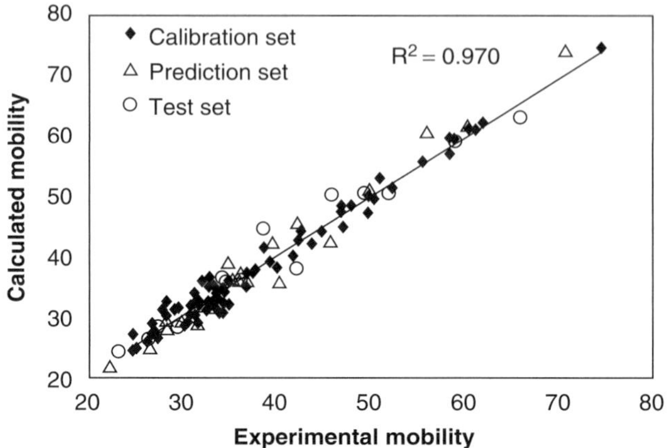

FIGURE 14.6. Plot of the calculated mobility against the experimental values for CART–ANFIS model.

acids. In addition, the results indicated that the CART–ANFIS model is superior over the mechanistic model and shows astonishing improvements for absolute average relative deviation (AARD) of calibration and prediction sets. The values of 3.78 and 4.81% for $AARD_c$ and $AARD_p$, respectively, should be compared with their counterpart values of 7.21 and 7.53% for the mechanistic model.

REFERENCES

1. Kasicka, V. (2003) *Electrophoresis*, **24**, 4013–4046.
2. Grossman, P.D., Colburn, J.C., and Lauer, H.H. (1989) *Anal Biochem*, **179**, 28–33.
3. Offord, R.E. (1966) *Nature (London)*, **211**, 591–593.
4. Compton, B.J. (1991) *J Chromatogr*, **599**, 357–367.
5. Adamson, N.J. and Reynolds, E.C. (1997) *J Chromatogr B*, **699**, 133–147.
6. Messana, I., Rossetti, D.V., Cassino, L., Misiti, F., Giardina, B., and Castagnola, M. (1997) *J Chromatogr B*, **699**, 149–171.
7. Kasicka, V. (1999) *Electrophoresis*, **20**, 275–279.
8. Cifuentes, A. and Poppe, H. (1997) *Electrophoresis*, **18**, 2362–2376.
9. Jalali-Heravi, M., Shen, Y., Hassanisadi, M., and Khaledi, M.G. (2005) *Electrophoresis*, **26**, 1874–1885.
10. Janini, G.M., Metral, C.J., Issaq, H.J., and Muschic, G.M. (1999) *J Chromatogr A*, **848**, 417–433.
11. Jalali-Heravi, M., Shen, Y., Hassanisadi, M., and Khaledi, M.G. (2005) *J Chromatogr A*, **1056**, 58–68.
12. Cross, R.F. and Granham, N.F. (2001) *Chromatographia*, **54**, 639–646.

13. Ma, W., Luan, F., Zhang, H., Zhang, X., Liu, M., Hu, Z., and Fan, B. (2006) *Analyst (Cambridge, UK)*, **131**, 1254–1260.
14. Yu, K. and Cheng, Y. (2007) *Talanta*, **71**, 676–682.
15. Li, Q., Dong, L., Jia, R., Chen, X., Hu, Z., and Fan, B.T. (2002) *Comput Chem*, **27**, 297–303.
16. Metral, C.F., Janini, G.M., Muschik, G.M., and Issaq, H.J. (1999) *High Resolut Chromatogr*, **22**, 373–378.
17. Greooman, P.D., Colburn, J.C., and Lauer, H.H. (1989) *Anal Biochem*, **179**, 28–33.
18. Wasburn, M.B., Wolters, D., and Yates, J.R., III (2001) *Nat Biotechnol*, **19**, 242–247.
19. Janini, G.M., Metral, C.J., and Issaq, H.J. (2001) *J Chromatogr A*, **924**, 291–306.
20. Minitab Release 12. http://www.minitab.com (accessed December 2, 1999).
21. Taft, R.W., Jr. and Newman, M.S. (ed.) (1956) *Organic Chemistry*, John Wiley and Sons, New York.
22. Pardon, J.R., Carrasco, R., and Pellon, R.F. (2002) *J Pharm Pharmaceut Sci*, **5**, 258–265.
23. Simon, H. (1983) Why should machines learn? In *Machine Learning: An Artificial Intelligence Approach* (eds. P. Michalski, J. Carbonell, and T. Mitchell), Tioga Press, Palo Alto, CA, pp. 25–38.
24. Wu, C.H. (1997) *Comput Chem*, **21**, 237–256.
25. Sun, Z., Rao, X., Peng, L., and Xu, D. (1997) *Protein Eng*, **10**, 763–769.
26. Cifuetes, A. and Poppe, H. (1994) *J Chromatogr A*, **680**, 321–340.
27. Kevin, D.A. (1999) *J Chromatogr A*, **856**, 443–463.
28. Cheng, Y. and Yuan, H. (2006) *Anal Chim Acta*, **565**, 112–120.
29. Jalali-Heravi, M. and Shahbazikhah, P. (2008) *Electrophoresis*, **29**, 363–374.
30. Breiman, L., Friedman, J.H., Olshen, R.A., and Stone, C.J. (1984) *Classification and Regression Trees*, Wadsworth, Monterey.
31. Loukas, Y.L. (2001) *J Med Chem*, **44**, 2772–2783.
32. Lavrac, N. (1999) *Artif Intell Med*, **16**, 3–23.
33. Deconinck, E., Hancock, T., Coomans, D., Massart, D.L., and Vander Heyden, Y. (2005) *J Pharm Biomed Anal*, **39**, 91–103.
34. Massart, D.L., Andeginst, B.G.M., Buydens, L.M.C., De Jong, S., Lewi, P.J., and Smeyers-Verbeke, J. (1997) *Handbook of Chemometrics and Qualimetrics Part A*, Elsevier, Amsterdam.
35. Wronski, M. (1993) *J Chromatogr A*, **657**, 165–173.

CHAPTER 15

ASSESSMENT OF SOLUTE–MICELLE INTERACTIONS IN ELECTROKINETIC CHROMATOGRAPHY USING QUANTITATIVE STRUCTURE–RETENTION RELATIONSHIPS

EDGAR P. MORAES,[1] FERNANDO G. TONIN,[2] LUÍS G. DIAS,[3] JOÃO P.S. FARAH,[1] and MARINA F.M. TAVARES[1]
[1]Institute of Chemistry, University of Sao Paulo, SP, Brazil
[2]Department of Food Engineering, Faculty of Zootechny and Food Engineering, University of Sao Paulo, SP, Brazil
[3]Department of Chemistry, Faculty of Philosophy, Sciences and Language of Ribeirão Preto (FFCLRP), University of Sao Paulo, SP, Brazil

CONTENTS

15.1. INTRODUCTION

Quantitative structure–retention relationships (QSRR) is a term first coined by R. Kaliszan in 1987 (1), that encompasses statistically derived relationships

Chemometric Methods in Capillary Electrophoresis. Edited by Grady Hanrahan and Frank A. Gomez

between retention parameters and descriptors characterizing the solute molecular structure (2–4). QSRR studies have found numerous applications in many scientific and industrial domains as compiled comprehensively by recent reviews (4–8).

Historically, QSRR has its foundation in the efforts of physical organic chemists who rationalize solute substituent effects on reaction rates and equilibria. The most notorious of all, the Hammett equation (9), inaugurates the "linear free energy relationships" (LFER), where the logarithm of the reaction equilibrium constant, K, is a linear function of the substituent constant (σ), an arbitrarily derived parameter based on the ionization of benzoic acid derivatives in water, as follows:

$$\log \mathrm{K} = f_{linear}(\sigma) \qquad \text{(Eq. 15.1)}$$

A specific subset of the broader class of thermodynamic LFERs is known under the acronym LSER, "linear solvation energy relationships." The LSER paradigm invokes explicitly the processes in which the solute transfer between two phases takes place and evolved from the work of Abraham (7, 10), who built on the pioneer work of Hammet and later on Kamlet and Taft (11, 12), enabling the extension of LFER from the realm of atomic properties to the realm of intermolecular interactions.

In the LSER formalism, the LFER equilibrium constant K becomes the partition coefficient P. The solvation process consists in the transfer of a given solute from the gaseous phase into a condensed phase and is described by the following hypothetical steps: (i) solvent cavitation, (ii) solute transfer to the cavity, (iii) launching of specific interactions between solute and solvent at the cavity surface, and finally (iv) the Born work, that is, the reversible work of charging the polarizable solute in nonhomogeneous dielectric medium. Conceptually, solute solvation is thus explained by three types of intermolecular interactions: hydrophobic (the cavity formation-dispersive interactions, V term), polar (dipolarity–polarizability interactions, E and S terms), and specific (hydrogen bond interactions, A and B terms), as represented by Equation 15.2:

$$\log \mathrm{P} = f_{linear}(\mathrm{V}, \mathrm{E}, \mathrm{S}, \mathrm{A}, \mathrm{B}) \qquad \text{(Eq. 15.2)}$$

QSRR is simply an LSER in the sense that solute retention parameters obtained in a given flow regimen are always representative of the equilibrium condition. In QSRR, descriptors of solute structure and properties can be incorporated in the model to give the general expression:

$$\log \mathrm{k} = f_{linear}(\text{descriptors}) \qquad \text{(Eq. 15.3)}$$

The benefits of estimating P from k measurements are unprecedented. First of all, a truly estimate of the ΔG associated with the solute transfer between the two condensed phases is provided. Second, in a single run, it is possible to simultaneously generate precisely measured retention factors for a large set of solutes, all submitted to rigorously the same environmental conditions. The application of QSRR to organized media presents additional aspects: micelle structure, size, and shape can be easily altered by the medium properties. Therefore, by modulating the electrolyte composition with additives (organic solvents being the most effective), a multitude of new dispersed phases are devised for solute interaction. Micellar electrokinetic chromatography (MEKC) becomes thus a unique technique to study solute–micelle interactions, via QSRR, the focus of the present chapter.

15.2. BUILDING QSRR FROM MEKC DATA

The search for a linear correlation between log k and solute descriptors (Eq. 15.3) allows one to establish in a qualitative and quantitative manner which intermolecular forces govern the phenomenon under investigation. Building QSRR thus demands the use of refined chemometric tools for variable selection, criteria to detect and eliminate outliers, and, ultimately, data validation procedures.

Figure 15.1 depicts an schematic representation of the steps underlying QSRR development in MEKC. In general terms, once the solute set is defined, retention data in a given electrophoretic system are acquired. In the sequence, solute descriptors are generated and combined statistically with retention

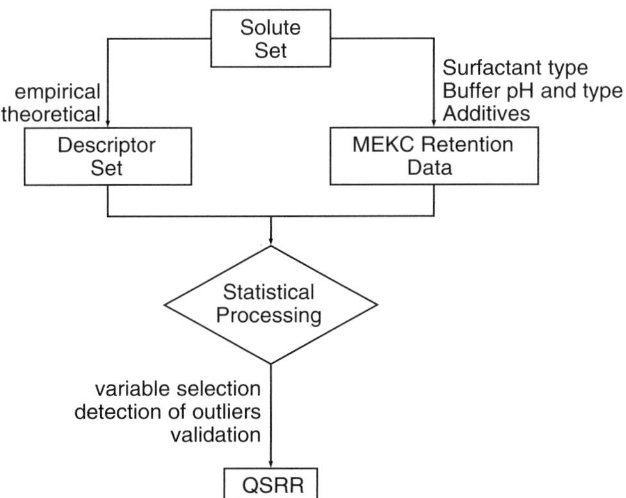

FIGURE 15.1. Schematic representation of the steps underlying QSRR development in micellar electrokinetic chromatography.

factors in a multivariate linear regression (MLR). Each building block of Figure 15.1 deserves a more detailed commentary.

15.2.1. Solute Set

So far the most important block in Figure 15.1 is the solute set because it is the solute structure that defines both descriptor set and optimal electrolyte composition. Depending upon the solute structure, the presence of certain functional groups, acid base properties, solubility in the aqueous versus micellar phases, etc., a proper electrolyte might be selected. The relevance and implications of the solute set selection will be clearly recognized by the literature examples of section 15.3.

15.2.2. MEKC Retention Data

In MEKC, a variety of electrolyte systems (phosphate and tetraborate buffers at extreme pH, among others) and separation carriers (sodium octyl-, decyl-, or dodecyl- sulfate [SDS], lithium dodecyl sulfate [LDS], tris(hydroxymethyl)aminomethane dodecyl sulfate, sodium dodecyl sulfonate, alkyltrimethylammonium bromide salts [alkyl: dodecyl, DDAB; tetradecyl, TTAB; hexadecyl or cetyl, CTAB], lithium perfluorooctanesulfonate [LPFOS], sodium N-dodecanoyl-N-methyltaurine, polyoxyethyleneglycol dodecyl ether [Brij35], and bile salts [cholate, deoxycholate, taurocholate, taurodeoxycholate], among others) modified by additives (organic solvents, cyclodextrins, etc.) may be contemplated to alter the separation selectivity. Depending upon the relative magnitude of the electroosmotic flow (EOF) and separation carrier velocities, three elution modes (normal, reversed, and restricted) are devised. For each MEKC mode, a corresponding distinct equation for calculating retention factor as a function of migration time (solute, micelle and EOF) applies (13).

Therefore, the practical evaluation of retention factors relies strongly on precise measurements of solute and micelle migration times as well as the EOF time, preferably all at the same run. A proper choice of EOF and micelle markers is not always trivial. Organic solvents and the usual refractive index baseline disturbances when UV detection is employed can be used to flag EOF time (14, 15). The measurement of time at the baseline deflection must be consistent, either at the beginning, middle, or terminal section. As micelle markers hydrophobic compounds such as Sudan III, polyaromatic hydrocarbons (e.g., anthracene), long-chain alkyl benzenes, and alkyl phenyl ketones have all been considered (14, 15). In these surfactant systems, adsorption-related distorted peaks often emerge, which compromise the precise determination of migration time of the micelle marker. Anyhow, triplicate injections of solutes and markers are mandatory and relative standard deviation (RSD) better than 1% should be pursued. Precision in the measurement of retention factors is important, especially when narrow migration windows are contemplated. Fuguet et al. examined the adequacy of EOF and micelle markers under the perspective of LSER for seven commonly used micellar systems finding methanol, acetonitrile and formamide as the best EOF markers whereas dodecanophenone was the most appropriate micelle marker (15).

15.2.3. Descriptor Set

Descriptors are atomic or molecular parameters or even molecular properties that contain information about the energy of each type of intermolecular interaction. They can be classified into two broad categories: empirical and theoretical. Empirical descriptors depend on experimental measurements; thus, they are available for a limited number of solutes (16). Theoretical descriptors are derived from the solute structure; they are usually based on *ab initio* or semiempirical quantum chemistry calculations or on the connectivity of atoms in the molecule. With the proper use of dedicated software, the number of structural descriptors that can be assigned to a given solute is practically unlimited. Comprehensive compilations of the literature (17, 18) register over 2000 known theoretical descriptors.

15.2.4. Statistical Processing

To obtain reliable QSRR, appropriate input data and stringent statistical analysis must be conducted. An important point to be emphasized here is that when QSRR are built from MEKC data, a physicochemical model for the solute–micelle interaction must be established before any statistical processing takes place. Therefore, considering that the intermolecular interactions responsible for solute retention are hydrophobic, polar, and specific in character, only the descriptors able to account for these interactions must be preselected.

Second, the nature and number of compounds in the solute set must be examined. If a large number of solutes with varying functional groups are under consideration, empirical descriptors covering the three types of intermolecular interactions might be selected and further processed. However, if a restricted number of solutes belonging to the same chemical class are considered, theoretical descriptors are then the most likely to explain solute retention. Furthermore, if the intermolecular interaction can be already rationalized for that particular set of solutes, descriptors of that interaction might be chosen. Otherwise, theoretical descriptors covering the three types of intermolecular interactions should be selected for statistical processing.

Once the descriptor set is selected, single or MLR of log k as a function of descriptors (Eq. 15.3) is performed. The next step is intrinsically related to any linear regression analysis. The variables (descriptors) must be inspected for variability and multicolinearity. For a given set of solutes, it is desirable for the descriptors to cover a reasonable numerical range with a uniform distribution of values within the range. The variability of the descriptor set can be evaluated by several ways. Histogram plots for each descriptor can readily inform about descriptor variability. No tendencies or clustering of values should be observed. A more elegant way to inspect descriptor variability is to calculate Shannon entropies (19). Large entropy values are desirable and reflect the amount of information a descriptor carries.

Equally important is to check the descriptor set for multicolinearity. Correlation between descriptor values results in unreliable MLR with overestimate goodness-of-fit parameters and poor predictive capability. Cross-correlation matrices provide information on descriptor multicolinearity. It is worth mentioning that when two descriptors, X and Z, are statistically correlated, it does not necessarily mean that physicochemically they are also redundant. Principal component regression or partial least squares regression can be used to address multicolinearity. Alternatively, the impact of descriptors X and Z on the QSRR should be inspected separately. It should be possible to select from the solute set those solutes with varying X values and constant Z values and vice versa.

There are other misconceptions regarding multicolinearity as postulated by Guyon and Elisseeff (20). For instance, if descriptor X presents a better correlation with log k than descriptor Z, it does not mean that X should be selected over Z. Furthermore, if a given descriptor X does not correlate with log k, it does not mean that X in combination with other descriptors will also present a poor fitting.

The inspection of outliers is the next step in the statistical processing. In the presence of outliers, least squares estimation is biased. Nowadays a number of robust MLR methods are available to treat data that contain outliers (21), least trimmed squares being one of the many popular alternatives. Interestingly, in many QSRR studies, it is precisely the outliers that are of physicochemical interest.

The final step on statistical processing and perhaps the most important it is the validation procedure. Depending on the solute set size, leave-one-out (LOO), bootstrap, or leave-group-out (LGO) procedures might be employed. LGO procedures may be quite informative when solutes of the same organic functionality are grouped and left out of the entire solute set at a time for testing. The predicted sum of squares (PreSS) and coefficient of determination (Q^2) are qualifying parameters for the validation procedure, whereas the number of datapoints (n), Fisher statistics parameter (F), standard error (S), and the coefficient of determination (R^2, and not simply the coefficient of correlation, R^2) qualify the regression.

Once a QSRR equation is built and validated, the ultimate evaluation of the real impact of each descriptor in the response (log k, Eq. 15.3) is given not by the magnitude of its coefficient but solely by its statistical importance. Therefore, the variance (S^2), the partial F (F_{partial}), and the random probability (*p value*) must be computed for each descriptor coefficient. Coefficients with small values of F_{partial}, large p values, and large variances should be disregarded.

15.3. THE PHILOSOPHY BEHIND QSRR

In general terms there are two lines of thought in the published literature regarding QSRR in MEKC. The first approach relies on the statistical treatment of a large set of solutes (LSS) with a gamut of chemical functional groups, representative of all possible intermolecular interaction types. The second approach models a restricted set of solutes (RSS), usually from the same chemical class, and possibly governed by a particular intermolecular interaction type. QSRR with both LSS and RSS can be built by statistical processing of a large set of descriptors (LSD) or a small set of descriptors (SSD). Any of the four combinations is possible with important consequences.

15.3.1. LSS

QSRR studies in MEKC involving LSS with LSD are scarce and not helpful in delineating solute–micelle interactions but are useful for predictive purposes (22). Typically, LSD are screened in quantitative structure activity relationships (QSAR) studies where the exact nature of the relationship between the solute structure and biological activity is hardly established by a physicochemical model. Just for reference, data mining procedures for LSD and statistical modeling aspects of QSRR using chromatographic data have been reviewed recently (23).

The literature of QSRR with LSS is dominated by a specific SSD, the LSER solute parameters V, E, S, A, and B, as defined in Equation 15.2. An extraordinary amount of attention has been paid to predict retention (24, 25) and to establish phase selectivity in MEKC using LSER (5, 7, 26–31). Attempts to classify and to contrast micellar phases with basis on the LSER coefficients have been pursued by many researchers (5, 26, 27, 29). Interesting approaches comprise the classification of micellar phases by the combined use of LSER parameters and retention indexes (32), the clustering of micellar systems by principal component analysis (26), the use of LSER parameters to compose vectors for characterization of lipophilicity scales (33), and, more recently, the establishment of micellar selectivity triangles (34, 35) in analogy to the solvent selectivity triangle introduced by Snyder to classify solvents and ultimately mobile phases in liquid chromatography.

Because of the ubiquitous use of UV detectors in capillary electrophoresis systems, the LSER studies derived from MEKC data, a subset of solutes within approximately 100 compounds with UV-absorbing properties (mostly benzene derivatives and compounds with carbonyl moieties) is usually selected (27, 29). Interestingly, the benzene derivatives of the solute set present an additional structural feature: the majority of the compounds exhibit organic multifunctionalities in the attempt to impart the necessary variability to the descriptor parameters.

The properties of a variety of surfactants (36–41), novel cationic phases (42–44), mixed micelles (45–49), microemulsions (50), vesicles (51–55), liposomes (56, 57), and synthetic polymers (58–65) have all be screened by LSER. Micelle structural modifications by differing head groups (66) and spacers (67, 68), chain lengths (69), and counterions (70), as well as the use of deuterated water buffers (71) and the addition of cyclodextrins (72) and organic solvents (73) to the micellar medium, have also been characterized by LSER studies.

Despite the fact that the blend of intermolecular interactions and their relative contribution are somehow different in many surfactant systems, the overall conclusion of the LSER studies cited above is rather astonishing: for neutral compounds, cavitation work (V term, positive sign) and the solute hydrogen bond basicity (B term, negative sign) are recognizably dominant factors describing the distribution of solutes in micellar aqueous systems. It is widely proclaimed by these studies that large nonpolar molecules tend to incorporate into the micellar phase relatively to the protic bulk aqueous solution of high dielectric constant. On the other hand, solutes that can accept proton via hydrogen bonding have their retention modulated, with increased solute basicity leading to decreased retention.

In order to investigate the scope of these findings in the interpretation of solute-micelle interactions, and to define a possible solubilization locus, Figure 15.2 was built using distribution data between water and SDS micelles as compiled by Sprunger et al. (74) (mostly derived from micellar liquid chromatography [MLC]) and Quina et al. (75) (including MLC and a variety of other techniques, such as absorption spectroscopy, calorimetry, potentiometry, and densitometry), and MEKC retention data as compiled by Kelly et al. (25) and by Poole and Poole (30). The resulting LSER statistics for each database is organized in Table 15.1. In Table 15.1, hexadecane–water partition database (16) was also included as a reference, the database from Sprunger et al. was further computed with and without outliers (set of large alkanes, decanol, and propionamide), and the database from Poole and Poole, originally published as log k, was converted to log P, using a phase ratio of 0.009 (25). A cross-correlation matrix was included in Table 15.1 for appreciation of the embedded multicolinearity of the descriptors associated with the MEKC solute set.

Figure 15.2a allows the evaluation of the representativity of the set of solutes used in MEKC for estimation of partition coefficients. Despite the limitations imposed by the use of UV detectors, the set of solutes used in MEKC predicts log P similarly to the set of solutes used by other techniques, which comprise solutes of a much richer blend of organic functionalities. In that sense, both solute databases are in fact rather complementary: the range of log P experimentally obtained from MEKC data reinforces the lower section of the graph ($1 < \log P < 4$), whereas the range derived from other techniques is more equally distributed throughout the entire range of log P, although it reinforces the upper section of the graph ($4 < \log P < 8$).

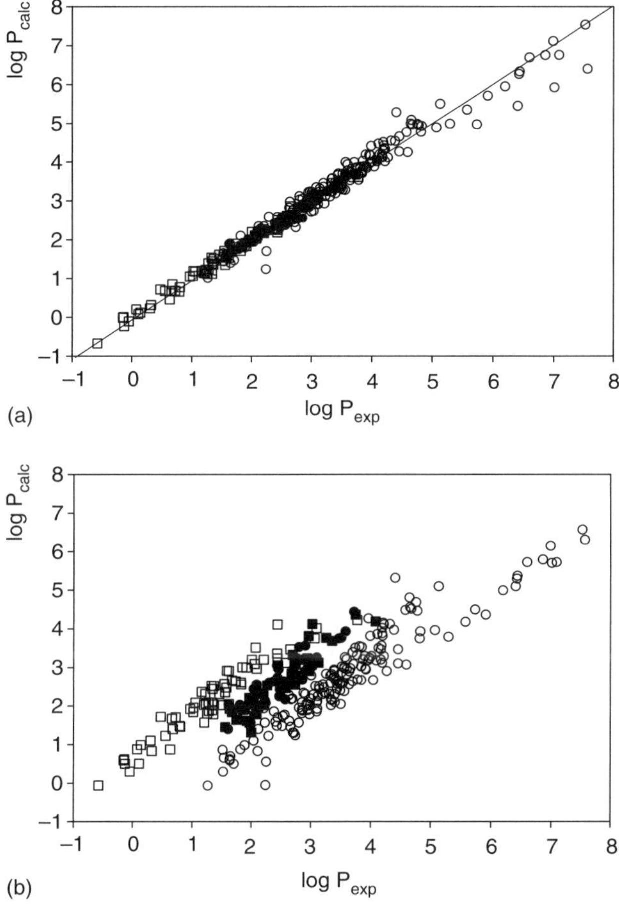

FIGURE 15.2. Prediction of partition coefficients for sodium dodecyl sulfate micellar systems using separate MLR for each database (a) and a single MLR for all databases (b). Databases: (○) data compiled by Sprunger et al. (74) and (□) by Quina et al. (75) for distribution between water and SDS; (●) data compiled by Kelly et al. (25) and (■) by Poole and Poole (30) using MEKC.

However, if a single MLR is fitted to the four databases altogether, the predicted log P discriminates the data sets as visualized in Figure 15.2b. As observed, the databases seem to be parallelly arranged. A better appreciation of the results of Figure 15.2b can derive from the overall quality of the MLR associated with each database and specifically with the magnitude, statistical relevance, and physicochemical meaning of the linear coefficients or system constants (Table 15.1). The linear coefficient of a QSRR (Eq. 15.3, when log k is computed) comprises the logarithm of the phase volume ratio, β (total volume of the micellar phase over the total volume of the aqueous phase). For chromatographic and electrophoretic data, retention factors, k, are

TABLE 15.1. Linear free energy relationships in SDS (as log P) using discrete databases

Database (as log P)	System Constant (p value)	v (p value; $F_{partial}$)	e (p value; $F_{partial}$)	s (p value; $F_{partial}$)	a (p value; $F_{partial}$)	b (p value; $F_{partial}$)
Hexadecane–water	0.080 ± 0.020	4.40 ± 0.03	0.67 ± 0.03	−1.61 ± 0.03	−3.59 ± 0.04	−4.85 ± 0.04
Abraham et al. (16)	(0.002)	(0; 29596)	(0; 522)	(0; 2279)	(0; 8072)	(0; 15737)
SDS–water	1.26 ± 0.08	3.39 ± 0.08	0.23 ± 0.08	−0.54 ± 0.10	−0.11 ± 0.08	−2.28 ± 0.12
Sprunger et al. (74)	(0)	(0; 1605)	(0.005; 8)	(2e-7; 29)	(0.2; 2)	(0; 375)
SDS–water[a]	1.27 ± 0.05	3.10 ± 0.06	0.42 ± 0.06	−0.54 ± 0.07	−0.11 ± 0.05	−1.96 ± 0.08
Sprunger et al. (74)	(0)	(0; 2337)	(0; 59)	(0; 65)	(0.04; 4)	(0; 594)
SDS–water	−0.61 ± 0.05	3.26 ± 0.08	0.36 ± 0.06	−0.64 ± 0.08	−0.07 ± 0.07	−1.82 ± 0.11
Quina et al. (75)	(0)	(0; 1522)	(0; 32)	(0; 64)	(0.3; 1)	(0; 292)
MEKC–SDS	0.13 ± 0.12	2.98 ± 0.12	0.36 ± 0.09	−0.43 ± 0.11	−0.26 ± 0.06	−1.70 ± 0.11
Kelly et al. (25)	(0.28)	(0; 593)	(0.0002; 16)	(0.0004; 15)	(7e-5; 19)	(0; 236)
MEKC–SDS (as log P)	0.097 ± 0.071	2.99 ± 0.07	0.46 ± 0.05	−0.44 ± 0.05	−0.30 ± 0.05	−1.88 ± 0.08
Poole and Poole (30)	(0.18)	(0; 1881)	(0; 98)	(0; 69)	(0; 37)	(0; 537)
MEKC–SDS (as log k)	−1.82 ± 0.07	2.99 ± 0.07	0.46 ± 0.05	−0.44 ± 0.05	−0.30 ± 0.05	−1.88 ± 0.08
Poole and Poole (30)	(0)	(0; 1881)	(0; 98)	(0; 69)	(0; 37)	(0; 537)
All (SDS)	−0.16 ± 0.16	4.24 ± 0.19	0.06 ± 0.18	−0.78 ± 0.21	0.34 ± 0.17	−2.05 ± 0.26
	(0.32)	(0; 477)	(0.74; 0.1)	(0.0003; 13)	(0.05; 4)	(0; 61)

Databases	n	PreSS	Q^2	S^2	R^2_{adj}	F
Hexadecane–water	379	0.02	0.996	0.02	0.996	18282
Abraham et al. (16)						
SDS–water	171	0.08	0.94	0.07	0.95	653
Sprunger et al. (74)						
SDS–water[a]	165	0.03	0.97	0.03	0.98	1318
Sprunger et al. (74)						
SDS–water	65	0.02	0.98	0.02	0.98	603
Quina et al. (75)						
MEKC–SDS	50	0.01	0.96	0.01	0.96	255
Kelly et al. (25)						
MEKC–SDS as log P	38	0.006	0.98	0.005	0.99	577
Poole and Poole (30)						
MEKC–SDS as log k[b]	38	0.006	0.98	0.005	0.99	577
Poole and Poole (30)						
All (SDS)	324	0.58	0.53	0.57	0.68	141

v, e, s, a, and b are the coefficients of V, E, S, A, and B, respectively.
[a]Outliers excluded: alkanes, decanol, and propionamide.
[b]Cross-correlation matrix for the MEKC database. Data from Poole and Poole (30).

V	1.00				
E	0.19	1.00			
S	0.00	0.33	1.00		
A	0.29	0.22	0.03	1.00	
B	0.00	0.03	0.42	0.17	1.00

experimentally measured and usually converted to partition coefficients, P, by means of β, and used to generate LSER (Eq. 15.2). Therefore, data in Table 15.1 are LSER; that is, log P was used instead of log k (except for the Poole and Poole database, presented in both ways). Therefore, different values of β might be used to convert log k into log P leading to differing system constants. Indeed the system constants that can be statistically compared (small *p* value) vary form –0.61 to 1.27. Hexadecane–water database provided a system constant close to zero. The Quina et al. SDS database whose partition data were derived from several techniques (many providing direct measure of log P) provided the smallest value as opposed to the largest value of the Sprunger et al. database, whose partition data was mostly derived from MLC. The Poole and Poole QSRR based on log k provided a system constant of –1.82 ± 0.07, a value that can be used to estimate the system phase ratio (β = 0.015).

Other observations from Table 15.1 include the following: the overall MLR is very poor (n = 324 compounds, F = 141, $R^2_{\mathrm{adj}} = 0.68$, S^2 = 0.57), whereas the MLR of the hexadecane–water database is quite superior (n = 379 compounds, F = 18282, $R^2_{\mathrm{adj}} = 0.996$, S^2 = 0.02). When the outliers were rejected from the database of Sprunger et al., the LOO validation parameters improved (PreSS decreased and Q^2 increased).

If the statistical quality of the parameters V, E, S, A, B from Table 15.1 are now inspected, it is clear that they are not equally relevant. For instance, parameter A must be rejected from the statistical standpoint: its coefficient presents a small F_{partial}, *sine qua non* condition to rejection, and a large *p* value for some databases. This is also expected from the physicochemical standpoint: the coefficient of parameter A (solute hydrogen bond acidity) reflects the minor differences in hydrogen bond basicity of hydrated sulfate head groups of SDS micelle and the water molecule in the aqueous bulk. Other parameters should be inspected bearing the same statistical criteria in mind. Therefore, the analysis of Table 15.1 reveals that only parameters V and B are statistically significant, with the coefficient of V the most prominent in magnitude and statistical relevance.

This last observation deserves further consideration. Because any molecule exhibits a measurable volume, the parameter V is expected to be the most important variable in the MLR (Table 15.1) in comparison with other parameters, especially those representing specific interactions, properties that some of the solutes in the set might lack. If a three-compartment model can be invoked for the micelle structure (inner core, interface, and surface) as opposed to the Hartley model ("oil droplet," hydrophobic core encased by a hydrophilic region), remarkable differences in cavitation energy between the aqueous bulk and the micelle interface as well as between the aqueous bulk and the micelle inner core are anticipated. Thus, the parameter V coefficient in the MLR with the entire set of solutes is expected to be prominent as well. More importantly, the parameter V coefficient reflects an average behavior, that is, it is indicative of cavitation energy differences between a given micelle compartment and bulk, weighted by the population of solutes in each micelle compartment, according to Equation 15.4:

$$\nu = n_{\mathrm{surface}} \times \nu_{\mathrm{surface}} + n_{\mathrm{interface}} \times \nu_{\mathrm{interface}} + n_{\mathrm{core}} \times \nu_{\mathrm{core}} \qquad \text{(Eq. 15.4)}$$

where ν is the parameter V coefficient and n is the number of solutes populating that particular micelle compartment over the total number of solutes in the set. Note that ν_{surface} approaches zero, that is, differences in cavitation energy between the aqueous bulk and the micelle surface are negligible.

Equation 15.4 leads to important practical observations. If a solute set is overcrowded by compounds that incorporate in the micelle inner core, the corresponding term of Equation 15.4 ($n_{\mathrm{core}} \times \nu_{\mathrm{core}}$ in the example) prevails, imparting a large value for the parameter V coefficient. Therefore, the results of any study that rely on that particular set of solutes will be compromised. That is why the many literature studies aiming at establishing phase selectivity differences fail. Many surfactants, even with different headgroups, lengths, and counterions, seem to behave alike. The solute set used in MEKC is overcrowded by compounds that are believed to incorporate in the micelle at the interface region (29), precluding revelation of the actual existence of selectivity differences.

From the chemometric perspective, in order to evaluate properly the importance of a given variable in an MLR, that variable must be studied while the others held constant. Homologous series (alkanes, alkyl benzenes, alkyl phenyl ketones, etc.), in which all LSER parameters are constant, except the volume, are a particularly interesting class of compounds to establish the importance of the parameter V. It is well known since the late 1960s that a homologous series with increasing number of methylene groups exhibits a linear relationship between log k and the number of carbon atoms (76, 77); the linear coefficient of such regression is a characteristic of the homologous series functional group and depends on the phase ratio. Both slope and linear coefficient depend on the composition of the aqueous phase and the nature of the micellar phase. Thus, plots of log k or log P versus the McGowan volume for selected homologous classes of compounds, as depicted in Figure 15.3, can be informative of the solubilization site of that particular class.

In Figure 15.3a, the log P_{SDS} versus V results generated by the MEKC solute set (25, 30) were contrasted with MLC and other techniques data (74, 75), bracketed by alkanes and crown ethers, the most hydrophobic and hydrophilic homologous series components of the entire database, respectively. For comparison purposes, hexadecane–water partition data for alkanes were also included (16). Figure 15.3b depicts the same plots for selected classes of compounds (MEKC solute set only) (25), and Table 15.2 was assembled with the corresponding statistical data.

If the slope of V can be used to estimate the degree of penetration into the micelle, Figure 15.3a indicates that alkanes are definitely the closest to the micelle inner core a homologous series can assess, with an approximate slope

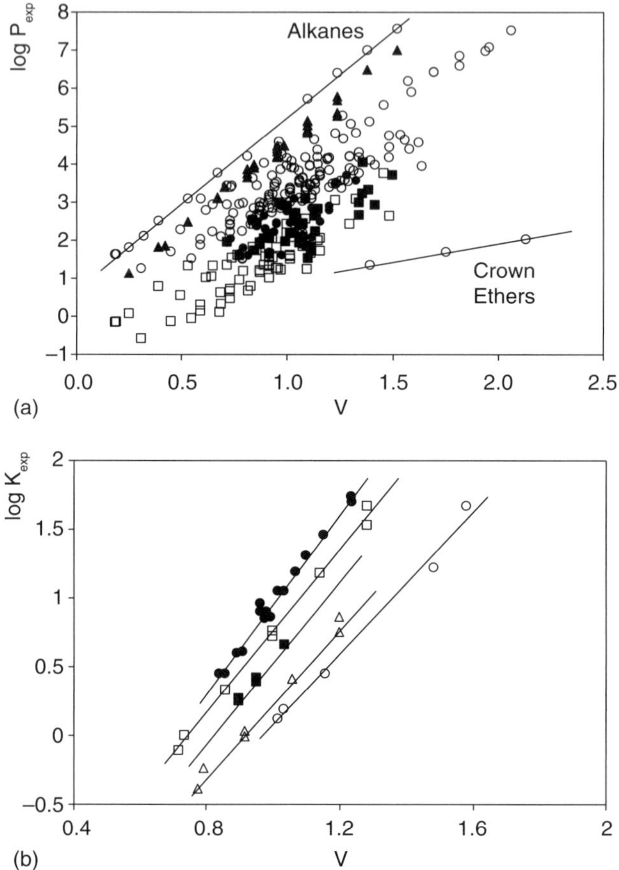

FIGURE 15.3. Distribution data between water and sodium dodecyl sulfate micelles as a function of the solute McGowan volume for the entire database (a) and for categorized classes of solutes (b), data from Reference 25. Database labels in (a) as in Figure 15.2. (▲) hexadecane–water partition data for alkanes, from Reference 16. Labels in (b): (□) alkyl benzenes; (○) alkyl phenyl ketones; (△) alkyl phenols; (●) halo benzenes; and (■) halo phenols.

of 4.6, and crown ethers are the farthest away, possibly at the interface region near the micelle surface (approximate slope of 0.92). Notice that for reference, the coefficient of the V parameter for the LSER from hexadecane–water data presented a value of 4.40 (Table 15.1), hexadecane being a well-characterized single organic phase, although at the interface hexadecane–water, regions of varying hydrophobicity might be postulated as well. On the other hand, the MEKC solute set seems to inhabit the micelle interface (slopes from 2.60 to 3.29, Table 15.2), confirming what was postulated previously in the literature (5, 7, 8, 29). An interesting observation from Figure 15.3a is that the solutes presenting the largest volumes (V > 1.5) do not seem to occupy the same

TABLE 15.2. Retention factors (as log k) as a function of McGown volume for categorized classes of solutes using MEKC databases for SDS

Series	*n*	Intercept (*p* value)	Slope (*p* value)	PreSS	Q^2	S^2	R^2_{adj}	*F*
Halobenzene	16	−2.32 ± 0.11 (6e-12)	3.29 ± 0.11 (4e-14)	0.003	0.98	0.003	0.98	911
Alkyl benzene	8	−2.23 ± 0.08 (1e-7)	2.99 ± 0.08 (2e-08)	0.004	0.99	0.002	0.995	1459
Halo phenol	5	−2.37 ± 0.14 (0.0004)	2.93 ± 0.14 (0.0003)	0.0004	0.98	0.0003	0.990	417
Alkyl phenol	7	−2.46 ± 0.12 (5e-6)	2.72 ± 0.12 (3e-6)	0.005	0.98	0.003	0.99	517
Alkyl phenyl ketone	5	−2.52 ± 0.20 (0.001)	2.60 ± 0.16 (0.0005)	0.02	0.95	0.007	0.98	263

Data compiled from Reference 25.

hydrophobic micelle locus alkanes do, demystifying the premise that large solutes solubilize into the micelle inner core.

A close inspection of Figure 15.3b shows that not only homologous series (alkyl benzenes, alkyl phenols, and alkyl phenyl ketones) but also other non-homologous series (halo benzenes and halo phenols) were considered. Figure 15.3b indicates that alkyl benzenes are the most hydrophobic homologous series MEKC can provide (slope of 2.99, Table 15.2); however, halo benzene derivatives also show a linear relationship between the experimental log P and V with a slope indicative of deeper penetration into the SDS micelle (slope of 3.29). The same is true when alkyl phenol (slope of 2.72) and halo phenols (slope of 2.93) are contrasted. A possible explanation relies on the opposed mesomeric effects halogen atoms and alkyl groups exert in the benzene ring, modifying its hydrogen bond basicity (B parameter). For reference, benzene, ethylbenzene, and chlorobenzene have B values of 0.14, 0.15, and 0.07, respectively, whereas phenol, 3-methylphenol, and 4-chlorophenol have B values of 030, 0.34, and 0.20, respectively (16). The larger the hydrogen bond basicity of the benzene derivative, the less hydrophobic regions of the micelle interface it populates because of its ability to perform interactions via hydrogen bond augments.

The MEKC behavior of surfactants of differing types versus SDS for the alkyl benzene homologous series is demonstrated in Figure 15.4. Because the data were normalized, slopes of one indicate similarity to SDS. As it can be observed from Figure 15.4, from all surfactants under consideration, only sodium deoxycholate (SDC, slope of 1.14 ± 0.04) and LPFOS (slope of 0.696 ± 0.009) exhibited distinctive behavior, despite the observations above concerning the restraints the solute set impart when phase selectivity is contrasted. The structure of SDC micellar aggregates is quite unusual being described by a helical model (78). The helix has the lateral surface covered by

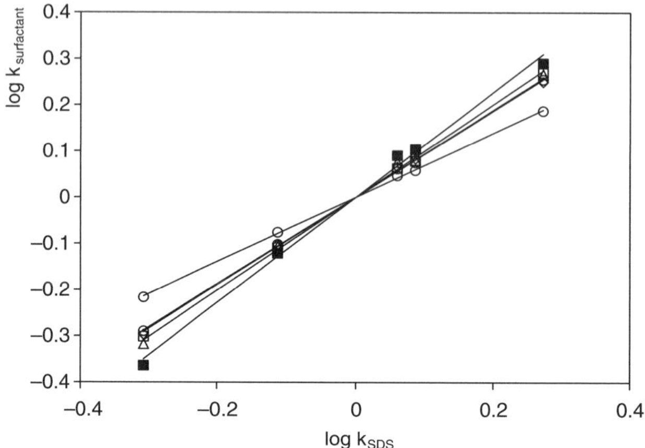

FIGURE 15.4. Comparison of MEKC surfactants of different types versus SDS for alkyl benzene homologous series. Surfactants: (●) LPFOS, (■) SDC, (□) LDS, (△) SC, (○) TTAB, and (◇) CTAB. Data compiled from Reference 27.

deoxycholate anions with the hydroxyl and the angular methyl groups protruding toward the inside and outside of the helix, respectively, and the nonpolar face of the deoxycholate anions oriented toward the aqueous medium. Therefore, it is not unexpected that the behavior of SDC toward solutes in MEKC is so different from SDS spherical micelles. LPFOS has also a few notable structural differences from SDS: its monomer comprises a fully fluorinated octyl chain and a sulfonate headgroup.

An interesting example of the selectivity changes LPFOS exerts in mixed micelle systems was published by Fuguet et al. (46). In that work, the behavior of a relatively large set of solutes in electrolytes prepared by increasing amounts of LPFOS in LDS was studied. Table 15.3 compiles the corresponding log k versus V statistics for alkyl benzene and alkyl phenyl ketone series (data from Reference 46). For both homologous series as observed in Table 15.3, an increase in the slope with the molar fraction increase of LPFOS is evident, although it is more pronounced for alkyl benzenes, again confirming that alkyl benzenes penetrate the micelle deeper than do alkyl phenyl ketones. While the slope reflects differences in cavitation work, the intercept of such plot contains information on the phase ratio, as stated earlier (77). The contributions of other intermolecular interactions (E, S, A, B parameters) are also part of the intercept; however, because homologous series are considered here, these parameters remain constant. Therefore, the variation of the intercept within a homologous series contains information on the "effective" phase ratio, that is, the volume of micellar phase that is available to a given solute, not the total volume occupied by the micellar phase, related to the volume of aqueous phase, as commonly defined in MEKC.

Figure 15.5 shows the variation of the intercept (as antilog) with the increase of molar fraction of LPFOS (X_{LIPFOS}) in LSD for alkyl benzene and alkyl

TABLE 15.3. Retention factor (as log k) as a function of McGown volume for alkyl benzene and alkyl phenyl ketone homologous series in mixed micelle systems LDS/LPFOS

X_{LIPFOS}	Intercept (p value)	Slope (p value)	PreSS	Q^2	S^2	R^2_{adj}	F
Alkyl benzene homologous series; alkyl: H, methyl, ethyl, n-propyl, di-methyl, and n-butyl							
0.00	−2.11 ± 0.03 (3e-7)	2.22 ± 0.03 (2e-7)	0.0002	0.999	0.0002	0.999	5081
0.25	−2.22 ± 0.04 (7e-7)	2.59 ± 0.04 (3e-7)	0.0004	0.998	0.0003	0.999	4245
0.50	−2.28 ± 0.04 (6e-7)	2.89 ± 0.04 (2e-7)	0.0004	0.999	0.0003	0.999	5267
0.75	−2.31 ± 0.05 (1e-6)	3.06 ± 0.05 (4e-7)	0.0006	0.998	0.0005	0.998	4047
1.00	−2.34 ± 0.06 (2e-6)	3.18 ± 0.06 (6e-7)	0.001	0.996	0.0006	0.998	3131
Alkyl phenyl benzene homologous series; alkyl: ethyl, n-propyl, n-butyl, and n-pentyl							
0.00	−1.78 ± 0.06 (0.001)	1.96 ± 0.05 (0.0005)	0.0007	0.993	0.0002	0.998	1838
0.25	−1.90 ± 0.05 (0.0006)	2.11 ± 0.04 (0.0003)	0.0004	0.996	0.0001	0.999	3254
0.50	−2.14 ± 0.08 (0.002)	2.37 ± 0.07 (0.0008)	0.002	0.989	0.0005	0.997	1201
0.75	−2.30 ± 0.09 (0.002)	2.50 ± 0.08 (0.0009)	0.002	0.988	0.0006	0.997	1109
1.00	−2.4 ± 0.1 (0.002)	2.55 ± 0.08 (0.001)	0.002	0.987	0.0007	0.997	984

Data compiled from Reference 46.

phenyl ketone homologous series (Table 15.3, data from Reference 46). As it can be observed from Figure 15.5, in LDS (X_{LIPFOS} = 0), the effective phase ratio or the micellar volume available to alkyl phenyl ketones is much larger than that available to alkyl benzenes. However, as the molar fraction of LPFOS increases, and the resulting mixed micelle shrinks (a large dodecyl sulfate surfactant is replaced by a smaller fluorinated octanesulfonate surfactant), the micellar volume available to alkyl phenyl ketones decreases abruptly. In LPFOS micelles, both series assess identical micellar volume.

Another interesting work published recently in the literature addresses the issue of solute localization dependence on phase selectivity. The influence of the length of flexible hydrophobic (67) and hydrophilic (68) spacers of anionic dimeric surfactants was studied by MEKC in contrast to SDS. A set of 41 solutes was categorized into nonhydrogen bond, hydrogen bond acceptors, and hydrogen bond donors. Although the authors chose to perform an LSER

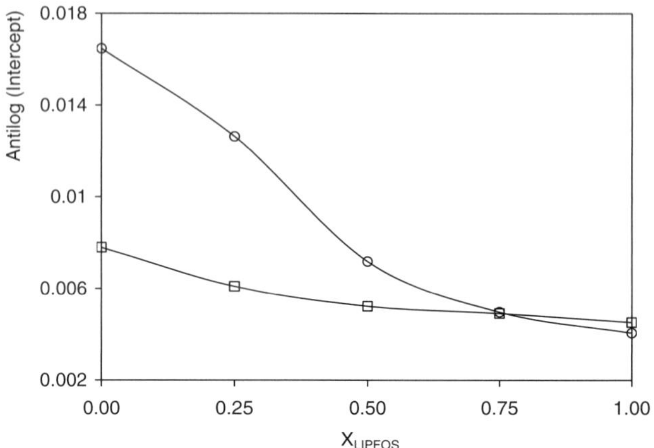

FIGURE 15.5. Effective phase ratio as a function of molar fraction of LPFOS in LDS for alkyl benzene and alkyl phenyl ketone homologous series. Data compiled from Reference 46.

with the entire set of solutes, log k_{DS} versus log k_{SDS} plots clearly demonstrated the distinct behavior of the categorized solutes, corroborating the idea that solutes of different hydrogen bond capabilities occupy different micelle loci.

15.3.2. RSS

QSRR studies involving RSS are much more informative of the solute–micelle interactions. Here the separation of solutes belonging to the same chemical class, with small or even subtle structural differences, is contemplated. Optimization of the separation to result in baseline resolution of all solutes under investigation is mandatory. If the solutes are separated in a particular optimized MEKC electrolyte system, it means that the micelle is able to sense the solutes' subtle structural differences. Therefore, the challenge is to search for an appropriate set of descriptors capable of explaining these differences. Considering that a specific micelle compartment is being assessed by the solute, descriptors of similar characteristics, representative of the same kind of intermolecular interaction, must be screened and selected.

Although the separation of RSS has been addressed by many authors, unraveling the fundamental aspects of MEKC theory consolidating the phenomenological models of solute migration (13), there are not many representative examples of QSRR studies involving RSS. Indeed, QSRR studies involving RSS with LSD are rare and require massive statistical processing (79); as pointed out before, when LSD are screened, the results are not always meaningful, leading to a straightforward interpretation of solute–micelle interactions. Examples of QSRR studies involving RSS with SSD comprise the MLC separation of amines (phenethylamines and antihistamines) in mixed SDS–pentanol systems (80) and the MEKC separation of flavonoids in SDS electrolytes modified by solvents (81). Both studies show no statistically relevant dependence of log k or solute mobility on the solute McGown volume. In another study conducted in our group involving 18 flavonoids, it was found that log k varied inversely with the flavonoid volume (82). These findings altogether suggest the micelle surface as a possible site of interaction for protonated amines and undissociated flavonoids.

15.4. CONCLUSIONS

All the results discussed in this chapter helped to delineate a unique physicochemical model for micelle structure and solute–micelle interactions. The SDS micelle is viewed as an entity composed of numerous compartments of distinct hydrophobicities; at least three of them can be readily defined: the inner core, the interface, and the surface. The interface can be further divided into a number of levels. The MEKC solute set used in the LSER studies predominantly occupies the interface distributing themselves into these levels according to their hydrophobicity. Therefore, the characterization of phase selectivity depends strongly on the composition of the solute set. By knowing the preferential locus of the solute series into the micelle, it is possible to promote changes in that specific locus and thus that must be the only way to alter and to contrast phase selectivity. Moreover, meaningful studies on solute–micelle interactions by QSRR can only derive from an RSS assessing a particular locus in the micelle and, of course, reliable chemometry.

ACKNOWLEDGMENTS

The authors wish to acknowledge the Fundação de Amparo à Pesquisa do Estado de São Paulo (Fapesp 04/08503-2; 04/08931-4) and the Conselho Nacional de Pesquisa e Desenvolvimento (CNPq 300595/2007-7) of Brazil for financial support and fellowships.

REFERENCES

1. Kaliszan, R. (1987) *Quantitative Structure-Chromatographic Retention Relationships*, Wiley, New York.
2. Kaliszan, R. (1997) *Structure and Retention in Chromatography. A Chemometric Approach*, Harwood Academic, Amsterdam.
3. Kaliszan, R. (2000) Recent advances in quantitative structure-retention relationships, in *Separation Methods in Drug Synthesis and Purification* (ed. K. Valko), pp. 503–530, Elsevier, Amsterdam.
4. Kaliszan, R. (2007) *Chem Rev*, **107**, 3212–3246.
5. Poole, S.K. and Poole, C.F. (2008) *J Chromatogr A*, **1182**, 1–24.

6. Héberger, K. (2007) *J Chromatogr A*, **1158**, 273–305.
7. Vitha, M. and Carr, P.W. (2006) *J Chromatogr A*, **1126**, 143–194.
8. Poole, C.F. and Poole, S.K. (2002) *J Chromatogr A*, **965**, 263–299.
9. Hammett, L.P. (1937) *J Am Chem Soc*, **59**, 96–103; (1935) *Chem Rev*, **17**, 125–136.
10. Abraham, M.H., Ibrahim, A., and Zissimos, A.M. (2004) *J Chromatogr A*, **1037**, 29–47.
11. Kamlet, M.J. and Taft, R.W. (1976) *J Am Chem Soc*, **98**, 377–383.
12. Taft, R.W. and Kamlet, M.J. (1976) *J Am Chem Soc*, **98**, 2886–2894.
13. Pyell, U. (2006) *Electrokinetic Chromatography. Theory, Instrumentation and Applications*, Wiley, Chichester.
14. Lin, C.E. (2004) *J Chromatogr A*, **1037**, 467–478.
15. Fuguet, E., Ràfols, C., Bosch, E., and Rosés, M. (2002) *Electrophoresis*, **23**, 56–66.
16. Abraham, M.H., Chadha, H.S., Whiting, G.S., and Mitchell, R.C. (1994) *J Pharm Sci*, **83**, 1085–1100.
17. Karelson, M., Lobanov, V.S., and Katritzky, A.R. (1996) *Chem Ver*, **96**, 1027–1044.
18. Todeschini, R. and Consonni, V. (2000) *Handbook of Molecular Descriptors*, Wiley, Weiheim.
19. Godden, J.W., Stahura, F.L., and Bajorath, J. (2000) *J Chem Inf Comput Sci*, **40**, 796–800.
20. Guyon, I. and Elisseeff, A.J. (2003) *Machine Learning Res*, **3**, 1157–1182.
21. Rousseeuw, P.J. and Leroy, A.M. (1987) *Robust Regression and Outlier Detection*, Wiley, New York.
22. Golmohammadi, H. and Fatemi, M.H. (2005) *Electrophoresis*, **26**, 3438–3444.
23. Put, R. and Heyden, Y.V. (2007) *Anal Chim Acta*, **602**, 164–172.
24. Liu, H.X., Yao, X.J., Liu, M.C., Hu, Z.D., and Fan, B.T. (2006) *Anal Chim Acta*, **558**, 86–93.
25. Kelly, K.A., Burns, S.T., and Khaledi, M.G. (2001) *Anal Chem*, **73**, 6057–6062.
26. Fuguet, E., Ràfols, C., Bosch, E., Abraham, M.H., and Rosés, M. (2006) *Electrophoresis*, **27**, 1900–1914.
27. Fuguet, E., Ràfols, C., Bosch, E., Abraham, M.H., and Rosés, M. (2002) *J Chromatogr A*, **942**, 237–248.
28. Trone, M.D. and Khaledi, M.G. (2000) *J Chromatogr A*, **886**, 245–257.
29. Poole, C.F., Poole, S.W., and Abraham, M.H. (1998) *J Chromatogr A*, **798**, 207–222.
30. Poole, S.K. and Poole, C.F. (1997) *Analyst*, **122**, 267–274.
31. Vitha, M.F., Dallas, A.J., and Carr, P.W. (1997) *J Colloid Interface Sci*, **187**, 179–183.
32. Muijselaar, P.G., Claessens, H.A., and Cramers, C.A. (1997) *Anal Chem*, **69**, 1184–1191.
33. Ishihama, Y. and Asakawa, N. (1999) *J Pharm Sci*, **88**, 1305–1312.
34. Fu, C. and Khaledi, M.G. (2009) *J Chromatogr A*, **1216**, 1891–1900.

35. Fu, C. and Khaledi, M.G. (2009) *J Chromatogr A*, **1216**, 1901–1907.
36. Fuguet, E., Ràfols, C., Bosch, E., and Rosés, M. (2003) *Langmuir*, **19**, 55–62.
37. Fuguet, E., Ràfols, C., and Rosés, M. (2003) *Langmuir*, **19**, 6685–6692.
38. Yang, S., Bumgarner, J.G., and Khaledi, M.G. (1996) *J Chromatogr A*, **738**, 265–274.
39. Yang, S., Bumgarner, J.G., Kruk, J.G., and Khaledi, M.G. (1996) *J Chromatogr A*, **721**, 323–335.
40. Yang, S.Y. and Khaledi, M.G. (1995) *J Chromatogr A*, **692**, 301–310.
41. Yang, S.Y. and Khaledi, M.G. (1995) *Anal Chem*, **67**, 499–510.
42. Schnee, V.P. and Palmer, C.P. (2008) *Electrophoresis*, **29**, 761–766.
43. Schnee, V.P. and Palmer, C.P. (2008) *Electrophoresis*, **29**, 767–776.
44. Schnee, V.P. and Palmer, C.P. (2008) *Electrophoresis*, **29**, 777–782.
45. Bailey, D.J. and Dorsey, J.G. (2001) *J Chromatogr A*, **919**, 181–194.
46. Fuguet, E., Ràfols, C., Bosch, E., Rosés, M., and Abraham, M.H. (2001) *J Chromatogr A*, **907**, 257–265.
47. Fuguet, E., Ràfols, C., Torres-Lapasió, J.R., García-Álvarez-Coque, M.C., Bosch, E., and Rosés, M. (2002) *Anal Chem*, **74**, 4447–4455.
48. Rosés, M., Ràfols, C., Bosch, E., Martínez, A.M., and Abraham, M.H. (1999) *J Chromatogr A*, **845**, 217–226.
49. Khaledi, M.G., Bumgarner, J.G., and Hadjmohammad, M. (1998) *J Chromatogr A*, **802**, 35–47.
50. Abraham, M.H., Treiner, C., Rosés, M., Ràfols, C., and Ishihama, Y. (1996) *J Chromatogr A*, **752**, 243–249.
51. Agbodjan, A.A. and Khaledi, M.G. (2003) *J Chromatogr A*, **1004**, 145–153.
52. Bui, H.H. and Khaledi, M.G. (2002) *J Colloid Interface Sci*, **253**, 397–401.
53. Schuster, S.A. and Foley, J.P. (2005) *J Sep Sci*, **28**, 1399–1408.
54. Pascoe, R.J. and Foley, J.P. (2003) *Electrophoresis*, **24**, 4227–4240.
55. Pascoe, R.J. and Foley, J.P. (2002) *Electrophoresis*, **23**, 1618–1627.
56. Burns, S.T., Agbodjan, A.A., and Khaledi, M.G. (2002) *J Chromatogr A*, **973**, 167–176.
57. Burns, S.T. and Khaledi, M.G. (2002) *J Pharm Sci*, **91**, 1601–1612.
58. Shi, W. and Palmer, C.P. (2002) *Electrophoresis*, **23**, 1285–1295.
59. Peterson, D.S. and Palmer, C.P. (2001) *Electrophoresis*, **22**, 3562–3566.
60. Tellman, K.T. and Palmer, C.P. (1999) *Electrophoresis*, **20**, 152–161.
61. Shamsi, S.A., Iqbal, R., and Akbay, C. (2005) *Electrophoresis*, **26**, 4138–4152.
62. Akbay, C. and Shamsi, S.A. (2004) *Electrophoresis*, **25**, 635–644.
63. Leonard, M.S. and Khaledi, M.G. (2002) *J Sep Sci*, **15**, 1019–1026.
64. Akbay, C., Agbaria, R.A., and Warner, I.M. (2005) *Electrophoresis*, **26**, 426–445.
65. Fujimoto, C. (2001) *Electrophoresis*, **22**, 1322–1329.
66. Trone, M.D. and Khaledi, M.G. (2000) *Electrophoresis*, **21**, 2390–2396.
67. Van Biesen, G. and Bottaro, C.S. (2007) *J Chromatogr A*, **1157**, 437–445.
68. Van Biesen, G. and Bottaro, C.S. (2008) *J Chromatogr A*, **1157**, 171–178.
69. Trone, M.D. and Khaledi, M.G. (2000) *J Microcolumn Sep*, **12**, 433–441.

70. Trone, M.D., Mack, J.P., Goodell, H.P., and Khaledi, M.G. (2000) *J Chromatogr A*, **888**, 229–240.
71. Greenaway, M., Okafo, G., Manallack, D., and Camilleri, P. (1994) *Electrophoresis*, **15**, 1284–1289.
72. Filipic, S., Nikolic, K., Krizman, M., and Agbaba, D. (2008) *QSAR Comb Sci*, **27**, 1036–1044.
73. Liu, Z., Zou, H., Ye, M., Ni, J., and Zhang, Y. (1999) *J Chromatogr A*, **863**, 69–79.
74. Sprunger, L., Acree, W.E., Jr., and Abraham, M.H. (2007) *J Chem Inf Model*, **47**, 1808–1817.
75. Quina, F., Alonso, E., and Farah, J.P.S. (1995) *J Phys Chem*, **99**, 11708–11714.
76. Tanford, C. (1969) *The Hydrophobic Effect*, Academic Press, New York.
77. Colin, H. and Guiochon, G. (1980) *J Chromatogr Sci*, **18**, 54–63.
78. Esposito, G., Giglio, E., Pavel, N.V., and Zanobi, A. (1987) *J Phys Chem*, **91**, 356–362.
79. Liang, H.R., Vuorela, H., Vuorela, P., Hiltunen, R., and Riekkola, M.-L. (1998) *J Liq Chromatogr Rel Technol*, **21**, 625–643.
80. Gil-Agustí, M., Estece-Romero, J., and Abraham, M.H. (2006) *J Chromatogr A*, **1117**, 147–155.
81. Wang, S., Xue, C., Liu, M., and Hu, Z. (2004) *J Chromatogr A*, **1033**, 153–159.
82. Tonin, F.G., Jager, A.V., Micke, G.A., Farah, J.P.S., and T avares, M.F.M. (2005) *Electrophoresis*, **26**, 3387–3396.

CHAPTER 16

CHEMOMETRICAL ANALYSIS OF CHEESE PROTEOLYSIS PROFILES BY CAPILLARY ELECTROPHORESIS: PREDICTION OF RIPENING TIMES

NATIVIDAD ORTEGA,[1] SILVIA M. ALBILLOS,[2] and MARÍA D. BUSTO[1]
[1]Department of Biotechnology and Food Science, University of Burgos, Burgos, Spain
[2]Institute of Biotechnology IMBIOTEC, León Scientific Park, León, Spain

CONTENTS

16.1. INTRODUCTION

The animal origin and quality of milk play a very important role in the production of all types of cheese, affecting both cheese yield and properties of cheese

Chemometric Methods in Capillary Electrophoresis. Edited by Grady Hanrahan and Frank A. Gomez

(1–4). Another important characteristic of this dairy product, also evaluated in quality control, is the ripening time. Significant qualitative and quantitative information concerning the animal origin of cheese can be obtained from protein analysis (5–7). Furthermore, proteolysis is recognized as one of the most complex biochemical events, and possibly the most important, for flavor and texture development during cheese ripening (8–10).

Different protein-based methods have been reviewed for species identification in milk and dairy products, and for characterization of cheese maturity, such as electrophoretic, chromatographic, and immunological techniques (11, 12). In addition to new developments in these techniques, the interdisciplinary and dynamic nature of milk product analysis is being enhanced by the application of disciplines already used to analyze other foodstuffs. Among them, capillary electrophoresis (CE), polymerase chain reaction, and isotope ratio mass spectrometry are just gaining popularity for solving dairy authenticity problems (13–15).

CE, with its high resolving power, rapid method development, easy sample preparation, and low operational cost, is reported to be an excellent technique for resolving caseins (including different genetic variants), peptides derived from them, and whey proteins (16–19). Peptide profiles obtained by CE supplement the information obtained by reversed-phase high performance liquid chromatography (RP-HPLC) (17, 20). The application of CE to the assessment of proteolysis in milk and different cheese types has acquired an enormous importance in recent years. Reviews on the application of CE to this field can be found in papers by Otte et al. (21) and Recio et al. (22).

CE also suffers from several weaknesses as an analytical technique (e.g., adsorption of charged species to the capillary wall, presence of Joule heating). Hence, it is important to be able to determine optimal conditions in CE method development (23). Various chemometric-based techniques including multivariate experimental design and response surface methodology have been devised to help optimize the performance of a system (23–26).

In addition, because of the complexity of proteolytic patterns during cheese ripening, the amount of data generated from such analyses and their interpretation becomes both large and complicated. For this reason, researchers working on proteolysis during cheese ripening need methods for objective evaluation and data reduction and interpretation in addition to the traditional visual examination of the proteolytic profiles (27–30). In this sense, multivariate statistical techniques can be used to better understand the complexity of proteolysis during cheese ripening, and even to predict the ripening time (6, 31–33). This approach has been used to identify cheese types (34, 35), to differentiate cheese within a type (driven more recently by the desire to protect "Appellation of Origin" or "Protected Designation of Origin" cheese) (36–39), to broadly group cheese according to maturity (40), to develop indices of maturity (35), to determine the effect of manufacturing process alterations on cheese properties (mainly degree of proteolysis) (41), and finally, to predict cheese properties (42).

The use of multivariate analysis (multiple linear regression [MLR], principle component regression [PCR], and partial least square [PLS]) to predict the ripening time has received great attention in recent years (Table 16.1). García-Ruiz et al. (31) applied MLR, PCR, and PLS in order to calculate the ripening time of commercial Manchego cheese based on physicochemical and proteolysis parameters. These authors found that PLS regression yielded the best prediction for ripening time. The equation proposed by these authors was improved by the prediction model described by Poveda et al. (43), as it included a higher number of samples for the calibration and it also reduced the number of variables that took part in the predictive equation (pH, a_w, and dry matter). Recently, Alvarenga et al. (44) reported that when MLR was used to correlate a combination of chemical, color, and rheological parameters, the prediction of the ripening periods suffered an estimation error of as low as 1.74 d.

Within this context, the aim of this work was to apply chemometric experimental designs for the optimization of casein separation by CE using a neutral capillary and to build a multivariate model for the reliable prediction of cheese

TABLE 16.1. Multivariate regression methods used to predict cheese ripening time

Cheese	Analytical Method	Statistical Method	Reference
Manchego (ewe's milk)	Physicochemical (a_w, pH, TN) and proteolysis parameters (WSN, WSN/TN, N-PTA/TN)	MLR, PCR, PLS	(31)
Manchego (ewe's milk)	a_w, pH, and DM	PLS	(43)
Terrincho (ewe's milk)	Chemical parameters (moisture, acidity, pH, a_w) and physical parameters (color and texture)	MLR	(59)
Terrincho (ewe's milk)	RP-HPLC (α_{s1}-CN and α_{s1}-I peptide)	MLR	(60)
Bovine/ovine cheese	CE (pH 4.6-insoluble fraction)	PLS, PCR	(32)
Ovine cheese	CE (pH 4.6-insoluble fraction)	PLS, PCR	(33)
Caprine cheese	CE (ethanolwater protein fraction)	PLS	(61)
Ovine milk	CE (ethanolwater protein fraction)	PLS	(61)
Serpa (ewe's milk)	Instrumental and color parameters	MLR	(44)
Ragusano (cows' milk)	Amino acid and peptide analysis	PLS, PLSDA	(62)

CE = capillary electrophoresis; DM = dried matter; N-PTA = phosphotungstic acid-soluble nitrogen; MLR = multiple linear regression; PLS = partial least squares; PLSDA = partial least squares discriminant analysis; PCR = principal component regression, RP-HPLC = reversed-phase high performance liquid chromatography; TN = total nitrogen; WSN = water-soluble nitrogen.

ripening time from peak areas of caseins and peptides separated by the CE method.

16.2. EXPERIMENTAL

16.2.1. Origin and Preparation of Samples

16.2.1.1. Milk and Cheese Samples. Milk samples (from cows and ewes) and cheeses were supplied by Quesos Frías, S.A. (Burgos, Spain). Two cheese types were investigated in detail: one type manufactured from raw ewe's milk and another type made from a combination of cow's and ewe's milk.

16.2.1.2. Isolation of Caseins. Isoelectric caseins were obtained by precipitation from whole milk or from 5g of homogenized cheese in 30mL of water by adding 2M HCl to pH 4.6 followed by centrifugation at 3500rpm for 15min. To isolate the casein fractions completely from whey and eliminate the remaining fat, it was washed once with 1M sodium acetate buffer (pH 4.6) and three times with dichloromethane/1M sodium acetate buffer (pH 4.6) (1:1, v/v). The casein fractions obtained were lyophilized and stored at −20°C.

16.2.1.3. Sample Preparation for CE. To dissociate the caseins, all samples were dissolved in a sample buffer containing 8M urea and 10mM dithiothreitol at pH 8, and left for at least 1h at room temperature before filtration (0.22μm Millex-GV_{13}, Millipore) and CE analysis. The isoelectrically precipitated casein from milk and cheese and the purified casein standards were dissolved at 10mg/mL. To most samples, 1μL of additional tripeptide Lys-Tyr-Lys (50mg/mL) was added per 50μL of sample as a reference compound.

16.2.2. Equipment and Capillary Electrophoretic Conditions

CE experiments were performed on a Bekcman P/ACE System 2200, equipped with an autosampler, a temperature-controlled fluid-cooled capillary cartridge, an automatic injector, a power supply able to deliver up to 30kV, and a UV detector. A System Gold Software data system version 810 was used for instrument control and for data acquisition and analysis. The separations were performed using a neutral capillary (eCAP Neutral Capillary, Beckman Instruments) of 45cm (33cm to the detector window) × 50μm internal diameter. This capillary utilizes a secondary layer of polyacrylamide to generate a hydrophilic surface.

All experiments were carried out in the cationic mode (anode at the inlet and cathode at the outlet). The sample introduction was achieved by pressure injection for 5s at 0.5psi. The run buffer was 0.32M citric acid/0.02M sodium citrate with 6M urea (pH 3.0) containing 0.055% (w/v) hydroxypropyl methyl cellulose (HPMC). During sample analysis, constant voltage was applied and the separation temperature was kept at 21 °C (or at a different temperature if stated in the text) with circulating coolant surrounding the capillary. The capillary was rinsed sequentially between successive electrophoretic runs, with 0.1 M HCl (2 min) and ionized water (2 min), and the rinse buffer (pH 3.0) contained 0.32 citric acid, 0.020 M sodium citrate, 6 M urea, and 0.042 M 3-morpholinopropane-sulfonic acid (5 min). For all experiments, detection was carried out at 214 nm (data collection rate 5 Hz). The first electropherogram in a series was always discarded. All solutions were based on highly purified water (Milli Q grade). Buffer solutions were filtered through 0.45 μm HAWP and 0.22 μm GSWP filters (Millipore) before used.

The detector response linearity (peak area vs. concentration) was evaluated by preparing five calibration samples using a 1:1:1 mixture of α_s-casein (CN), β-CN, and κ-CN (each solution was injected three times). The calibration range was 2–30 mg/mL for α_s- and β-CN and 2–10 mg/mL for κ-CN.

16.2.3. Experimental Design: Optimization of Casein Separation by CE

The first step of the procedure was to establish the criteria that define the quality of the analysis. The criteria typically used in CE include the values of resolution, efficiency, and run time required to achieve the best separation in the shortest analysis time (45). In our case the key response to determine the optimal conditions was the resolution (*Rs*) calculated using the following equation:

$$Rs = \frac{2(tm_2 - tm_1)}{(w_1 + w_2)} \qquad \text{(Eq. 16.1)}$$

where *tm* and *w* are the migration time and the peak width, respectively. A minimum resolution (*Rs*) of 1.5 was chosen.

Preliminary experiments were carried out to screen the appropriate parameters and to determine the experimental domain. From these experiments, two electrophoretic factors were investigated: the running voltage (X_1) and the temperature (X_2). Their influence was evaluated according to a 3^2 full factorial design with four replicates at the central point. The range and levels of the studied variables are given in Table 16.2. The experimental results of the experimental design were fitted to a second-order polynomial equation:

$$y = b_0 + \sum_{i=1}^{k} b_i X_i + \sum_{i=1}^{k} b_{ii} X_i^2 + \sum_{i}^{i<j}\sum_{j} b_{ij} X_i X_j + e \qquad \text{(Eq. 16.2)}$$

where y is the dependent variable (response variable) to be modeled, X_i and X_j are the independent variables (factors), b_o, b_i, b_{ii}, and b_{ij} are regression coefficients, and e is the error. The model was simplified by dropping terms that

TABLE 16.2. Independent variables and levels for the 3^2 full factorial design

Independent Variables	Symbol	Coded Variable Levels		
		Low (−1)	Medium (0)	High (1)
Running voltage (kV)	X_1	18.50	24.50	30.00
Temperature (°C)	X_2	23.0	34.0	45.0

were not statistically significant ($p > 0.05$) by analysis of variance (ANOVA). The lack of fit test was used to determine whether the constructed models were adequate to describe the observed data (46). If the F-test for the model is significant at the 5% level (i.e., <0.05), there is evidence that the model has some power to explain the variation in the response. The R^2 statistic indicates the percentage of the variability of the optimization parameter that is explained by the model (47). Three-dimensional surface plots were drawn to illustrate the main and interactive effects of the independent variables on the dependent ones.

16.2.4. Multivariate Analysis of CE Data

Multivariate analysis techniques were applied to peak areas obtained by CE to evaluate the ripening time of the cheese. Data were autoscaled prior to model calculations. This normalization involved the subtraction of the mean and then the division of each value of a given variable by the standard deviation of all the values for this variable over the entire sample collection period (48). After normalization, all variables had the same weight because they had a mean of zero and unitary variance.

16.2.4.1. Principal Component Analysis. Principal component analysis (PCA) was used to reduce the dimensionality of the data obtained from the peptide profile. The analysis of the principal component (PC) scores gives evidence of sample grouping in the PC space according to similarities in their characteristics (cheese ripening time), while the examination of the PC loadings considers the influence of the original variables (peak area of casein and peptide) in the sample arrangement.

16.2.4.2. Multivariate Regression Methods. The main goal of this study was to build a multivariate model for the reliable prediction of a property of interest $\mathbf{y}$ (cheese ripening time) from a number of predictor variables, $\mathbf{x}_1$, $\mathbf{x}_2$ ... (peak area of casein and peptide obtained by CE). This model should describe the measured $\mathbf{x}$ and $\mathbf{y}$ data of the calibration set (cheese samples at different ripening time). In particular, in this research, the PCR and PLS methods were evaluated.

Full cross-validation was applied to all the regression models. Cross-validation is a strategy for validating calibration models based on systematically removing groups of samples in the modeling, and testing the removed samples in the model based on the remaining samples; only one sample at a time is left out in full cross-validation. The regression models were evaluated using the correlation coefficient (r^2) and the root-mean-square error of cross-validation (RMSECV) as the term indicating the prediction error of the model. The RMSECV is defined by:

$$\text{RMSECV} = \left(\sum_{i=1}^{n} \left(t_i - t_{(i)} \right)^2 / n \right)^{1/2} \qquad \text{(Eq. 16.3)}$$

where t_i is the real ripening time for the ith sample of the standard cheese, $t_{(i)}$ is the predicted ripening time obtained with the model constructed without the ith sample, and n was the number of standard cheese used in the calibration model. Statistical analysis of experimental data was performed using the Q-PARVUS 3.0 package (49).

16.3. RESULTS AND DISCUSSION

16.3.1. Application of Factorial Design and Response Surface Methodology to the Analysis of Caseins by CE using a Neutral Capillary

Various chemometric experimental designs have been employed for the optimization of CE methods. These include central composites, fractional factorials, Plackett–Burman, simplex, and overlapping resolution mapping (24). By far, central composite design is the most widely used method for the optimization of CE separations, as it offers the possibility of evaluating the curvature of the data and fitting the experimental points to response surfaces. A central composite design and response surface method was applied by our research group in a previous work (50) to optimize the bovine casein separation by capillary zone electrophoresis (CZE) using a fused-silica capillary. Nevertheless, it is known that one of the major problems in CE analysis of proteins is the adsorption of proteins and peptides to the negatively charged fused-silica surface, which leads to distorted peak shapes and poor separation (51). In fact, in our previous research, the electropherogram obtained under the optimized conditions showed that resolution of α_s-CN and β-CN had been achieved, but κ-CN and β-CN had not been separated (50). To solve this problem, neutral polymers like HPMC have been used as dynamic coating in CZE (50, 52).

A different approach used to solve the fused-silica capillary disadvantages is the development of coatings covalently bonded to the inner surface of the capillary wall (51). In the present research, a neutral capillary that utilizes a polyacrylamide layer covalently linked onto the inner wall to generate a hydrophilic surface was selected. The hydrophilic layer has high viscosity and it is therefore capable of suppression of the electroosmotic flow (EOF). The disadvantage of suppressed EOF capillaries in the analysis of proteins and peptides is that they must be sufficiently charged at the working pH of the buffer electrolyte in order to achieve separation at a reasonable time (51). To meet this requirement, in this study the electrophoretic separation of the caseins was carried out at pH 3.0, because at this pH value caseins are positively charged. Other experimental factors that play an important role in the electrophoretic mobility in neutral capillaries are the temperature and voltage applied. Thus, Castagnola et al. (53) suggested that the best separations are achieved at an acceptable low temperature and high voltage, but every experimental scheme may require a particular compromise. In this case, where factor interactions are found to be relevant, multivariate experimental designs or multivariate sequential optimization methods should be used for a proper optimization (54).

Taking into account these previous experiences, in the present research a three-level full factorial design (3^2) was used to evaluate the influence of running voltage (X_1) and temperature (X_2) on the separation of caseins. This design required nine runs. The experimental matrix included four extra experiments at the central level of the design to obtain an estimation of the experimental error. Thus, the entire design required 13 runs. The individual runs of the design were carried out in a randomized sequence. Randomization offers some assurance that uncontrolled variation of factors, other than those studied, will not influence the estimation. The measured response was the resolution of the main peak of κ-CN and β-CN B. In Table 16.3, the experimental matrix and response factors are detailed.

Figure 16.1 shows the graphical representation (Pareto plot) of the "size effect" of each of the parameters investigated upon resolution of the peaks

TABLE 16.3. Experimental design and results according to the 3^2 full factorial design

Run	Variable Level[a]		Resolution (*Rs*)	
	X_1	X_2	Experimental	Predicted
1	+1 (30.00)	−1 (23)	1.234	1.246
2	0 (24.25)	0 (34)	1.019	1.013
3	0 (24.25)	−1 (23)	1.258	1.284
4	−1 (18.50)	+1 (45)	0.545	0.513
5	−1 (18.50)	0 (34)	1.042	1.112
6	+1 (30.00)	0 (34)	0.885	0.856
7	−1 (18.50)	−1 (23)	1.301	1.263
8	+1 (30.00)	+1 (45)	0.000	0.018
9	0 (24.25)	+1 (45)	0.280	0.294
10	0 (24.25)	0 (34)	1.001	1.013
11	0 (24.25)	0 (34)	1.012	1.013
12	0 (24.25)	0 (34)	1.020	1.013
13	0 (24.25)	0 (34)	1.052	1.013

[a]Numbers in parenthesis represent actual experimental amounts. The X_1 and X_2 are running voltage (kV) and temperature (°C).

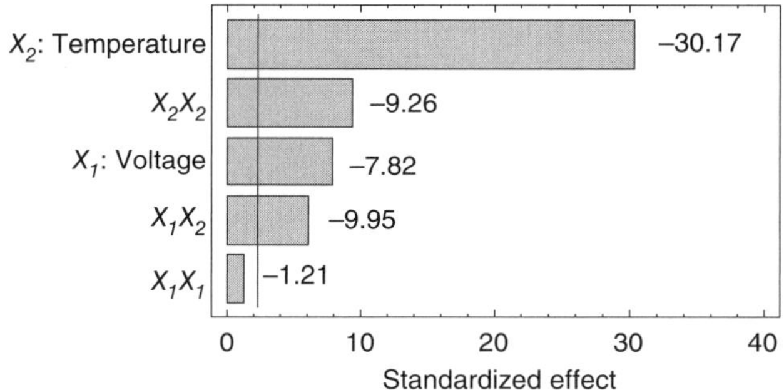

FIGURE 16.1. Pareto chart for the effect of voltage (X_1) and temperature (X_2) on the resolution between κ-CN and β-CN B. Experimental data and conditions are shown in Table 16.3.

TABLE 16.4. Analysis of variance

Source	SS	d.f.	MS	F Value	p Value
X_1: voltage	0.0986	1	0.0986	272.49	0.0001
X_2: temperature	1.4682	1	1.4682	4059.08	0.0000
X_1X_1	0.0024	1	0.0024	6.49	0.0635
X_1X_2	0.0571	1	0.0571	157.92	0.0002
X_2X_2	0.1382	1	0.1382	381.96	0.0000
Lack-of-fit	0.0098	3	0.0033	9.07	0.0295
Pure error	0.0014	4	0.0004	—	—
Total	1.8155	12	—	—	—

$R^2 = 0.9938$; R^2_{adj}: 0.9893; standard error of estimate: 0.0190; mean absolute error: 0.0234.
SS = sum of squares; d.f. = degrees of freedom; MS = mean square.

κ-CN and β-CN B. In this treatment a parameter is deemed to have a significant influence if the size effect is greater than 2. The analysis of the overall data set indicated that the most significant factor was the temperature (X_2), although the voltage (X_1) exerted a statistically significant effect, as did the interaction X_1X_2.

ANOVA was important in determining the adequacy and significance of the quadratic model. ANOVA summary is shown in Table 16.4. The fitness of the model was expressed by the R^2 value, which is 0.9938, indicating that 99.38% of the variability in the response can be explained by the model. The adjusted R^2 value of 0.9893 suggested that the model was statistically significant.

A 3^2 full factorial design provides sufficient data for the fitting of a second-degree expression. In this sense, the following second order polynomial equation explains the data obtained

$$y = -1.13 + 8.50\,10^{-2} X_1 + 1.30\,10^{-1} X_2 - 1.89\,10^{-3} X_1X_2 - 1.85\,10^{-3}(X_2)^2 \quad \text{(Eq. 16.4)}$$

where y represents the experimental response, and X_i the independently evaluated factors (in coded variables, X_1 = running voltage and X_2 = temperature).

The response surface (Fig. 16.2) was used to determine the local optimal conditions that maximize the resolution. Optimal conditions were found to be 25.1 kV and 21 °C. Under these conditions the separation of α_s-CN, β-CN, and κ-CN was achieved as shown in Figure 16.3. Identification of peaks was based on the results previously obtained with a fused-silica capillary (50).

In order to validate the feasibility and validity of the method of analysis developed, linearity and precision were assessed as described below. Furthermore, the results obtained with the neutral capillary were compared with those obtained previously by our research group using a fused-silica capillary (50).

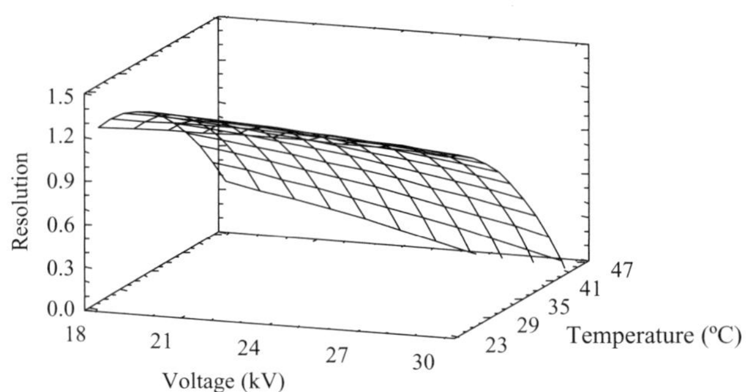

FIGURE 16.2. Response surface plot for the resolution between κ-CN and β-CN B as a function of voltage and temperature.

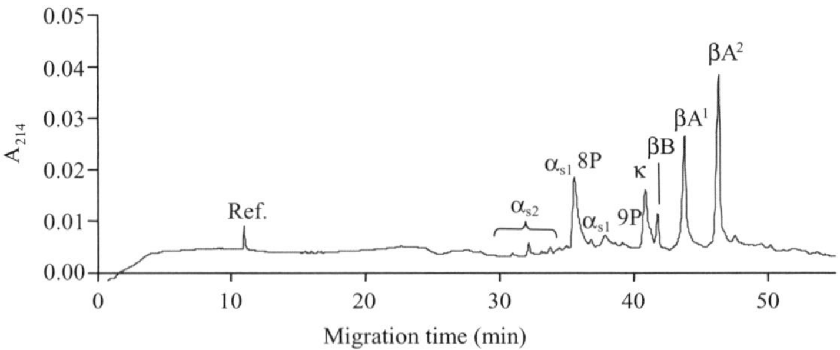

FIGURE 16.3. Capillary electrophoresis separation of a 1:1:1 mixture of α_s-CN, β-CN, and κ-CN. Separation was performed using a neutral capillary at 21 °C and 25.1 kV (~50 μA), and the run buffer was 50 mM phosphate with 6 M urea and 0.05% HPMC, pH 3.0. Peaks: $\alpha_{s2} = \alpha_{s2}$-CN; α_{s1} 8P = α_{s1}-CN 8P; α_{s1} 9P = α_{s1}-CN 9P; κ = κ-CN; βB = β-CN B; βA^1 = β-CN A^1; βA^2 = β-CN A^2.

16.3.1.1. Linearity. The detector response linearity (peak area vs. concentration) was evaluated by preparing calibration curves using a 1:1:1 mixture of α_s-CN, β-CN, and κ-CN. The results demonstrated that the correlation between casein concentration and resulting peak area in the electropherogram was linear, with correlation coefficient (r^2) values of 0.9706–0.9899 and 0.9823–0.9988 for the fused-silica and the neutral capillary, respectively (Table 16.5). In all cases the correlation coefficients were slightly worse for the fused-silica capillary. It is noteworthy that a correlation between α_{s2}-CN concentration and resulting peak area in the electropherograms was only achieved with the neutral capillary. In addition, the calibration ranges for α_s-CN and β-CN were broader for the neutral capillary (2–30 mg/mL) than for the fused-silica capillary (2–10 mg/mL).

16.3.1.2 Precision. Precision was determined by measuring repeatability of migration time, relative migration time, peak area, normalized peak area, and peak height (Table 16.6). The repeatability (within-day precision) of the method was determined by performing replicate injections (n = 10) of a 1:1:1 mixture of α_s-CN, β-CN, and κ-CN at 5 mg/mL each. The relative standard deviation (RSD) for migration times was always less or equal to 1.2% and 0.4% for the fused-silica capillary and the neutral capillary, respectively. The RSD values of the peak area were less satisfactory, ranging between 4.20% and 6.18% (fused-silica capillary) and 1.40% and 4.86% (neutral capillary, with the exception of α_{s2}-CN). The high RSD value for α_{s2}-CN (7.36%) might be due to the fact that this casein was separated into at least four peaks, however this value was similar to the one reported by Heck et al. (19) (5.7%) and lower than that obtained by Chen and Zhang (55) (11.29%). In conclusion, acceptable levels of precision were obtained for both methods in terms

TABLE 16.5. Regression data for the calibration curves of the method assessed with a 1:1:1 mixture of purified caseins

	Fused-Silica Capillary[a]			Neutral Capillary[b]		
	Slope	Intercept	R^2	Slope	Intercept	R^2
α_{s1}-CN 8P	5.11 ± 0.03	−1.87 ± 0.45	0.9823	1.08 ± 0.08	−0.25 ± 0.24	0.9988
α_{s1}-CN 9P	1.43 ± 0.01	0.05 ± 0.01	0.9878	0.34 ± 0.03	−0.42 ± 0.18	0.9939
α_{s2}-CN	—	—	—	0.12 ± 0.01	0.19 ± 0.05	0.9856
β-CN B	0.65 ± 0.01	0.67 ± 0.07	0.9899	0.15 ± 0.05	0.05 ± 0.03	0.9982
β-CN A^1	4.00 ± 0.12	3.14 ± 1.10	0.9813	0.92 ± 0.08	0.27 ± 0.11	0.9956
β-CN A^2	6.93 ± 0.48	1.65 ± 0.20	0.9883	1.67 ± 0.07	−0.87 ± 0.65	0.9949
κ-CN	2.52 ± 0.40	2.68 ± 0.19	0.9706	0.45 ± 0.06	0.11 ± 0.05	0.9823

[a]Data from Ortega et al. (50). Concentration range: 2–10 mg/mL.
[b]Concentration range: 2–30 mg/mL (α_s-CN and β-CN) and 2–10 mg/mL (κ-CN).

TABLE 16.6. Method precision given as RSD values in %

Capillary	Parameter	α_{s2}-CN	α_{s1}-CN 8P	α_{s1}-CN 9P	β-CN B	β-CN A^1	β-CN A^2	κ-CN
Fused-silica[a]	Migration time (t_m)	—	1.04	0.95	1.06	1.13	1.17	1.03
	Relative t_m[b]	—	0.63	0.54	0.65	0.71	0.76	0.63
	Peak area (*pa*)	—	6.06	6.18	4.51	5.44	5.63	4.20
	Normalized *pa*[c]	—	6.83	5.71	4.24	5.00	5.07	4.26
	Peak height	—	6.21	4.88	4.97	3.95	4.78	3.93
Neutral	Migration time (t_m)	0.39	0.39	0.41	0.37	0.37	0.38	0.36
	Relative t_m[b]	0.55	0.55	0.57	0.53	0.52	0.53	0.52
	Peak area (*pa*)	7.36	1.40	4.02	3.21	4.89	1.74	1.48
	Normalized *pa*[c]	7.26	1.31	4.56	3.00	4.69	1.47	1.34
	Peak height	7.37	1.41	4.02	3.21	4.89	1.74	1.56

[a]Data from Ortega et al. (50).
[b]Relative to the reference compound.
[c]Peak area divided by migration time.

of repeatability, although the neutral capillary showed the best results for all parameters studied.

16.3.2. Analysis of Bovine and Ovine Casein by CE Using a Neutral Capillary: A Comparative Study

Once the analytical methodology was optimized it was applied to the analysis of casein extracts from milk. Figure 16.4 shows the electrophoretic profiles of bovine and ovine milk. Identification of caseins was established by comparing the migration times of standard proteins for cow's milk and comparing electropherograms from previous reports (56–58) for ewe's milk. Very similar patterns were obtained when bovine and ovine caseins were analyzed using a fused-silica capillary and a neutral capillary (Fig. 16.4). The only clearly visible difference between electropherograms obtained with both capillaries was the narrower peaks for the neutral capillary. Peak broadening was likely caused by a nonspecific interaction between casein and the charged inner surface of the fused-silica capillary. Furthermore, the neutral capillary provides a better resolution than the fused-silica capillary in the separation of a blend of bovine and ovine milk. As can be seen in the electropherogram obtained with the fused-silica capillary (Fig. 16.4) κ-CN (C4), β-CN B (C5), and β-CN A^1 (C6) for bovine milk showed the same migration time as α_{s1}-CN III (E4), κ-CN (E5), and β-CN (E6) for ovine milk, respectively, while using the neutral capillary only bovine β-CN A^1 (C6) was overlapped with ovine β-CN (E6).

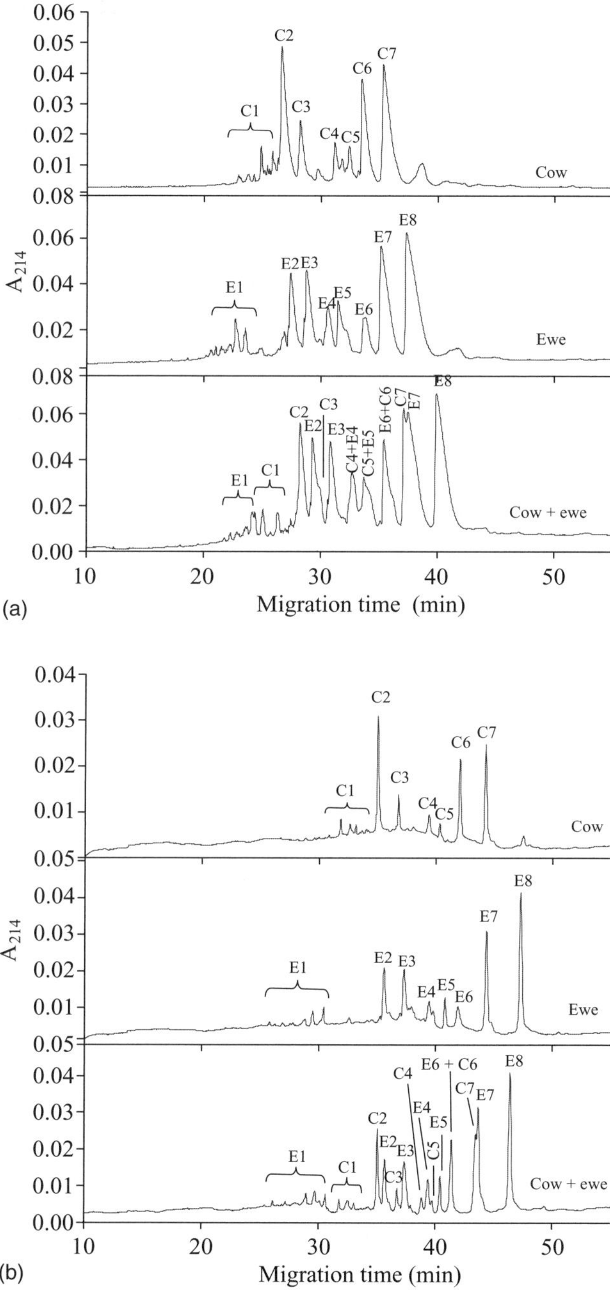

FIGURE 16.4. CE analysis of cow's and ewe's milk and a 1:1 mixture of both milks. Separations were performed in a fused-silica capillary at 18.5 kV (separation conditions described by Ortega et al. [50]) (a), and a neutral capillary at 25.1 kV (~50 mA) (b). Cow's milk: C1 = α_{s2}-CN, C2 = α_{s1}-CN 9P, C3 = α_{s1}-CN 8P, C4 = κ-CN, C5 = β-CN B, C6 = β-CN A^1, C7 = β-CN A^2. Ewe's milk: E1 = α_{s2}-CN, E2 = α_{s1}-CN I, E3 = α_{s1}-CN II, E4 = α_{s1}-CN III, E5 = κ-CN, E6 = β-CN, E7 = β_2-CN, E8 = β_1-CN.

16.3.3. Chemometrical Analysis of Proteolysis during Ripening of Ewe's Milk Cheese and Milk Mixture Cheese

The proteolytic process during the ripening of commercial cheese, such as ewe's milk cheese and cheese made from cow's and ewe's milk, was analyzed by CE using the neutral capillary. Figure 16.5 shows the electropherograms of the pH 4.6-insoluble fraction of ewe's milk cheese after 139 d of ripening period and cheese manufactured with mixtures of cow's and ewe's milk after 167 d of ripening. The peaks were indicated on the electropherograms with serial numbers (in order of migration time) followed by the letter *e* (ewe's milk cheese) or *m* (milk mixture cheese).

Totals of 21 and 16 peaks were visually recognized and matched in the electropherograms obtained from the ovine and bovine/ovine cheese, respectively (Table 16.7). The identification of the peaks corresponding to intact casein and the peptide release during the cheese ripening has been previously described by our research group (32, 33).

Because of the complexity of proteolytic patterns during cheese ripening, chemometrics has recently been proposed as an objective approach for the evaluation of proteolytic profiles and data interpretation.

16.3.3.1. Application of Multivariate Regression Methods (PLS and PCR) to Predict the Ripening Time of Ewe's Milk and Milk Mixture Cheese. The methods that we selected to analyze the CE peak data were PLS and PCR. In a preliminary analysis, PLS regression was applied to the calibration samples of ovine and mixture cheeses with ripening times of 0 to 139 d ($n = 12$) and 0 to 167 d ($n = 14$), respectively, using the areas of the specified peaks (Table 16.7) as the predictor variables. A first model, with the whole data set of the standard cheeses, indicated the peaks with a low modeling power of variance (peaks 11*e*, 13*e*, and 20*e* [ewe's milk cheese] and peaks 1*m*, 2*m*, 6*m*, 12m, 13*m*, and 16*m* [milk mixture cheese]). Table 16.8 shows the results when PLS and PCR regressions were applied to the area of peaks selected (18 and 10 for the ovine and mixture cheese, respectively). These results include the number of components selected by cross-validation (a), the determination coefficient (R^2), and the RMSECV. The RMSECV was used as a diagnostic test for examining the errors in the predicted maturation time of the cheese samples (31). It indicates both precision and accuracy of prediction.

In all regression methods the percentage of the explained variance for the models was >96%, yielding good correlation (R^2 values > 0.989) between the observed and calculated ripening times. For each cheese, the values of RMSECV obtained with the PLS and PCR models were similar, and it was possible to predict the ripening time of ovine cheese and milk mixture cheese with an error lower than 4 and 8 d, respectively. Similar values have been reported when these cheeses were analyzed using a fused-silica capillary (32, 33). These values can be considered to be relatively low. In fact, García-Ruiz et al. (31) and Poveda et al. (43) obtained values of 10.3 and 11.9 d, respec-

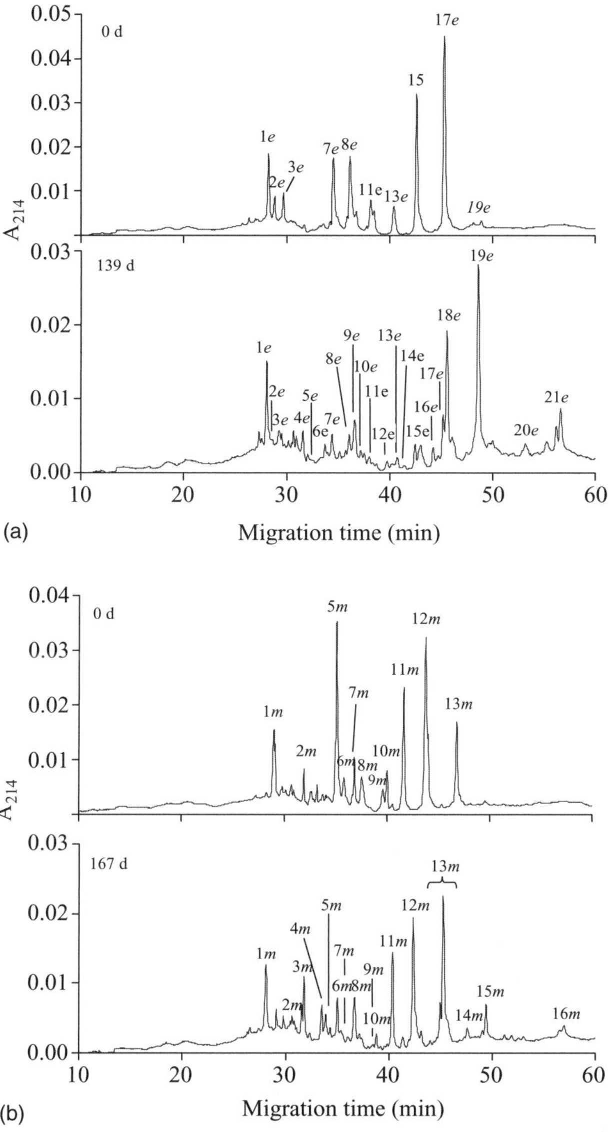

FIGURE 16.5. CE analysis of the pH 4.6-insoluble fraction of ewe's milk cheese (0-day-old and 139-day ripening time) (a) and cheese made from cow's and ewe's milk (0-day-old and 167-day ripening time) (b). Separations were performed in a neutral capillary at 25.1 kV (~50 μA). Other conditions are described under Materials and Methods. Peak identification is given in Table 16.7.

TABLE 16.7. Peak visually recognized and matched in the electropherograms of the pH 4.6-insoluble fraction of ewe's milk cheese (Fig. 16.5a) and cheese made from cow's and ewe's milk (Fig. 16.5b)

Peak[a]	Casein or Peptide	Peak[a]	Casein or Peptide[b]
1*e*	p-κ-CN	1*m*	p-κ-$CN_{(C)}$ + p-κ-$CN_{(E)}$ + α_{s2}-$CN_{(E)}$
2*e*	α_{s2}-CN	2*m*	α_{s2}-$CN_{(C)}$
3*e*	α_{s2}-CN + peptide	3*m*	α_{s2}-$CN_{(E)}$ + γ-CN
4*e*	γ_2-CN	4*m*	γ-CN
5*e*	P[c]	5*m*	α_{s1}-$CN_{(C)}$ 8P
6*e*	Peak 6*e*	6*m*	α_{s1}-$CN_{(E)}$ I
7*e*	α_{s1}-CN I	7*m*	α_{s1}-$CN_{(C)}$ 9P
8*e*	α_{s1}-CN II	8*m*	α_{s1}-$CN_{(E)}$ II + γ CN
9*e*	γ_3-CN	9*m*	α_{s1}-$CN_{(E)}$ III
10*e*	P[c]	10*m*	β-$CN_{(C)}$ B
11*e*	α_{s1}-CN III	11*m*	β-$CN_{(C)}$ A^1 + β-$CN_{(E)}$
12*e*	Peak 12*e*	12*m*	β-$CN_{(C)}$ A^2 + β_2-$CN_{(E)}$
13*e*	β-CN	13*m*	β_1-$CN_{(C)}$ + P[c] + α_{s1}-I-CN
14*e*	P[c]	14*m*	Peak 14*n*
15*e*	β_2-CN	15*m*	Peak 15*n*
16*e*	P[c]	16*m*	P[c]
17*e*	β_1-CN		
18*e*	Peak 18*e*		
19*e*	α_{s1}-I-CN		
20*e*	P[c]		
21*e*	P[c]		

[a]*e* = ewes' milk cheese; *m* = mixture milk cheese.
[b](C) = cow; (E) = ewe.
[c]P = peptides from the action of plasmin on caseins.

TABLE 16.8. Partial least squares regression (PLS) and principal components regression (PCR) results for the prediction of the ripening times of ewe's milk cheeses and cheeses made from cow's and ewe's milk

	Ovine Cheese[a]		Bovine/Ovine Cheese[b]	
	PLS	PCR	PLS	PCR
a[c]	4	4	5	5
(R^2)[d]	0.9981	0.9975	0.9929	0.9890
RMSECV[e]	4.1	3.6	7.8	7.5
% var[f]	99.24	96.18	97.46	97.17[g]

[a]Data from Albillos et al. (33).
[b]Data from Albillos et al. (32).
[c]Number of components selected by cross-validation.
[d]Determination coefficient.
[e]Root-mean-square error of prediction (within the day) of cross-validation.
[f]Percentage of explained variance.
[g]Data not published.

tively, using PLS regression to calculate ripening time in standard Manchego cheeses based on some physicochemical parameters and secondary proteolysis indices.

Figure 16.6 depicts the relationship between the ripening times predicted by cross-validation and real ripening times for the ewe's milk cheeses. Similar behavior was observed for the mixture milk cheeses. The fit for the prediction of the period between 0 and 139 d of ripening of the ewe's milk cheese and between 0 and 167 d for the cheese made from cow's and ewe's milk was good as shown by the values of r^2 obtained: 0.9982 and 0.9929, respectively.

16.3.3.2. Ripening Dynamic Using PCA Approach. Considering that peptide profiles generally lead to a large amount of data to be processed, PCA

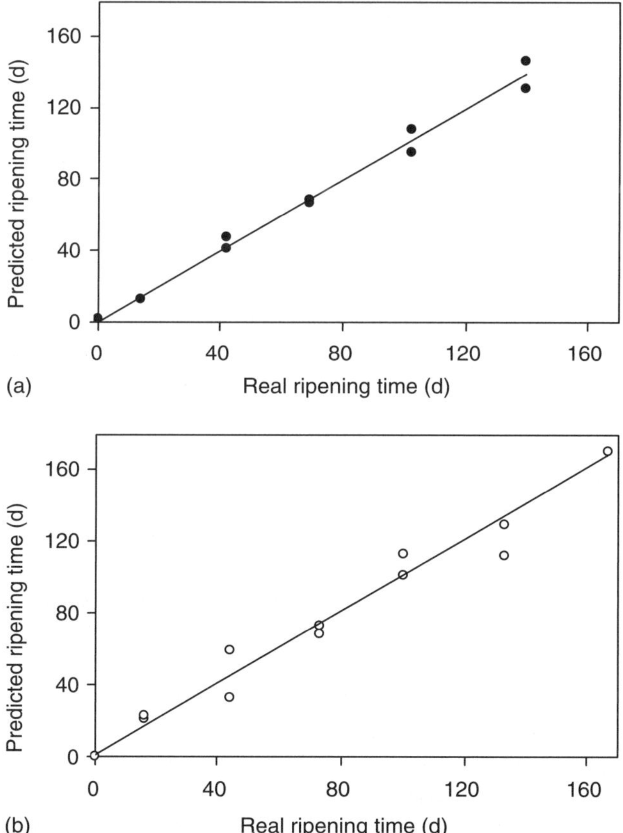

FIGURE 16.6. Correlation between the ripening times predicted by cross-validation using PCR regression and real ripening times for ewe's milk cheeses (a) and cheese made from a mixture of cow's and ewe's milk (b) analyzed by CE using a neutral capillary.

can be a useful tool to dimension and to examine data variation. The extent to which PCA is able to determine the maturity of a ripening cheese (to distinguish between cheese with different ripening times) when used to analyze data from CE of the pH 4.6-insoluble fraction has been described by our research group (32, 33). Accordingly, in order to establish the relationships between the different variables and to detect the most important causes of variability, PCA was applied to the area of peaks selected (except the peaks with a low modeling power of variance) for two batches of ewe's milk cheese ripened for 0, 14, 42, 69, 102, and 139 d. Four PCs were obtained, accounting for 96.69% of the total variance. PC1 explained 71.5% of the total variance and the peaks that correlated best with this PC and their factor loadings were ovine α_{s1}-casein I (peak 7*e*) (–0.266), α_{s1}-casein II (peak 8*e*) (–0.267), β_2-casein (peak 15*e*) (–0.269), β_1-casein (peak 17*e*) (–0.276), and peak 21*e* (0.268).

The peaks best correlated with PC2 (which accounted for 14.5% of the total variance) and their factor loadings were γ-casein peak 5*e* (–0.521), peak 6*e* (–0.407), peak 12*e* (0.346), and peak 16*e* (0.272). Therefore, these results indicate that PC1 was correlated with intact ovine caseins, while PC2 was associated with hydrolysis products released during cheese ripening. The distribution of the samples in the plane defined by PC1 and PC2 showed that the samples appeared separated according to their ripening times (33): cheeses with the lowest ripening time (0 and 14-day-old cheeses) located in the right side of PC had higher contents of nonspecific non-degraded casein (α_{s1}-CN I, α_{s1}-CN II, β_1-CN, and β_2-CN) than samples of 102 and 139 d of ripening, located in the left side of this PC. Furthermore, PC2 distinguished the cheeses at 42 d of ripening from the cheeses at 69 d of ripening. Similar results were obtained when a fused-silica capillary was used to analyze the casein fraction (33). However, in this case PC2 distinguished better between cheeses of 42 and 69 d of ripening time than PC2 of the equivalent results of the PCA analysis applied to data obtained from the neutral capillary.

As mentioned previously, a total of 16 peaks was identified when cheeses manufactured from mixture of milk (with 0, 16, 44, 73, 100, 133, and 167 d of ripening) were analyzed by CE using a neutral capillary. Areas of peaks selected (except peaks with a low modeling power of variance) were analyzed by PCA and two principal components were obtained, accounting for 84.3% of the total variance. Thus, the dimensionality of the data was reduced from 10 variables to two uncorrelated PCs with 15.7% loss of variation. PC1 explained 72.2% of the total variance and was strongly correlated to bovine α_{s1}-CN 8P (peak 5*m*), bovine α_{s0}-CN 9P (peak 7*m*), and bovine β-CN B (peak 10*m*) (positive values). PC2 was correlated to peak 8*m* (ovine α_{s1}-CN II + γ-CN) (positive values) and peak 11*m* (bovine β-CN A^1 + ovine β-CN) (negative values). A biplot showing the projection of the samples (scores) and the variables (loadings) on the plane of the first and the second PC is given in Figure 16.7. It can be observed that samples appear separated according to their ripening time from right to left, although with the PC1 no separation occurred between samples of cheese at 44 and 73 d of ripening. PC2, which accounted

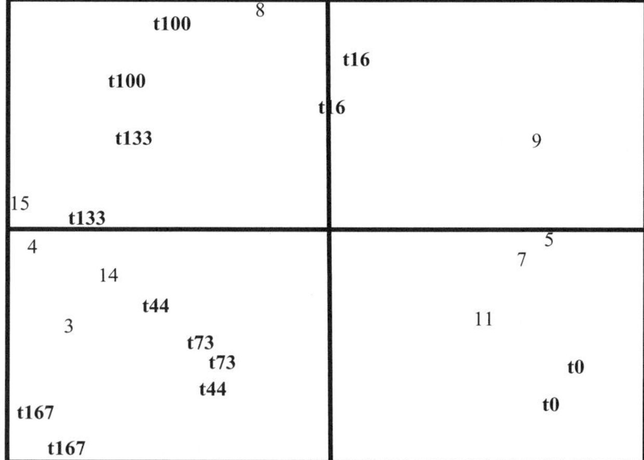

FIGURE 16.7. Biplot on the plane of the first and second eigenvectors. Training samples are represented by ***ti*** (where *i* is the ripening time). The numbers represent the original variables, plotted according to their respective loadings.

for only 12.1% of the variation, distinguished better than PC1 between cheeses at 100 and 133 d of ripening.

16.3.4. Concluding Remarks

Factorial design and response surface methodology have been used successfully for the optimization of a capillary electrophoresis method, using a neutral capillary, applied to the separation of caseins. By combining the electrophoretic profiles and multivariate regression analysis, PLS and PCR, it was possible to predict the ripening times of commercial cheese within approximately 4–8 d. In conclusion, the chemometrical strategy described in this chapter shows that it is a proven powerful tool to achieve adequate separation of bovine and ovine caseins by CE and to obtain information on the biochemical process of proteolysis during the ripening of cheese manufactured with milk from different origins.

REFERENCES

1. Ulberth, F. (2003) Testing the authenticity of milk and milk products, in *Dairy Processing. Improving Quality* (ed. G. Smit), CRC Press Lt., Boca Raton, Florida, p. 208.
2. Summer, A., Franceschi, P., Bollini, A., Formaggioni, P., Tosi, F., and Mariani, P. (2003) *Vet Res Commun*, **27**, 663–666.
3. Fox, P. and McSweeney, P.L.H. (2004) Cheese: An overview, in *Cheese—Chemistry, Physics and Microbiology. Vol 1. General Aspect* (eds. P.F. Fox, P.L.H. McSweeney, T.M. Cogan, and T.P. Guinee), Elsevier Academic Press, Amsterdam, p. 1.

4. Karoui, R. and Baerdemaeker, J. (2007) *Food Chem*, **102**, 621–640.
5. Molina, E., Ramos, M., and Martin Alvarez, P.J. (1995) *Z Lebesnm Unters Forsch*, **201**, 331–335.
6. Herrero-Martínez, J.M., Simó-Alfonso, E.F., Ramis-Ramos, G., Gelfi, C., and Righetti, P.G. (2000) *J Chromatogr A*, **878**, 261–271.
7. Veloso, A.C.A., Teixeira, N., and Ferreira, I.M.L.V.O. (2002) *J Chromatogr A*, **967**, 209–218.
8. Fox, P.F. (1989) *J Dairy Sci*, **72**, 1379–1400.
9. Fox, P.J., Law, J., McSweeney, P.L.H., and Wallace, J. (1993) Biochemistry of cheese ripening, in *Cheese: Chemistry, Physics and Microbiology* (ed. P.F. Fox) Chapman & Hall, London, p. 389.
10. Visser, S. (1993) *J Dairy Sci*, **76**, 329–350.
11. Upadhyay, V. K., McSweeney, P.L.H., Magboul, A.A.A., and Fox, P.F. (2004) Proteolysis in cheese during ripening, in *Cheese—Chemistry, Physics and Microbiology. Vol 1. General Aspect* (eds. P.F. Fox, P.L.H. McSweeney, T.M. Cogan, and T.P. Guinee) Elsevier Academic Press, Amsterdam, p. 391.
12. Ramos, M. and Juárez, M. (1986) *Int Dairy Fed Bull*, **202**, 175–190.
13. de la Fuente, M.A. and Juárez, M. (2005) *Crit Rev Food Sci Nut*, **45**, 563–585.
14. Mayer, H.K. (2005) *Int Dairy J*, **15**, 595–604.
15. Crittenden, R.G., Andrew, A.S., LeFourmour, M., Young, M.D., Middleton, H., and Stockmann, R. (2006) *Int Dairy J*, **17**, 421–428.
16. de Jong, N., Visser, S., and Olieman, C. (1993) *J Chromatogr A*, **652**, 207–213.
17. Otte, J., Zakora, M., Kristiansen, K.R., and Qvist, K.B. (1997) *Lait*, **77**, 241–257.
18. Molina, E., Martín-Álvarez, J., and Ramos, M. (1999) *Int Dairy J*, **9**, 99–105.
19. Heck, J.M.L., Olieman, C., Schennink, A., van Valenberg, H.J.F., Visker, M.H.P., Meuldijk, R.C.R., and van Hooijdonk, A.C. (2008) *Int Dairy J*, **18**, 548–555.
20. Molina, E., de Frutos, M., and Ramos, M. (2000) *J Dairy Res*, **67**, 209–216.
21. Otte, J., Ardö, Y., Weimer, B., and SØrensen, J. (1999) *Bull Int Dairy Fed*, **337**, 10–16.
22. Recio, I., Ramos, M., and López-Fandiño, R. (2001) *Electrophoresis*, **22**, 1489–1502.
23. Hanrahan, G., Montes, R., and Gomez, F.A. (2008) *Anal Bioanal Chem*, **390**, 169–179.
24. Altria, K.D., Clark, B., Filbey, S.D., Nelly, M.A., and Rudd, D.R. (1995) *Electrophoresis*, **16**, 2143–2148.
25. Siouffi, A.M. and Phan-Tan-Luu, R. (2000) *J Chromatogr A*, **892**, 75–106.
26. Sentellas, S. and Saurina, J. (2003) *J Sep Sci*, **26**, 875–885.
27. Pripp, A.H., Rehman, S.-U., McSweeney, P.L.H., and Fox, P.F. (1999) *Int Dairy J*, **9**, 473–479.
28. Pripp, A.H., Rehman, S.-U., McSweeney, P.L.H., Sørhaug, T., and Fox, P.F. (2000) *Int Dairy J*, **10**, 25–31.
29. Pripp, A.H., Stepaniak, L., and Sørhaug, T. (2000) *Int Dairy J*, **10**, 249–253.
30. Coker, C.J., Crawford, R.A., Jonhston, K.A., Singh, H., and Creamer, K.K. (2005) *Int Dairy J*, **15**, 631–643 .

31. García-Ruiz, A., Cabezas, L., Martín-Alvárez, P.J., and Cabezudo, D.Z. (1998) *Lebensm Unters Forsch*, **206**, 382–386.
32. Albillos, S.M., Busto, M.D., Perez-Matos, M., and Ortega, N. (2005) *J Agric Food Chem*, **53**, 6094–6099.
33. Albillos, S.M., Busto, M.D., Perez-Matos, M., and Ortega, N. (2006) *J Agric Food Chem*, **54**, 8281–8287.
34. Smith, A. M. and Nakai, S. (1990) Classification of cheese varieties by multivariate analysis of HPLC profiles. *Can Inst Food Sci Technol J*, **23**, 53–58.
35. Coker, C.J. (2003) Objective differentiation of cheese type and maturity. PhD Dissertation, Riddet Centre and Institute of Food Nutrition and Health, College of Science, Massey University, Palmerston North, New Zealand.
36. Pillonel, L., Albrecht, B., Badertscher, R., Chamba, J.F., Bütikofer, U., Tabacchi, R., and Bosset, J.O. (2003) *Ital J Food Sci*, **15**, 49–62.
37. Pillonel, L., Badertscher, R., Bütikofer, U., Casey, M., Dalla Torre, M., Lavanchy, P., Meyer, J., Tabacchi, R., and Bosset, J.O. (2003) *Eur Food Res Technol*, **215**, 260–267.
38. Pillonel, L., Bosset, J.O., Bütikofer, U., Tabacchi, R., and Schlichtherlecerny, H. (2005) Int Dairy J, **15**, 557–562.
39. Pillonel, L., Tabacchi, R., and Bosset, J.O. (2003) *Mitt Lebensm Hyg*, **94**, 60–69.
40. Pham, A.-M. and Nakai, S. (1984) *J Dairy Sci*, **67**, 1390–1396.
41. Amantea, G.F., Furtula, V.N., Choi, H.Y., Laleye, L.C., and Nakai, S. (1995) Assessment of accelerated cheese ripening by reverse-phase HPLC, in *Chemistry of Structure-Function Relationships in Cheese* (eds. E.L. Malin and M.H. Tunick) Plenum Press, New York, p. 113.
42. Noël, Y., Ardö, Y., Pochet, S., Hunter, A., Lavanchy, P., Luginnbühl, W., LeBars, D., Polychroniadou, A., and Pellegrino, L. (1998) *Lait*, **78**, 511–519.
43. Poveda, J.M., García, A., Martín-Alvarez, P.J., and Cabezas, L. (2004) *Food Chem*, **84**, 29–33.
44. Alvarenga, N., Silva, P., Rodriguez Garcia, J., and Sousa, I. (2008) *J Dairy Res*, **75**, 233–239.
45. de Frutos, M., Molina, E., and Amigo, L. (1996) *Milchwissenschaft*, **51**, 374–378.
46. Montgomery, D.C. (1991) *Diseño y Análisis de Experimentos*; Iberoamericana, Mexico.
47. Haaland, P.D. (1989) *Experimental Design in Biotechnology*, Marcel Dekker, New York.
48. Garrido Frenich, A., Jouan-Rimbaud, D., Massart, D.L., Kuttatharmmakul, S., Martinez Galera, M., and Martinez Vidal, J. (1995) *Analyst*, **120**, 2787–2792.
49. Forina, M., Lanteri, S., and Armanino, C. (2000) Q-PARVUS Release 3.0. An extendable package of programs for data explorative analysis, classification and regression analysis, http://parvus.unige.it (accessed July 10, 2009).
50. Ortega, N., Albillos, S.M., and Busto, M.D. (2003) *Food Control*, **14**, 307–315.
51. Rodriguez, I. and Li, S.F.Y. (1999) *Anal Chim Acta*, **383**, 1–26.
52. Lindner, H., Helliger, W., Sarg, B., and Meraner, C. (1995) *Electrophoresis*, **16**, 604–610.

53. Castagnola, M., Messana, I., and Rossetti, D.V. (1996) Capillary zone electrophoresis for the analysis of peptide, in *Capillary Electrophoresis in Analytical Biotechnology* (ed. P.G. Righetti), CRC Press, Boca Raton, Florida, p. 239.
54. Sentellas, S. and Saurina, J. (2003) *J Sep Sci*, **26**, 1395–1402.
55. Chen, F.T.A. and Zang, J.H. (1992) *J OAAC Int*, **75**, 905–909.
56. Cattaneo, T.M.P., Nigro, F., and Greppi, G.F. (1996) *Michwissenschaft*, **51**, 616–619.
57. Recio, I., Pérez-Rodríguez, M.L., Ramos, M., andAmigo, L. (1997) *J Chromatogr A*, **768**, 47–56.
58. Recio, I., Amigo, L., Ramos, M., and López-Fandiño, R. (1997) *J Dairy Res*, **64**, 221–230.
59. Pinho, O., Mendes, E., Alves, M.M., and Ferreira, M.P.L.V.O. (2004) *J Dairy Sci*, **87**, 249–257.
60. Ferreira, I.M.P.L.V.O., Veiros, C., Pinho, O., Veloso, A.C.A., and Peres, A.M. (2006) *J Dairy Sci*, **89**, 2397–2407.
61. Herrero-Martínez, J.M., Simó-Alfonso, E.F., Ramis-Ramos, G., Gelfi, C., and Righetti, P.G. (2000) *Electrophoresis*, **21**, 633–640.
62. Fallico, V., McSweeney, P.L.H., Siebert, K.J., Horne, J., Carpino, S., and Licitra, G. (2004) *J Dairy Sci*, **87**, 3138–3152.

PART IV

TRANSFORMATION TECHNIQUES

CHAPTER 17

TRANSFORMATION TECHNIQUES FOR CAPILLARY AND MICROCHIP ELECTROPHORESIS

TAKASHI KANETA
Department of Applied Chemistry, Graduate School of Engineering and Division of Translational Research, Center of Future Chemistry, Kyushu University, Fukuoka, Japan

CONTENTS

17.1 INTRODUCTION

Resolution and sensitivity are essential to the collection of analytical chemical data with accuracy and precision. It is well known that mathematical transformation techniques enhance the resolution and sensitivity of spectroscopic methods. Fourier transform (FT), cross correlation (CC), and Hadamard transform (HT) techniques allow for high resolution and high sensitivity of infrared spectroscopy (IR), fluorometry, nuclear magnetic resonance

Chemometric Methods in Capillary Electrophoresis. Edited by Grady Hanrahan and Frank A. Gomez

spectroscopy (NMR), and mass spectrometry. In addition, the application of the CC technique to chromatographic separation was first proposed by Izawa et al. (1). Subsequently, in the 1970s, CC techniques were applied to gas chromatography (GC) (2) and high performance liquid chromatography (HPLC) (3). Application of these mathematical transformation techniques to capillary electrophoresis (CE) was first demonstrated by Smit et al. (4), who employed a CC technique to improve the signal-to-noise (S/N) ratio of analytes separated by CE. Similarly, we applied HT to CE (5). Currently, high resolution in FT-IR and NMR requires the use of the FT technique. Advances in micromachining technology have accelerated the use of mathematical transformations in elecrophoretic separations.

Application of FT to electrophoresis was proposed by Manz et al. (6), who utilized a Shah convolution of the fluorescent signal during separation on a microchip. In addition, CC (7) and HT (8, 9) have also been applied to electrophoresis on a microchip. Recent developments in the use of these techniques in chromatography and electrophoresis have been reviewed by Kaljurand and Smit (10).

Among the mathematical transformation techniques, CC and HT substantially improve the S/N ratio in CE separations. It should be noted that Shah convolution Fourier transform (SCOFT) detection yielded results that differ from those obtained with either CC or HT. Although fundamental studies have demonstrated the potential of SCOFT (6, 11), both the advantages and disadvantages remain to be determined. Nevertheless, SCOFT apparently has the potential to improve the resolution and/or sensitivity of microchip electrophoresis (ME). In this chapter, we discuss the principles, instrumentation, and performance of CE and ME methods using CC, HT, and FT techniques.

17.2. CROSS CORRELATION AND HT

17.2.1. Theory

In CE, as well as in other separation techniques such as GC and HPLC, the input signal corresponds to the introduction of a sample as a single plug. The output signal from the detector is digitized at a constant frequency, resulting in an electropherogram. Thus, the digitized data set is represented as a time function, $y(t)$. If one uses a pseudo-random binary sequence (PRBS) as a function, $x(t)$, to sample a narrow sample input, the cross correlation, $\phi_{xy}(\tau)$, is represented by (12):

$$\phi_{xy}(\tau) = \frac{1}{N}\sum_{t=1}^{N} x(t-\tau)\,y(t) \qquad \text{(Eq. 17.1)}$$

where N is the number of data points in the input and output signals, and τ is the delay. After recording the electropherogram obtained by multiple injections according to a PRBS function, $\phi_{xy}(\tau)$ is plotted against τ, resulting in a correlogram that is similar to the electropherogram obtained as a result of a single impulse injection. If a circular matrix constituting the PRBS is employed, Equation 17.1 can be represented by a simple calculation as follows:

$$[Y]=[X]\times[E] \qquad \text{(Eq. 17.2)}$$

where $[X]$ is the circular matrix, $[E]$ is the electropherogram obtained from the single impulse input, and $[Y]$ is the data set of the electropherogram represented by $y(t)$ in Equation 17.1. We can see that Equation 17.2 is identical to the case of HT, in which only the matrix employed for successive injections is derived from a Hadamard matrix. In HT, the matrix, $[X]$, is a cyclic S matrix that is obtained by deletion of both the first row and column of a Hadamard matrix and by substitution of "–1" with "1" and "1" with "0" (13). In this case, the correlogram or data set that is transformed using an inverse HT is given by multiplying by the inverse matrix of $[X]$,

$$[E]=[X]^{-1}\times[Y] \qquad \text{(Eq. 17.3)}$$

where $[X]^{-1}$ is the inverse matrix of $[X]$. Comparing Equations 17.1 and 17.3, $\phi(\tau)$ represents each element in the vector of $[E]$. The significant improvement in the S/N ratio of either the correlogram or the transformed data is known as the multiplex or Fellgett advantage. The improvement factor is determined by the order of the length in the PRBS, N. The theoretical improvement factor for CC-CE is given by

$$\frac{I}{\sqrt{N}} \qquad \text{(Eq. 17.4)}$$

where I is the number of sample injections in the PRBS. For HT-CE, the theoretical improvement factor is represented by

$$\frac{N+1}{2\sqrt{N}} \approx \frac{\sqrt{N}}{2} \qquad \text{(Eq. 17.5)}$$

In general, I in Equation 17.4 is equal to $(N + 1)/2$, hence, the improvement factor in CC techniques is the same as that for HT, as written in Equation 17.5. Figure 17.1 shows a schematic illustration of CE based on multiple inputs, such as CC-CE and HT-CE. In conventional CE, the input is a single pulse formed by a small plug. Conversely, multiple input techniques are applied in CC- and HT-CE. That is, the sample is introduced into a separation channel according to the PRBS, as shown in Figure 17.1, resulting in an electropherogram modulated by the PRBS. The correlogram or transformed data is calculated by multiplying the inverse matrix, $[X]^{-1}$, with the modulated electropherogram, as represented by Equation 17.3.

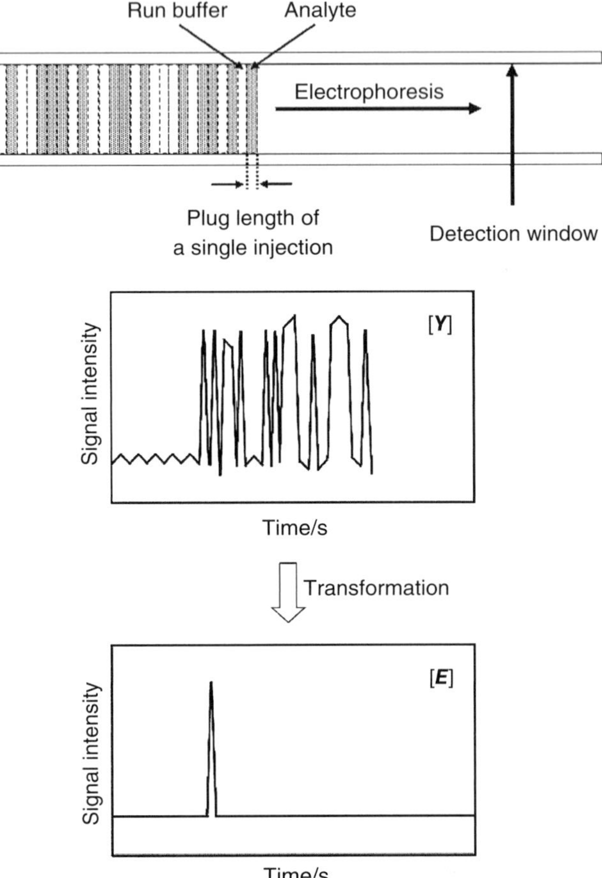

FIGURE 17.1. Schematic illustrations of PRBS injection and the expected results. References 4 and 15 should be referred to for the details.

17.2.2. Instrumentation for Cross Correlation and HT Electrophoresis

The key technology required for CE based on multiple sample injections is the injection device. In the first demonstration of CC-CE by Van der Moolen et al., a capillary was glued in a glass rod, in which a small hole had been drilled perpendicular to the capillary (4). The size of the hole (30 μm) was less than the inner diameter of the capillary (75 μm). A running buffer solution and a sample solution were electrokinetically introduced into the capillary through the hole. In a subsequent study, Van der Moolen et al. developed the microchip injection device (14) shown in Figure 17.2. Conversely, in the first demonstration of HT-CE, we employed an optically gated injection method for successive, high-precision injections (5, 15). We also reported an electrokinetic injection device for use in HT-CE that was constructed from a laser-

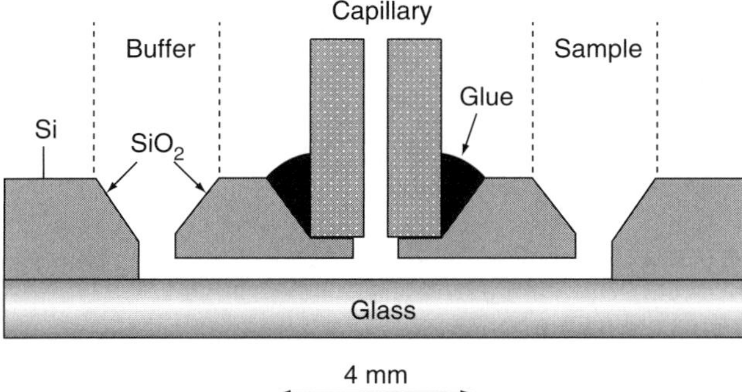

FIGURE 17.2. Design of the microchip injection device for CC-CE. Reproduced with the permission of the American Chemical Society (14).

drilled capillary (16, 17). The optical gating provides rapid and precise injections of discrete sample plugs due to the rapid response of the optical shutter. However, this injection method is applicable only to fluorescence detection. On the other hand, electrokinetic injection devices are applicable to any type of detector, including absorbance, fluorescence, and electrochemical detectors, although the time required to switch the electric potential is much greater than the need for modulation of an optical shutter. In ME, electrokinetic injection is generally employed for both the CC (7) and HT techniques (8, 9, 18). Table 17.1 shows the injection method, detector, and mathematical technique employed in CE and ME.

17.2.3. Fundamental Studies and Applications of Cross Correlation and HT Electrophoresis

In the first demonstration of CC-CE, Smit et al. observed a considerable reduction in the detection limit that was in agreement with the theoretically predicted values shown in Equation 17.5. They used the clock periods of 10 s (4) and 5 s (14), which correspond to the injection time for one element of the PRBS, depending on the injection device. In CC- and HT-CE, the sampling time of the detector signal is typically adjusted such that it is equal to the clock period. The sampling time of the detector signal is usually less than 1 s in CE experiments. Thus, clock periods of 5–10 s are too long to obtain electropherograms with high-resolution separation. A long clock period would be caused by slow switching of the high potential between the sample and buffer reservoirs. A relay is usually employed to switch the high voltage. In CC-CE, the time needed for the relay to signal the switch is on the order of 10 ms (14). As a result, the high voltage is insulated for less than 100 ms. To decrease the injection errors, the clock period must be much longer than the time required for switching the relay. Consequently, a clock period of several seconds might be necessary in CC-CE, although it is uncertain whether the clock time was

TABLE 17.1. Summary of electrophoretic methods based on sample injection according to PRBS

Method	Injection method	Detection	Reference
CC-CE	Electrokinetic injection on an injection device connected with a glass rod	Absorbance	(4)
CC-CE	Electrokinetic injection on a microchip injection device connected with a capillary	Absorbance	(14)
CC-CE	Electrokinetic injection assisted with the pressure of compressed air	Absorbance	(19)
CC-ME	Electrokinetic injection on a microchip	LIF	(7)
HT-CE	Optically gated injection	LIF	(5), (22), (23), (26), (27), (29), (31)
HT-CE	Electrokinetic injection with a laser-fabricated capillary	Absorbance, LIF	(16), (17)
HT-CE	Electrokinetic injection with a Tee connector	Absorbance	(30)
HT-CE	Electrokinetic injection assisted with the pressure of compressed air	Absorbance	(19), (28)
HT-CE	Pressure-assisted capillary injection	Absorbance	(21)
HT-ME	Electrokinetic injection on a microchip	LIF	(8), (9), (18), (25)
HT-ME	Electrokinetic injection on a microchip	Electrochemical	(24)

optimized in previous studies. Conversely, the relay used in our studies of HT-CE (16) can be switched in 3 ms. In this case, the clock period was reduced to 1 s in both HT-CE and HT-ME.

The time required to achieve the maximum voltage (rise time) and to return to the minimum voltage (fall time) depends on the magnitude of the voltage applied to the separation column. It is expected that both the rise time and fall time decline when the value of the high voltage is reduced. Thus, the clock period can be reduced in ME, as the applied voltage in ME is ~10-fold less than that in CE. Fister III et al. successfully demonstrated CC-ME, in which the clock period was reduced to 0.25 s (7). We also found that 0.5 s was the optimal clock period for HT-ME combined with a laser-induced fluorescence (LIF) detection system equipped with a compact Nd:YAG laser (18).

Pressure-assisted sampling devices have been developed and applied to CC-CE (19) and HT-CE (20). Kaljurand et al. have constructed an automated electrokinetic injection device that had no relay (21). In the sampling device, either the sample or the buffer solution was introduced into a T-shaped channel by air pressure. A capillary and an electrode were connected downstream from the T-shaped channel. Thus, either the sample or the buffer solution was injected into the capillary electrokinetically by pressurized injection of the solution. The pressure-assisted sampling devices allowed clock periods of 3.75 s (19) and 1 s (20), which are comparable with those required by our electrokinetic injection device (16).

In the first demonstration of HT-CE, we employed an optically gated sample injection method that was used as a fast sample injection technique in CE (5). In this method, a capillary is filled with a sample solution containing fluorescent analytes. A high-power laser is split into two parts that are used as the gating and probe beams. The gating beam is focused on the capillary. The beam is either passed through or blocked by an optical shutter that can be modulated by a controller interfaced with a computer. When the shutter is open, the fluorescent analyte is photobleached as a result of the strong irradiation by the laser light. Thus, the sample is injected only when the shutter is closed. In the early studies of HT-CE, a clock period of 0.5 s was employed for modulation of analyte introduction. Braun et al. proposed fast HT-CE (22) that achieved rapid analysis by reducing both the length of the injection sequence (by ~50%) and the clock period (10–100 ms).

Several investigations of HT-CE have been attempted, as follows: a photolytic optical gating injection technique for caged fluorescent labels (23); other detection methods, including either an electrochemical detector (24) or an LIF detector using a charge-coupled device (CCD) camera (25); and a modified transformation technique (24). The key advantage of the CC and HT techniques is that the S/N is improved more rapidly compared with averaging techniques that require repeated runs. Figure 17.3 shows the inverse transformation obtained using HT-CE with different order matrices. As shown in Figure 17.3, the S/N ratio is enhanced as the order of the matrix increases. When the order equal to 2047 was employed, the limit of detection (LOD) for fluorescein was 500 fM, which corresponds to 27 molecules in a single injection volume (26). An enhancement in the S/N ratio was also obtained for a mixture, as illustrated in Figure 17.4, which shows the results for an amino acid mixture obtained using conventional CE and HT-CE (17).

Only glutamic acid is evident in the electropherogram obtained using conventional CE (Fig. 17.4a), while additional peaks for Rhodamine B isothiocyanate and phenylalanine are detectable in HT-CE (Fig. 17.4c). The transformed data shown in Figure 17.4a were calculated from the electropherogram obtained using the PRBS injection shown in Figure 17.4b. Unfortunately, the improvement factor for each analyte in the mixture sample was slightly less than that of a sample containing a single component. This is attributed to additional errors that result from overlapping of the peaks from different analytes. However, HT-CE significantly improves the S/N ratio even for mixture samples, as seen in Figure 17.4. The enhanced S/N ratio was observed

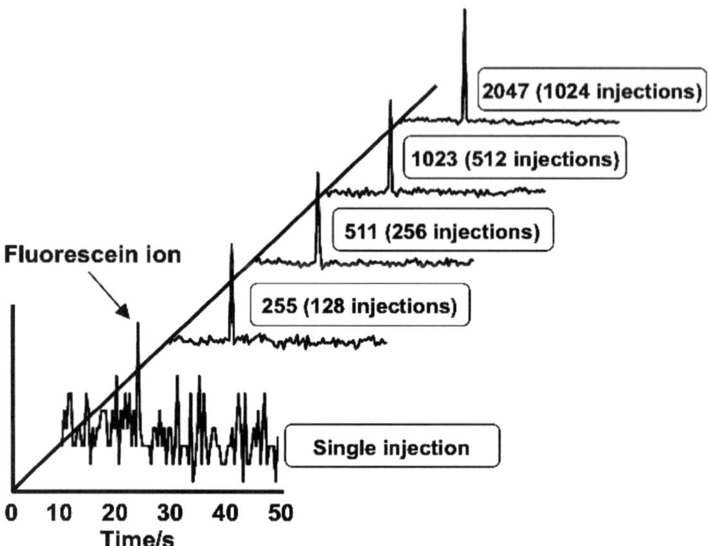

FIGURE 17.3. Electropherogram obtained by conventional single injection and transformed data obtained by HT-CE (optically gated injection) with different orders of the matrices. Analyte, sodium fluorescein (50 pM); injection time for a single segment, 0.5 s; buffer, 30 mM carbonate buffer (pH 9.3); laser power, 150 mW (gating beam 120 mW, probe beam 9 mW); wavelength, 488 nm; capillary, 25 µm inner diameter and 375 µm outer diameter; total length, 14 cm; effective length, 4.5 cm; and migration voltage, 10 kV.

in the separation of enantiomers (27) that have only small differences in electrophoretic mobility. Conversely, Seiman et al. suggested that HT-CE is unfavorably affected by the stacking phenomenon, but works well when analytes are present in sufficiently low concentrations (28). They concluded that additional theories are needed for the development of a mathematical procedure that permits a combination of the stacking method with HT-CE. In addition, a limitation of both the CC and HT techniques is that the major components of the sample interfere with the detection of minor species. According to our results (29), when the concentration of the major component was 100-fold greater than that of a minor component, detection of the minor component was difficult. However, minor species were detectable when the concentration of the major component was 20-fold.

Similar to conventional CE, HT-CE and HT-ME are applicable to quantitative analyses, as the calibration curve constructed from the transformed data shows good linearity even at concentrations less than the concentration limit of detection obtained using conventional CE (18, 29). HT-CE has been used in the analysis of actual samples. For instance, McReynolds et al. have successfully applied the HT-CE method with UV detection to the analysis of nitrates and nitrites in biological samples (30). We have also shown that the HT

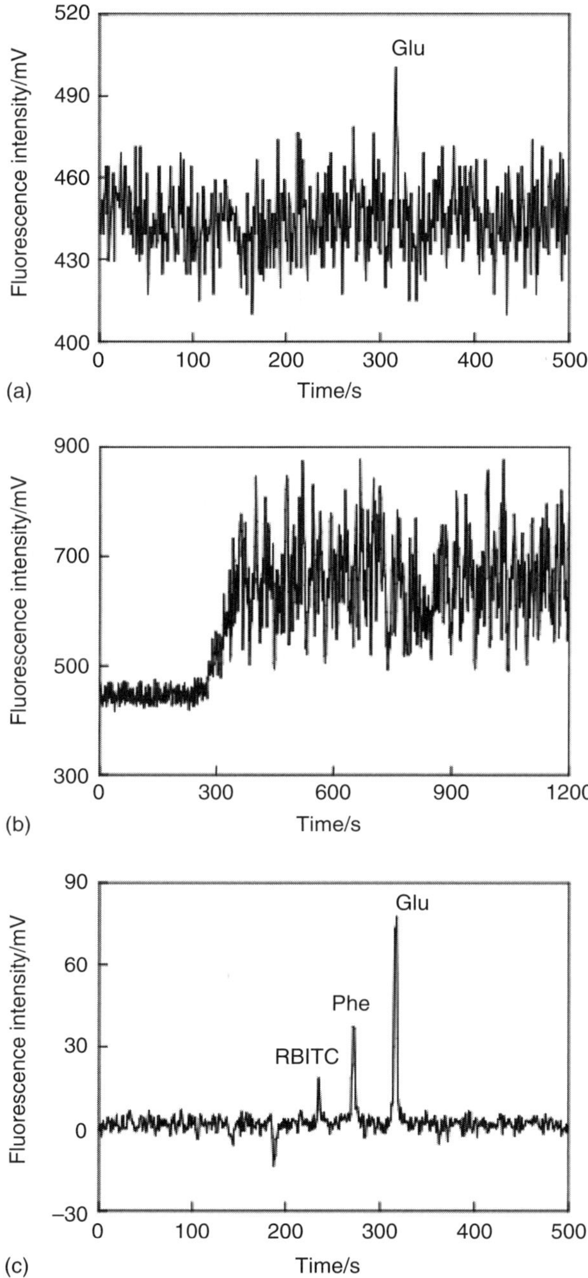

FIGURE 17.4. Electropherogram of a sample containing phenylalanine, glutamic acid, and free rhodamine B isothiocyanate (RBITC). (a) Single injection technique; (b) multiple sample injection according to the Hadamard sequence code; (c) inverse Hadamard transformed data. The concentrations of both phenylalanine and glutamic acid were 1.9 nM. Running buffer: borate–Tris (pH 9.0); effective length, 32 cm; total length, 60 cm; electric field, 150 V/cm. The order of the Hadamard matrix was 511. The injection period was set at 1.0 s. Reproduced with permission from Elsevier (17).

technique can be combined with micellar electrokinetic chromatography and the method is applicable to the determination of amino acids in beverages (31).

17.3. FT

17.3.1. Theory

The first successful use of an electrophoretic technique employing the FT is the SCOFT detection proposed by Crabtree et al. (6). In SCOFT, a slit positioned at a fixed distance is inserted into the separation channel of the microchip. The analyte ions are simultaneously separated and detected as they pass through the separation channel with the slit, resulting in an electropherogram with equally spaced peaks for each analyte. A schematic illustration of SCOFT detection is shown in Figure 17.5.

The data set obtained for the electropherogram, which is represented in the time domain, is converted into a data set in the frequency domain using an FT. When the number of the data points is N, FT of the data in the time domain yields N/2 + 1 complex points that are pairs of real and imaginary points in the frequency domain. These complex data are represented in terms of their magnitude as follows:

$$FT_{Mag} = \left(FT_{\mathrm{Re}}^2 + FT_{\mathrm{Im}}^2\right)^{1/2} \qquad \text{(Eq. 17.6)}$$

where FT_{Mag} is the magnitude, FT_{Re} are the real points, and FT_{im} are the imaginary points. The electropherogram is represented by plotting the magnitude in the frequency domain.

Another method for obtaining periodic signals during electrophoresis was proposed by Allen et al. (32). Multiple simultaneous separations were performed on a microchip with multiple separation channels that vary in distance from the injection port to the detection window. The separation channels are fabricated to be the square root of the linear increase in length, as the migration time for an analyte is proportional to the square of the channel length. The dependence of the migration time on the channel length is illustrated by the following relationship:

$$t = \frac{L}{v} = \frac{L}{E\mu} = \frac{L^2}{V\mu} \qquad \text{(Eq. 17.7)}$$

where t is the migration time, L is the channel length, v is the electrophoretic velocity, E is the electric field, μ is the electrophoretic mobility, and V is the applied voltage. Thus, if the differences of L^2 in the separation channels are constant, the migration times of an analyte are equally spaced in the electropherogram, that is, an electropherogram with periodic peaks is obtained. A

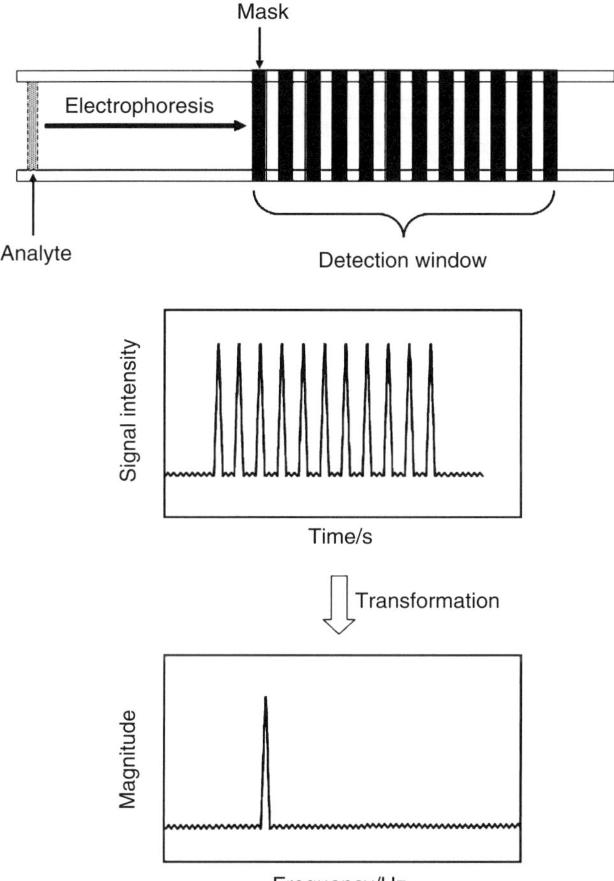

FIGURE 17.5. Schematic illustrations of multipoint detection for generation of periodic signals and the expected results in SCOFT. For the details, refer to Reference 6.

schematic illustration of the channels and the expected results is shown in Figure 17.6. The time-domain electropherogram is converted into frequency-domain plot by means of FT.

17.3.2. Instrumentation for FT Electrophoresis

In FT electrophoresis, well-designed microchips are employed to obtain an electropherogram with periodic peaks. Crabtree et al. (6) achieved SCOFT detection on a microchip with a Cr layer patterned on top of the separation channel. In their first report on SCOFT detection, fifty-five 300 μm-wide slits, spaced such that each slit is separated by 700 μm measured from its center, that is, 400 μm-wide detection windows, were aligned at 300 μm intervals. A

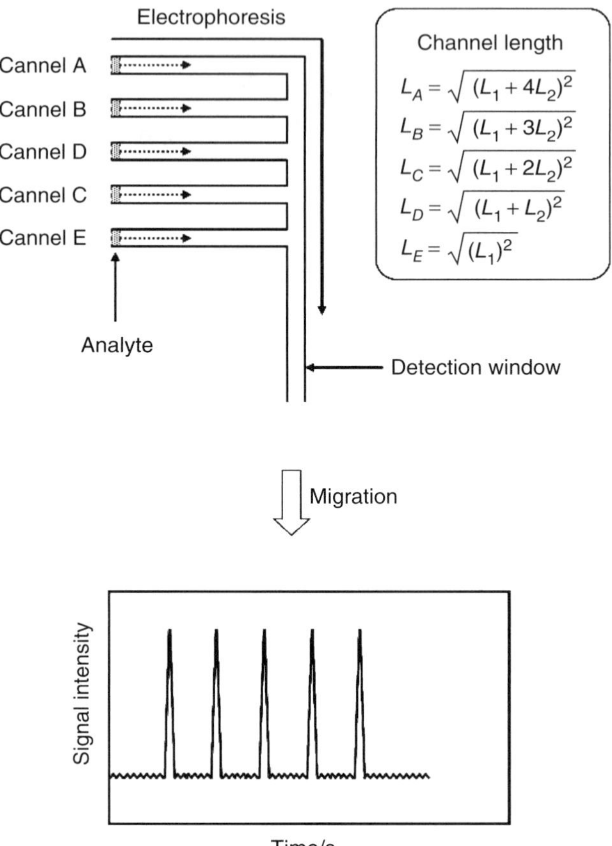

FIGURE 17.6. Schematic illustrations of simultaneous multichannel separation for generation of periodic signals and the expected result. For the details, refer to Reference 32.

laser beam was focused in the cylindrical shape using a convex lens so as to cover the 55 slits. A sample plug was injected into the separation channel and the fluorescence from the 55 slits was detected using a photomultiplier tube, resulting in 55 peaks for a single analyte.

McReynolds et al. modified the detection system for SCOFT, in which a CCD was employed for collection of fluorescence (33). In the system, a digital mask was used for generation of the Shah function, instead of the Cr-layer mask patterned on the microchannel. The CCD had a 1340×100 pixel imaging array. The length of the microchannel formed an image on the CCD with pixel dimensions of 1200×30. The 30 pixels perpendicular to the separation channel were binned into a single row and then, every 5 pixels along the separation channel were summed together before data readout, resulting in a 240×1-pixel image for each frame. The data set of a 240×1 image was multiplied by a Shah function consisting of alternating blocks of five 1's and five 0's. Thus, there were five pixels with fluorescence intensity followed by five pixels with no fluorescence. To generate the time domain signals, the sum of the total intensity for each frame was plotted against time.

In the case of multiple simultaneous separations (32), the sample solution was injected into separation channels with different lengths. The lengths of the separation channels were the square roots of the components of the following linear series: 5, 5.8, 6.6, 7.4, and 8.2. That is, the lengths were $\sqrt{5}\,(=2.24)$, $\sqrt{5.8}\,(=2.41)$, $\sqrt{6.6}\,(=2.57)$, $\sqrt{7.4}\,(=2.72)$, and $\sqrt{8.2}\,(=2.86)$ cm. The sample solution was introduced into the parallel channels (schematic illustration is shown in Fig. 17.6) simultaneously using pressure. Initially, the channels were filled with a migration buffer solution. Then, the sample solution was introduced into the separation channel at a pressure greater than that of the buffer solution. After the sample solution was introduced into the five separation channels, the entrance part of the five separation channels was filled with the buffer solution by decreasing the pressure of the sample solution. LIF detection was carried out using a CCD camera with only a single pixel for fluorescence detection. A photomultiplier tube can substitute for the camera.

17.3.3. Fundamental Studies in FT Electrophoresis

While the use of FT in the capillary format remains to be used for electrophoresis, FT has been employed in microchip electrophoretic separations. Figure 17.7 shows the results obtained using a microchip with detection windows consisting of 55 slits. In Figure 17.7, the resolution obtained in the frequency domain (Fig. 17.7c) was less than that seen in the time domain with single-point detection (Fig. 17.7a). Kwok and Manz reported that the S/N ratio was enhanced ~9-fold in the SCOFT detection (11). However, the analyte concentrations used in the SCOFT detection were much greater than those used in conventional LIF detection.

Some modifications of the detection technique have been attempted, for example, application of SCOFT to rear analysis (34), multiple-sample injection (35), and modified detection using a CCD camera (33). Unfortunately, simple inspection of the FTs of the electropherograms does not always yield useful information. As a result, some other mathematical techniques are applied to improve the resolution and/or sensitivity of FT-CE. For example, Eijkel et al. applied wavelet transformation to the results obtained using a Shah convolution detection system (36). Similarly, Allen et al. attempted a multivariate fit of a set of appropriate basis vectors to the FT results (32). The set of basis vectors was generated as a function of the mobility of the simulated analytes. The real component of the FT was compared with the set of basis vectors. Consequently, the best-fit coefficients were plotted against time, resulting in the electropherogram after FT. This type of data processing may enhance sensitivity and resolution. Generally, FT requires additional

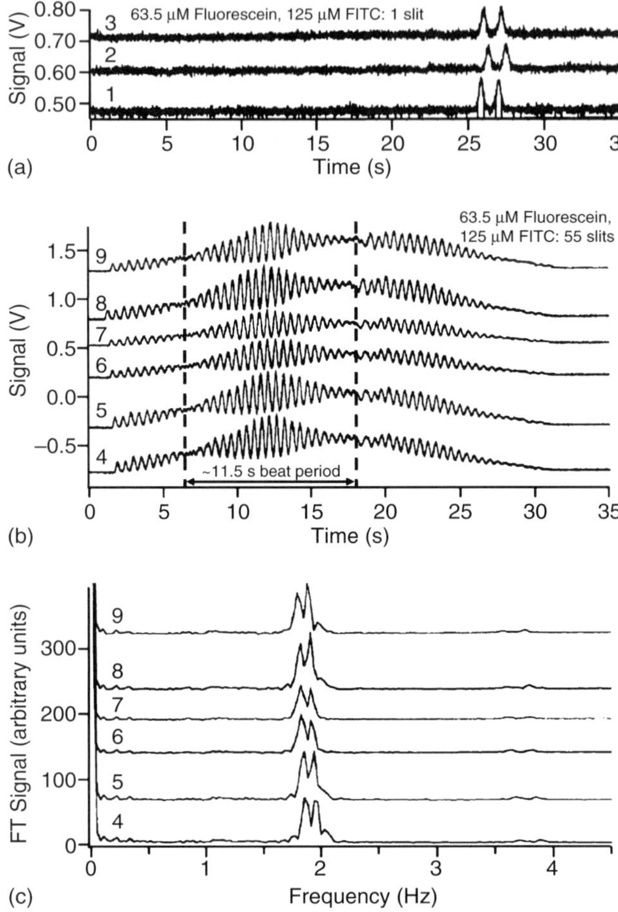

FIGURE 17.7. Two-component injection. (a) Single-point detection electropherograms of fluorescein and fluorescein isothiocyanate (FITC) show two resolved baseline peaks. (b) 55-point electropherograms generated for six injections. (c) Fourier transformations (FT) of the electropherograms of part (b): two fundamental frequencies at ~1.8 Hz are obvious, but harmonics at ~3.7 Hz are also visible. Reproduced with the permission of the American Chemical Society (6).

computational manipulation. In addition, how the method can be used in practical qualitative and quantitative analyses of actual samples remains to be determined. Thus, additional studies are required to demonstrate the usefulness of the FT technique in electrophoretic separation.

There are a few publications on the application of SCOFT to the measurement of particle velocity (36, 37). Briefly, the sample reservoir on a microchip was filled with a suspension of fluorescent microspheres, which migrate continuously under an applied potential. The time-domain signal obtained from the flowing microspheres was transformed using FT, resulting in a magnitude plot in the frequency domain. Unlike fluorescent molecules, the signal of the microspheres in the frequency domain showed a wide peak comprised of several spikes. The width of the peak may reflect the wide distribution of the charge-to-size ratio of the microspheres, although the distribution was not estimated from the experimental results. Thus, SCOFT could be applicable to precise measurement of the velocity of flowing particles, as the peak in the frequency domain reflects the results obtained using multiple detection, that is, each particle is detected at a large number of detection windows on a slit array.

17.4. CONCLUSIONS

Transformation techniques, including CC, HT, and FT, have been successfully combined with CE and ME techniques. Significant improvement in the S/N ratio was achieved for both capillary and microchip electrophoresis when CC and HT were employed. Fundamental studies of FT-ME and SCOFT have shown promising results, as demonstrated by enhancement in analyte sensitivity. These fundamental studies verified that these techniques work well theoretically and experimentally. However, how these techniques can be used in practical chemical analyses remains unclear. Therefore, further investigation should be aimed at discovery of important applications of these techniques, especially of FT and SCOFT.

REFERENCES

1. Izawa, K., Furuta, K., Fujiwara, T., and Suyama, T. (1966) *Ind Chim Belge*, **31**, 71.
2. Smit, H.C. (1970) *Chromatographia*, **3**, 515–518.
3. Lub., T.T., Smit, H.C., and Poppe, H. (1978) *J Chromatogr*, **49**, 721–733.
4. Van der Moolen, J.N., Louwerse, D.J., Poppe, H., and Smit, H.C. (1995) *Chromatographia*, **40**, 368–374.
5. Kaneta, T., Yamaguchi, Y., and Imasaka, T. (1999) *Anal Chem*, **71**, 5444–5446.
6. Crabtree, H.J., Kopp, M.U., and Manz, A. (1999) *Anal Chem*, **71**, 2130–2138.
7. Fister, J.C., Jacobson, S.C., and Ramsey, M. (1999) *Anal Chem*, **71**, 4460–4464.
8. Hata, K., Kichise, Y., Kaneta, T., and Imasaka, T. (2003) *Anal Chem*, **75**, 1756–1768.
9. Zhang, T., Fang, Q., and Fang, Z.L. (2003) *Chem J Chinese Univ*, **24**, 1775–1778.
10. Kaljurand, M. and Smit, H.C. (2005) *Chemometr Intell Lab Sys*, **79**, 65–72.
11. Kwok, Y.C. and Manz, A. (2001) *Analyst*, **126**, 1640–1644.
12. Annino, R. and Bullock, E.L. (1973) *Anal Chem*, **45**, 1221–1227.
13. Harwit, M. and Sloane, N.J.A. (1979) *Hadamard Transform Optics*, Academic Press, London.

14. Van der Moolen, J.N., Poppe, H., and Smit, H.C. (1997) *Anal Chem*, **69**, 4220–4225.
15. Kaneta, T. (2001) *Anal Chem*, **73**, 540A–547A.
16. Hata, K., Kaneta, T., and Imasaka, T. (2004) *Anal Chem*, **76**, 4421–4425.
17. Hata, K., Kaneta, T., and Imasaka, T . (2006) *Anal Chim Acta*, **556**, 178–182.
18. Hata, K., Kaneta, T., and Imasaka, T. (2009) *J Appl Phys*, **105**, 102018.
19. Kuldvee, R., Kaljurand, M., and Smit, H.C. (1998) *J High Resol Chromatogr*, **21**, 169–174.
20. Gao, L., Patterson, E.E., and Shippy, S.A. (2006) *Analyst*, **131**, 222–228.
21. Kaljurand, M., Ebber, A., and Somer, T. (1995) *J High Resol Chromatogr*, **18**, 263–265.
22. Braun, K.L., Hapuarachchi, S., Fernandez, F.M., and Aspinwall, C.A. (2006) *Anal Chem*, **78**, 1628–1635.
23. Braun, K.L., Hapuarachchi, S., Fernandez, F.M., and Aspinwall, C.A. (2007) *Electrophoresis*, **28**, 3115–3121.
24. Guchardi, R. and Schwarz, M.A. (2005) *Electrophoresis*, **26**, 3151–3159.
25. McReynolds, J.A. and Shippy, S.A. (2004) *Anal Chem*, **76**, 3214–3221.
26. Kaneta, T., Kosai, K., and Imasaka, T. (2003) *Anal Sci*, **19**, 1659–1661.
27. Kaneta, T., Nishida, M., and Imasaka, T. (2003) *Bunseki Kagaku*, **52**, 1193–1197.
28. Seiman, A., Kaljurand, M., and Ebber, A. (2007) *Anal Chim Acta*, **589**, 71–75.
29. Kaneta, T., Kosai, K., and Imasaka, T. (2002) *Anal Chem*, **74**, 2257–2260.
30. McReynolds, J.A., Gao, L., Barber-Singh, J., and Shippy, S.A. (2005) *J Sep Sci*, **28**, 128–136.
31. Hata, K., Kaneta, T., and Imasaka, T. (2007) *Electrophoresis*, **28**, 328–334.
32. Allen, P.B., Doepker, B.R., and Chiu, D.T. (2007) *Anal Chem*, **79**, 6807–6815.
33. McReynolds, J.A., Edirisinghe, P., and Shippy, S.A. (2002) *Anal Chem*, **74**, 5063–5070.
34. Kwok, Y.C. and Manz, A.J. (2001) *J Chromatgr A*, **2924**, 117–186.
35. Kwok, Y.C. and Manz, A.J. (2001) *Electrophoresis*, **22**, 222–229.
36. Eijkel, J.C.T., Kwok, Y.C., and Manz, A. (2001) *Lab Chip*, **1,** 122–126.
37. Kwok, Y.C., Jeffery, N.T., and Manz, A. (2001) *Anal Chem*, **73**, 1748–1753.

INDEX

Chemometric Methods in Capillary Electrophoresis. Edited by Grady Hanrahan and Frank A. Gomez